Bayer-Symposium II

New Aspects of Storage and Release Mechanisms of Catecholamines

Edited by

H. J. Schümann · G. Kroneberg

With 116 Figures

Springer-Verlag Berlin · Heidelberg · New York 1970

Professor Dr. Hans-Joachim Schümann, Direktor des Pharmakologischen Institutes am Klinikum Essen der Ruhruniversität Bochum, 4300 Essen-Holsterhausen

Professor Dr. Günter Kroneberg, Vorstand des Institutes für Pharmakologie der Farbenfabriken Bayer AG., 5600 Wuppertal-Elberfeld

The editors are indebted to Miss M. L. Chamberlain and Dr. K. Brandau for their valuable assistance

Bayer-Symposium II
held at Grosse Ledder near Cologne, Germany
October 9th—12th, 1969

ISBN-13: 978-3-642-46243-6 e-ISBN-13: 978-3-642-46241-2
DOI: 10.1007/ 978-3-642-46241-2

Contents

Welcome: G. Kroneberg 1

Opening Remarks: H. J. Schümann 2

Session 1

Origin and Axonal Transport of Adrenergic Nerve Granules

(Chairman: U. S. von Euler)

Studies on the Origin of the Granulated and Non-granulated Vesicles: A. Pellegrino de Iraldi and E. de Robertis. With 10 Figures 4

Discussion 17

The Effects of Drugs on Axonal Transport of Amine Storage Granules: A. Dahlström. With 8 Figures 20

Discussion 36

Invited Comment: L. B. Geffen 38

Discussion 41

Session 2a

Mechanism of Transmitter Release with Special Reference to the Problem of Exocytosis (I)

(Chairman: E. de Robertis)

Opening Remarks: E. de Robertis 44

Immunological Properties of Chromogranin: K. B. Helle. With 5 Figures 45

Discussion 56

Immunohistochemical Localization of Chromogranins in Sheep Sympathetic Neurons and their Release by Nerve Impulses: L. B. Geffen, B. G. Livett, and R. A. Rush. With 10 Figures 58

Discussion 72

Session 2b

Mechanism of Transmitter Release with Special Reference to the Problem of Exocytosis (II)

(Chairman: P. HOLTZ)

Quantal Aspects of the Secretion of Catecholamines and Dopamine-β-hydroxylase from the Adrenal Medulla: N. KIRSHNER and O. H. VIVEROS. With 1 Figure 78

Discussion 88

Some Aspects of the Quantal Release of the Adrenergic Transmitter: B. FOLKOW and J. HÄGGENDAL 91

Discussion 97

Some Further Aspects on the Release of the Adrenergic Transmitter: J. HÄGGENDAL. With 5 Figures 100

Discussion 109

Quantal or Graded Secretion of Adrenal Medullary Hormone and Sympathetic Neurotransmitter: L. STJÄRNE. With 7 Figures 112

Discussion 127

Session 3

Chemical Sympathectomy

(Chairman: P. HOLTZ)

Chemical Sympathectomy with 6-Hydroxydopamine: H. THOENEN, J. P. TRANZER, and G. HÄUSLER. With 10 Figures 130

Discussion 142

Session 4a

Effects of Drugs on Uptake and Release of Catecholamines (I)

(Chairman: H. BLASCHKO)

Effect of Some Metabolic Factors and Drugs on Uptake and Release of Catecholamines in vitro and in vivo: U. S. VON EULER. With 10 Figures 144

Discussion 159

Saturation of the Uptake of Noradrenaline as a Determinant of Pharmacological Effects: U. TRENDELENBURG, P. R. DRASKÓCZY, K.-H. GRAEFE, and H. HENNEMANN. With 2 Figures 161

Discussion 166

Cholinomimetic Drugs and Release of the Adrenergic Transmitter: E. MUSCHOLL. With 8 Figures 168

Discussion 186

Session 4b

Effects of Drugs on Uptake and Release of Catecholamines (II)

(Chairman: A. CARLSSON)

Membrane Effects of Catecholamine Releasing Drugs: D. PALM, H. GROBECKER, and I. J. BAK. With 8 Figures 188

Discussion 198

Effect of Angiotensin on Noradrenaline Release of the Isolated Rabbit Heart: H. J. SCHÜMANN. With 8 Figures 202

Discussion 209

Neuronal and Hormonal Control of Tyrosine Hydroxylase and Phenylethanolamine N-methyltransferase Activity: J. AXELROD, R. A. MUELLER, and H. THOENEN. With 4 Figures 212

Discussion 219

Session 4c

Effects of Drugs on Uptake and Release of Catecholamines (III)

(Chairman: J. AXELROD)

Effect of Drugs on Amine Uptake Mechanisms in the Brain: A. CARLSSON. With 7 Figures 223

Discussion 234

Release of Monoamines in the Central Nervous System: J. GLOWINSKI. With 3 Figures 237

Discussion 248

The Effect of Different Neuroleptic Drugs on the Monoamine Neurons of the Rat Brain: H. CORRODI 252

Discussion 256

Release of Catecholamines from the Hypothalamus by Drugs and Electrical Stimulation: A. PHILIPPU. With 10 Figures 258

Discussion 267

Session 5

(Chairman: H. J. Schümann)

General Discussion 270

Author Index 283

Subject Index 295

List of Participants

Dr. J. Axelrod, National Institute of Mental Health, Department of Health, Education and Welfare, Bethesda 14, Maryland 20014/USA

Prof. Dr. H. Blaschko, 24 Park Town, Oxford, England

Dr. K. Brandau, Farbenfabriken Bayer AG, Werk Elberfeld, Institut für Pharmakologie, 56 Wuppertal-Elberfeld, Friedrich-Ebert Straße 217, West-Germany

Prof. Dr. A. Carlsson, Department of Pharmacology, University of Göteborg, Göteborg, Sweden

Dr. H. Corrodi, AB Hässle, Fack Göteborg 5, Mölndal, Kärragatan 5, Göteborg, Sweden

Dr. A. Dahlström, Institute of Neurobiology, University of Göteborg, Faculty of Medicine, Medicinargatan 5, Göteborg, Sweden

Prof. Dr. D. Efron, National Institute of Mental Health, Bethesda 14, Maryland 20014, USA, and University of Baltimore

Prof. Dr. U. S. von Euler, Department of Physiology, Faculty of Medicine, Karolinska Institute, Stockholm, Sweden

Prof. Dr. B. Folkow, Department of Physiology, University of Göteborg, Göteborg, Sweden

Dr. L. B. Geffen, Department of Physiology, Monash University, Clayton, 3168, Victoria, Australia

Dr. J. Glowinski, Ministère de l'Education Nationale College de France, Laboratoire de Neurophysiologie, Generale Unite de Neuropharmacologie Biochimique, 11 Place Marcelin-Berthelot, Paris V^{e}, France

Dr. J. Häggendal, Department of Pharmacology, University of Göteborg, Göteborg, Sweden

Dr. K. B. Helle, Institute of Physiology, University of Bergen, Arstadviene 19, Bergen, Norway

Dr. F. Hoffmeister, Farbenfabriken Bayer AG, Werk Elberfeld, Institut für Pharmakologie, 56 Wuppertal-Elberfeld, Friedrich-Ebert-Straße 217, West-Germany

Prof. Dr. P. Holtz, Pharmakologisches Institut der Universität, 6 Frankfurt a.M., Ludwig-Rehn-Straße 14, West-Germany

Dr. L. Julou, Societé des Usines Chimiques, Rhone-Poulenc, 9, Quai Jules Guesde, Vitry s. Seine, France

Dr. N. Kirshner, Department of Biochemistry and Pathology, Duke University, Medical Center, Durham/North Carolina 27706, USA

Prof. Dr. G. Kroneberg, Farbenfabriken Bayer AG, Werk Elberfeld, Institut für Pharmakologie, 56 Wuppertal-Elberfeld, Friedrich-Ebert-Straße 217, West-Germany

Dr. D. R. Maxwell, May and Baker Ltd., Dagenham/Essex, England

Prof. Dr. E. Muscholl, Pharmakologisches Institut der Universität, 65 Mainz, Langenbeckstraße 1, West-Germany

Prof. Dr. D. Palm, Pharmakologisches Institut der Universität, 6 Frankfurt a.M., Ludwig-Rehn-Straße 14, West-Germany

Dr. A. Pellegrino de Iraldi, Inst. de Anatomia General y Embriologia, Facultad de Medicina, Universidad de Buenos Aires, Buenos Aires, Argentina

Prof. Dr. A. Philippu, Institut für Pharmakologie und Toxikologie der Universität, 87 Würzburg, Koellikerstraße 2, West-Germany

Prof. Dr. E. de Robertis, Inst. de Anatomia General y Embriologia, Facultad de Medicina, Universidad de Buenos Aires, Buenos Aires, Argentina

Prof. Dr. H. J. Schümann, Pharmakologisches Institut, Klinikum Essen der Ruhr-Universität, 43 Essen, Hufelandstraße 55, West-Germany

Dr. L. Stjärne, Department of Physiology, Faculty of Medicine, Karolinska Institutet, Stockholm 60, Sweden

Dr. H. Thoenen, Hoffmann-La Roche u. Co. AG, Abteilung für experimentelle Medicin, Basel 4002, Switzerland, Postfach

Prof. Dr. U. Trendelenburg, Institut für Pharmakologie und Toxikologie der Universität, 87 Würzburg, Koellikerstraße 2, West-Germany

Bayer-Symposium II, 1 (1970)

Welcome

G. Kroneberg

Ladies and Gentlemen!

I would like to welcome you in the name of Farbenfabriken Bayer and thank you for accepting the invitation to Grosse Ledder.

In recent years we have held several small scientific meetings in this place. The beautiful rural environment of the Bergisches Land with its calm and peaceful atmosphere, the temporary seclusion one enters, and the benefits of a "cuisine excellente" and a "bonne cave" have proved a particularly conducive background for scientific communication. But it was only last year that we institutionalized these meetings with a symposium about Immune Biology which is today followed by this symposium about the special problems of catecholamine research.

Apart from national and international congresses symposiums, too, often grow so extensive that the intensive discussion of details has to be neglected. We have therefore based these Bayer symposiums on the assumption that a small group of leading scientists from all over the world should be given an opportunity to lecture about and discuss specific and wellchosen problems of their particular field of pharmacological research.

The theme of this symposium, Storage and Release Mechanisms of Catecholamines, was chosen for two reasons. One is that this is indeed an important field of basic research immediately connected with applied pharmacology. The other, more personal reason is that my friend, Professor Schümann, and I have almost grown up with catecholamines in Professor Holtz's research laboratory. It is thus a particular honour and pleasure for us to have with us today Professor Blaschko, Professor von Euler and Professor Holtz who are pioneers in this field; and I think I am speaking on behalf of my younger colleagues when I give these three gentlemen an especially warm welcome.

The greatest poet of this country once said: "Tages Arbeit, abends Gäste, saure Wochen, frohe Feste". This (remark) it is hard to translate, but it means that somebody who has worked hard during the day may well be gay and enjoy his life in the evening.

It would certainly contribute to the wider meaning of this symposium if you, Ladies and Gentlemen, would practise this pleasant way of life here in Grosse Ledder. —

Bayer-Symposium II, 2—3 (1970)

Opening Remarks

H. J. Schümann

Ladies and Gentlemen!

More than a year ago my friend and colleague Prof. Kroneberg asked me to participate in the preparation and organization of a Bayer-Symposium on catecholamines, which should give a small number of participants the opportunity to present their recent data and — what is most important — to discuss the results in a group of experienced research workers free of time limitations. It was with great pleasure that I accepted this invitation, since for my own part I greatly welcome this kind of symposia in which the chosen field is of relatively small compass and can therefore be discussed with great intensity.

The organizing committee was very pleased to learn that most of the colleagues invited were able to accept this invitation and did not fear the long way from Australia, South America and from the United States of America. I should like to welcome you altogether cordially and especially the ladies Dr. Dahlström, Dr. Helle and Dr. Pellegrino de Iraldi.

I am very glad that also my teacher in Pharmacology Prof. Holtz is our guest. Under his guidance Dr. Kroneberg and I started to get the first experiences on the field of catecholamine research. — We are also very pleased to have Prof. Blaschko with us. In his laboratory in Oxford I spent a very stimulating and instructive time. Last but not least, I would like to welcome in particular Prof. von Euler as an old friend of the former Rostocker catecholamine group Holtz, Kroneberg, and Schümann.

I very much regret that Dr. Kroneberg, the initiator of this symposium who has taken a great deal of the burden of the organization, is not able to participate. Last sunday he had an accident and has now to stay in hospital. He asked me to give you his best regards and wishes a successful outcome of the meeting.

If you look at the program, you will notice that the first three sessions are devoted to more principal problems of the storage and release mechanisms of catecholamines, as there are: origin and axonal transport of adrenergic nerve granules, mechanism of transmitter release with special reference to the problem of exocytosis and finally chemical sympathectomy.

The following sessions are determined for the discussion of the effects of drugs and hormones on uptake and release mechanisms of catecholamines.

It was the aim of the organizing committee to plan the program in such a way that enough time is available for free discussion. I hope our arrangement will prove to be sufficient in this respect. May I ask the chairmen not to allow the speakers to exceed the time indicated. On the other hand I would like to emphasize that an interesting and stimulating discussion should not be stopped for reasons of the time schedule, because small alterations of this schedule will always be possible.

I would not like to finish my opening remarks without having given tribute and thanks to the Bayer AG, Leverkusen, which has in this way given us the opportunity to held this meeting at Grosse Ledder. I am quite sure that I expressed also your feelings. Finally, allow me, on behalf of every body, also to thank sincerely Dr. BRANDAU — a coworker of Prof. KRONEBERG — who did most of the organizing work.

I have now the pleasure to declare open the symposium on "new aspects of storage and release mechanisms of catecholamines" and to ask Prof. ULF VON EULER to take over the chairmanship of the first session.

Bayer-Symposium II, 4—17 (1970)

Session 1

Origin and Axonal Transport of Adrenergic Nerve Granules

Chairman: U. S. von Euler

Studies on the Origin of the Granulated and Nongranulated Vesicles*

AMANDA PELLEGRINO DE IRALDI and EDUARDO DE ROBERTIS

With 10 Figures

The compression or severance of an axon results in important ultrastructural changes at the regenerating segment. While in a normal myelinated axon only less than 1% of the volume is occupied by axially oriented material composed of neurotubules and neurofilaments (DE ROBERTIS and FRANCHI, 1953; THORNBURG and DE ROBERTIS, 1956) after sectioning the proximal stump becomes loaded with mitochondria, neurotubules and numerous clear and dense core (i.e. granulated) vesicles (for lit. see LUBINSKA, 1964; PELLEGRINO DE IRALDI and DE ROBERTIS, 1968).

The first observations of submicroscopic changes were made on regenerating axons in tissue culture; the growing tips were observed to be packed with microvesicular material (DE ROBERTIS and SOTELO, 1952). ESTABLE et al. (1957) found that after one day of the section of the guinea pig sciatic nerve, the proximal stump was filled with vesicular components of 200 to 700 Å and similar findings have been described by many investigators in different nerves and species. More recently the presence of granulated vesicles in the proximal stump of compressed adrenergic nerve fibers has been described by HÖCKFELT and DAHLSTRÖM (see DAHLSTRÖM and HÄGGENDAL, 1966; DAHLSTRÖM, 1969) and KAPELLER and MAYOR (1967) and correlated with the accumulation of noradrenaline observed with histochemical and biochemical methods.

These ultrastructural changes have been interpreted according to two main lines of thought. While some authors have preferred an explanation based on the stimulation of a local reaction caused by the nerve injury others have postulated a more mechanical explanation based on the flow of axonal components and of its damming at the site of compression.

* This wok has been supported by grants from the Consejo Nacional de Investigaciones, Cientificas y Técnicas, Argentina and U. S. Air Force (AF-AFOSR 67-0963 A).

Our investigations, made with electron microscopy and biochemical methods, deal mainly with the vesicular material concentrated at the proximal stump of regenerating axons. They show that granulated and non-granulated vesicles exist in adrenergic and non-adrenergic normal nerves and that they become highly concentrated in compressed nerves. Furthermore they suggest that a local reaction based in the dilatation, proliferation and pinching off of preexisting tubular structures, the neurotubules, is the main process involved in the formation of granulated and non-granulated vesicles. These neurotubules may integrate the vacuolar system of the axon. Some morphological and biochemical evidences suggesting that this axonal system may be related to the Golgi complex of the perikaryon will be presented.

Electron Microscope Observations

Our studies were carried out on the sciatic nerve of rats and cats and on the preganglionic fibres of the superior cervical ganglion of the rat (Pellegrino de Iraldi and de Robertis, 1968; Pellegrino de Iraldi and Rodriguez de Lores Arnaiz, unpublished). The electron microscope observations were made at different intervals between 6 hours and 12 days after compression. Tissue were fixed in glutaraldehyde followed by osmium tetroxide and processed for the electron-

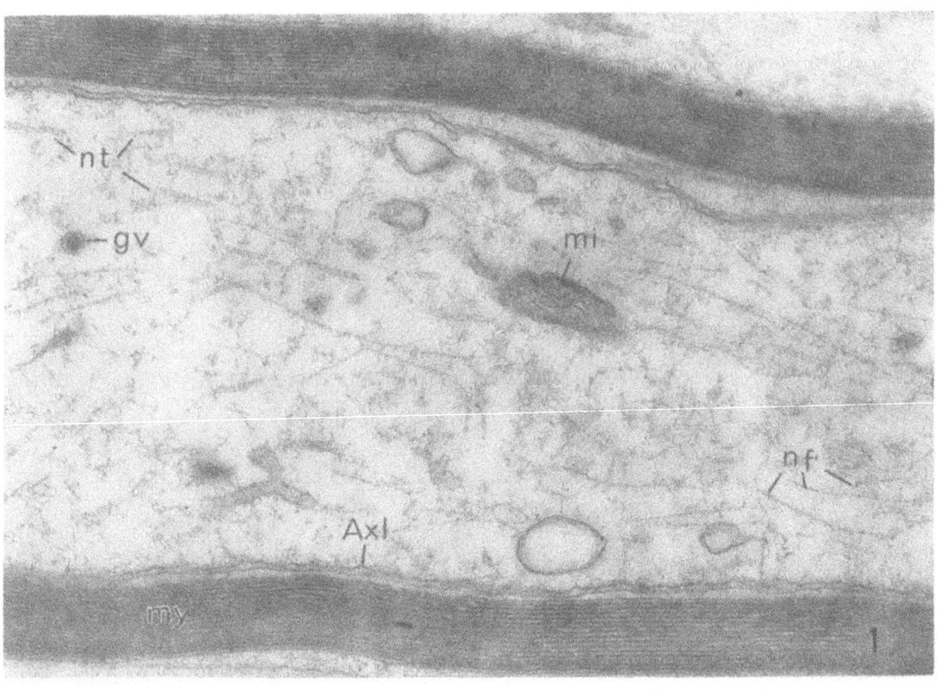

Fig. 1. Electron micrograph of a normal myelinated nerve fibre of the sciatic nerve of the rat (see description in the text). × 45.000. Reproduced from Z. Zellforsch. 87, 330—344 (1968). Legends for all figures: Axl, axolemma; ent, enlarged neurotubules; g, granules; gv, granulated vesicles; mi, mitocondrion; my, myelin sheath; nf, neurofilament; nt, neurotubule; t, tubular profiles; v, vesicle

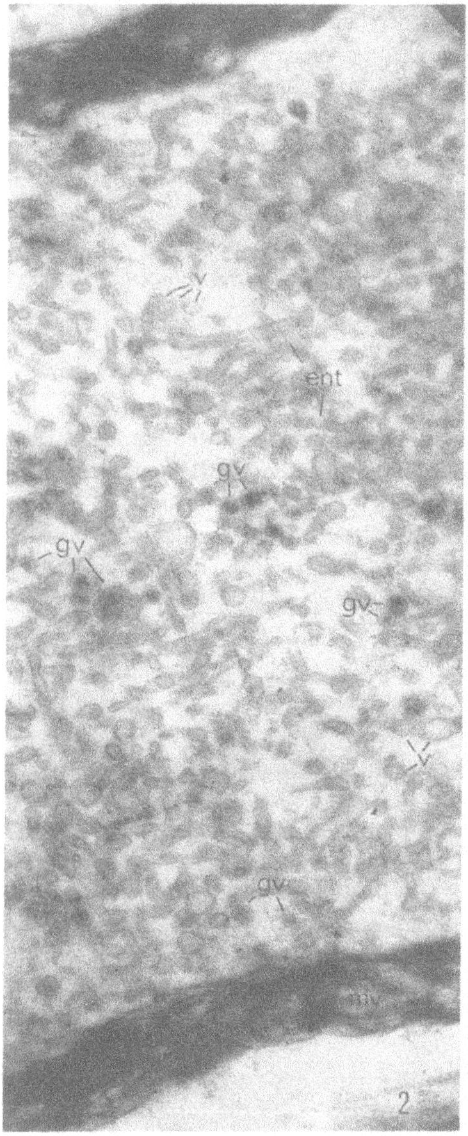

Fig. 2

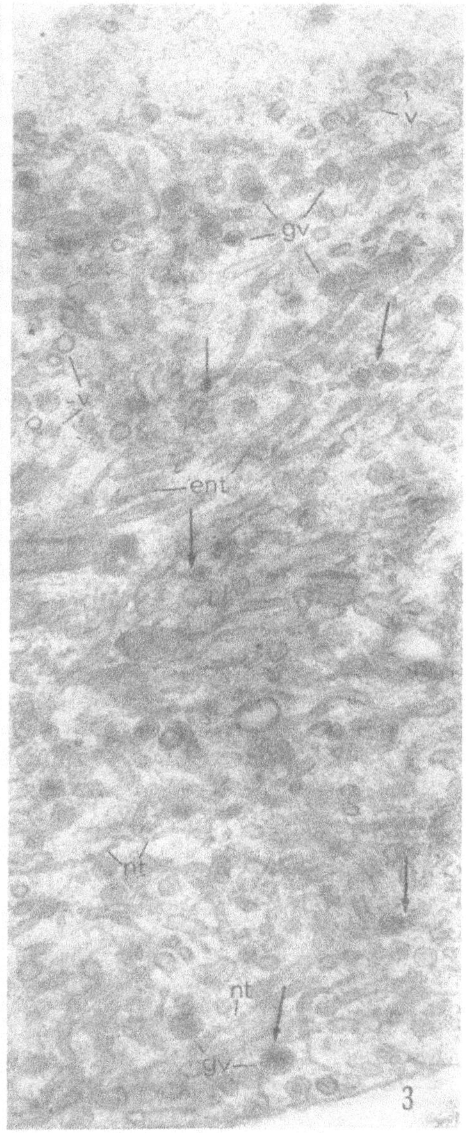

Fig. 3

Fig. 2. Mielynated fibre of the rat sciatic nerve 6 hours after compression. The axoplasm is filled with clear and granulated vesicles of about 750 Å in diameter and tubular profiles of 500—600 Å in diameter. × 60.000

Fig. 3. Unmyelinated fibre of the rat sciatic nerve 6 hours after compression. Many clear and granulated vesicles of about 880 Å in diameter and tubular structures of 500—600 Å tending to stack in parallel in many points, fill the axoplasm. Typical neurotubules may also be observed. With arrows are pointed the continuity of neurotubular structures with clear and granulated vesicles outer membranes. × 60.000

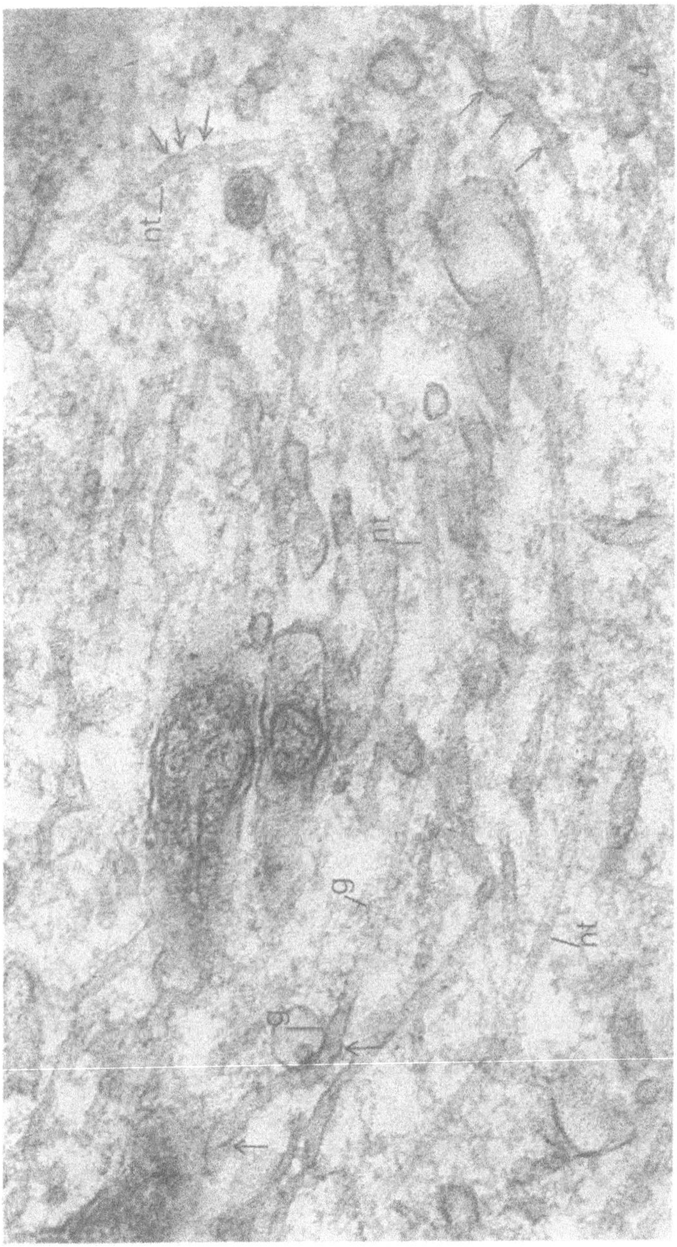

Fig. 4. Typical neurotubules with dilatations (arrows) and small granules inside. × 100.000

Fig 4 to 7. Electronmicrographs of axons from myelinated nerve fibres of the rat sciatic 6 hours after compression corresponding to a zone immediately above the illustrated in Figs. 2 and 3

microscope observation (see details in Pellegrino de Iraldi and de Robertis, 1968).

Under the electron microscope a normal myelinated fiber of the sciatic nerve of the rat looks rather empty showing a few axially oriented neurofilaments and neurotubules of 200 to 500 Å; there are also mitochondria and large clear vesicles considered as smooth endoplasmic reticulum. The presence of a few granulated vesicles of about 800 Å in diameter is also observed (Fig. 1). These vesicles are more abundant in unmyelinated fibres.

The changes observed in the axoplasm of the proximal stump were basically similar in all types of nerve fibres studied and varied according to the distance

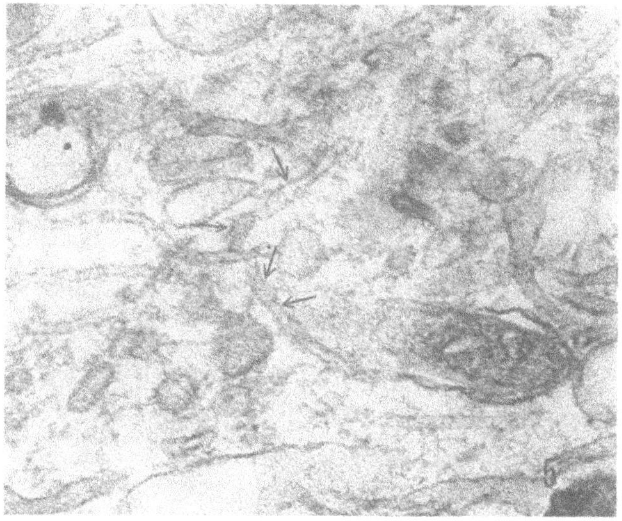

Fig. 5. Typical neurotubules with localized dilatations (arrows). × 100.000

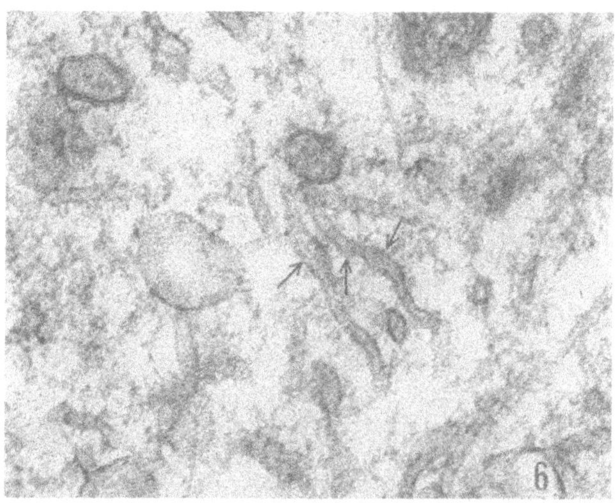

Fig. 6. Typical neurotubules some of them with localized dilatations (arrows). × 100.000

from the point of compression. In the immediate vicinity to this point there is a zone of few microns in which the axoplasm is completely amorphous. This is followed by a zone in which the axoplasm is filled with tubular profiles in different planes of section and with vesicles showing a clear content or a dense granule (Figs. 2 and 3). The diameter of granulated vesicles was about the same in the various types of fibres. In the myelinated axons of the rat sciatic nerve they varied between 550 and 1050 Å with a mean diameter of 750 Å. In the unmyelinated axons they were of 650 to 1250 Å with a mean diameter of 880 Å. In the preganglionic bundle of the superior cervical ganglion varied between 600 to 1500 Å with a mean diameter of 850 Å.

The tubular structures had a diameter which varied between 250 and 600 Å or more. While some of these are typical neurotubules axially oriented, straight and of uniform diameter others are convoluted, contain localized dilatations or are connected with vesicles and show a tendency to be stacked in parallel.

Morphological evidences suggesting a transformation of typical neurotubules (200 to 350 Å) into dilated tubular structures of about 500 Å and finally into

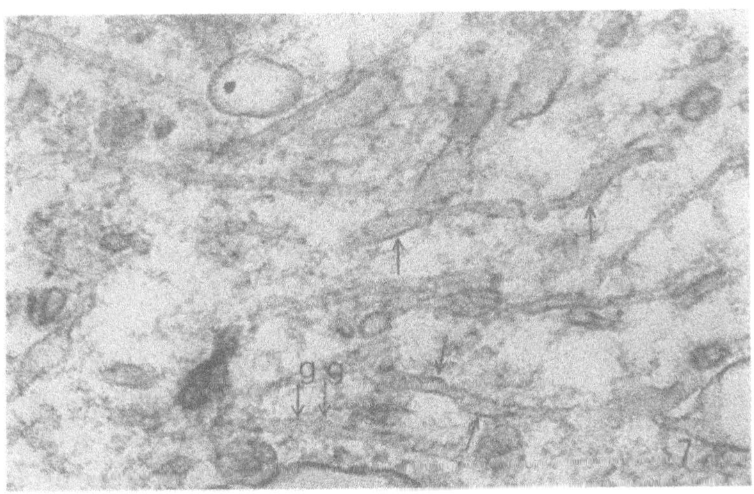

Fig. 7. Typical neurotubules some of which show localized dilatations (arrows). At the bottom of the figure one neurotubule with vesicular dilatations (arrows) and a small granule inside. × 100.000

vesicular elements of agranular and granular type (Fig. 4 to 7) were particularly well observed in a zone immediately above that previously illustrated in Figs. 2 and 3 and which contained predominantely vesicular elements.

After longer periods of compression in the cat sciatic, up to 12 days, similar changes were observed near the site of compression except for the presence of some lysosome-like bodies. Another interesting finding in these nerves was the connection of typical neurotubules with large vesicular structures.

The packing of tubular structures and vesicles sometimes resemble the organization of the Golgi complexes found in the perikaryon. There is a tendency for the tubular structures to aggregate in parallel and the vesicles seem to be pinched off from the ends of the tubules as in the case of the Golgi cisternae (Fig. 8).

Observations made on the distal stump of sciatic nerve, compressed for 6 hours, also revealed a considerable reaction with the appearance of vesicular structures but with abundant multivesicular bodies, small lamellar spheres and altered mito-

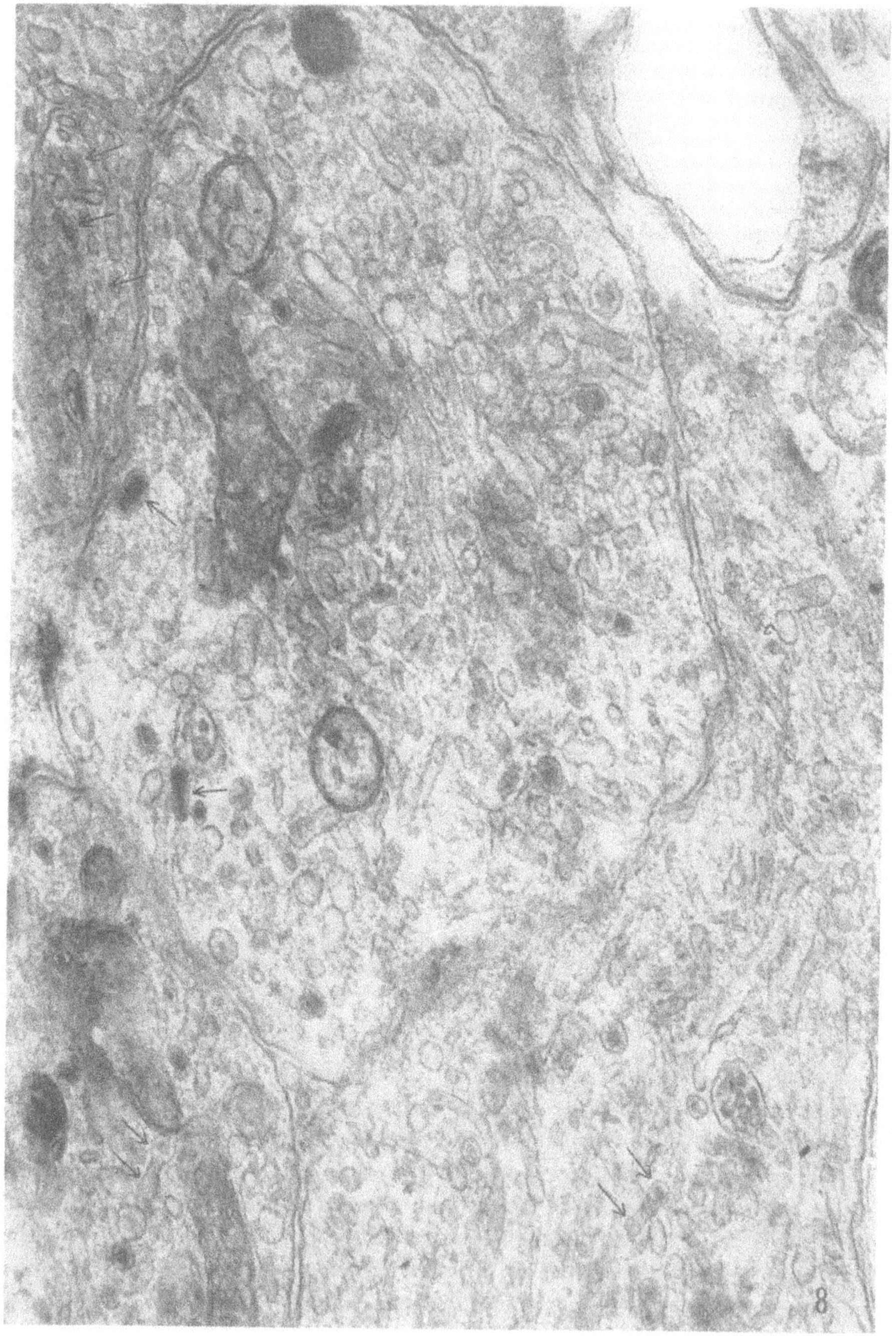

Fig. 8. Unmyelinated nerve fibre of the preganglionic bundle to the superior cervical ganglion of the rat 24 hours after compression. In the axon in the upper centre the enlarged neurotubules are organized in parallel and vesicles are apparently being formed from both ends of the tubules. Arrows point to elliptical shaped vesicles and to tubules containing granular material. Double arrows to granulated vesicles that are apparently in the process of pinching off from tubules. × 60.000. [Reproduced from Z. Zellforsch. **87**, 330—344, (1968)]

chondria. These changes suggest a predominant autophagic reaction of the axon, distally to the site of compression (PELLEGRINO DE IRALDI and DE ROBERTIS, 1968).

Biochemical Studies

After the work of NOVIKOFF and GOLDFISHER (1961) a number of investigators have shown that thiamine-pyrophosphatase is localized in the Golgi membranes and this enzyme is now used as a marker of the Golgi complex. These considera-

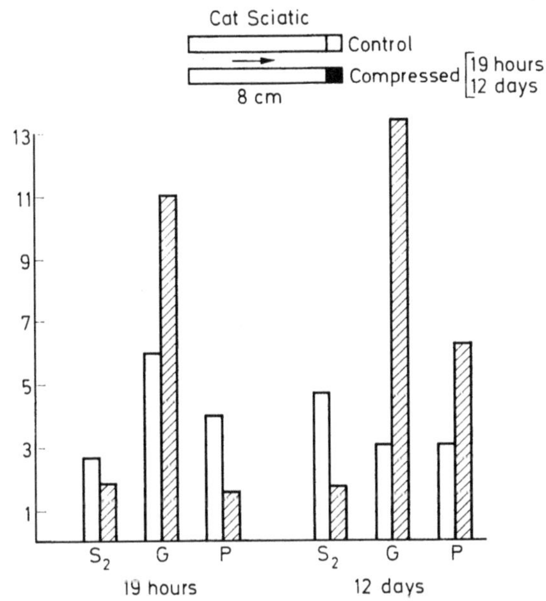

Fig. 9. Thiamino-pyrophosphatase (TTPase) activity in fractions of segments of the cat sciatic nerve 19 hours and 12 days after compression. The protein content and the enzyme activity were determined in the several fractions of pooled segments from 3 cat after gradient centrifugation (see more details in the text). The results per 5 mm length segment are expressed as the ratio between the treated and the control values. In the compressed nerves the specific activity of the Golgi fraction (G) increased 2 and 6 times, respectively 19 hours and 12 days after the compression. ▭ Protein $\frac{\text{Exper.}}{\text{Control}}$; ▨ TPPase $\frac{\text{Exper.}}{\text{Control}}$ (A. PELLEGRINO DE IRALDI and RODRIGUEZ DE LORES ARNAIZ, unpublished)

tions and the above mentioned finding of a Golgi-like structure in compressed nerve fibres led us to study the content of this enzyme in segments of 0.5 cm near the compression and at a distance from it. The findings were compared with enzyme determinations in the uncompressed contralateral nerve (Fig. 9).

The determination of the enzyme was made in the axoplasmic particles after homogenization and sedimentation of the large components of the nerve, which included: the myelin and connective tissue sheaths and the Schwann cells (PELLEGRINO DE IRALDI and RODRIGUEZ DE LORES ARNAIZ, to be published). An increase in the enzyme activity of 400% was observed at the axoplasm of the proximal stump 9 days after of the compression while at 9 cm above the site of the ligature the enzyme content was normal. The axoplasmic particles were further separated into three fractions with a gradient similar to that used by Schneider and Kuff (1954) to isolate the Golgi complex from the epididymis. The thiamine-pyrophosphatase was found to be highly concentrated in such a Golgi fraction. In the compressed nerves the specific activity increased 2 and 6 times respectively 19 hours and 12 days after the compression (Fig. 9). The electron microscope study of this fraction demonstrates that in the compressed nerves there was a considerable increase in tubular and vesicular material as compared with the control (Figs. 10 A and 10 B). This material has the same specific gravity as the Golgi complex and also a high content of the marker enzyme thiamino-pyrophosphatase.

Significance of the Vesicular Material in Regenerating Nerves

The discussion of the above mentioned morphological and biochemical findings will be concentrated upon two main points: a) the significance of the vesicular material present in the regenerating axonic stump and b) the probable origin of these vesicles.

The vesicular material which includes clear and granulated vesicles having a mean diameter of 850 Å is found in *all the axons of the regenerating stump*. Taking the case of the sciatic nerve the reaction is observed after compression in the myelinated, as well as in the unmyelinated axons, which include either efferent or afferent fibres of various nature (i.e. motor, sensitive, autonomus).

The granulated vesicles of the regenerating axons correspond to the medium size type of our classification (Pellegrino de Iraldi and de Robertis, 1965) and are very different than the small granulated vesicles of 500 Å in diametre present in peripheral adrenergic nerve endings (de Robertis and Pellegrino de Iraldi, 1961).

While the storage of some transmitter substances in such regenerating vesicles cannot be excluded, some preliminary observations done in our laboratory using the histochemical technique of Wood at the electron microscope level have revealed a negative reaction in them. This is at variance with what is observed in peripheral normal adrenergic endings in which the small and medium size granulated vesicles always show a positive reaction with glutaraldehydedichromate (Jaim Etcheverry and Zieher, 1968; Pellegrino de Iraldi and Gueudet, 1969).

One may speculate that some of the vesicles in regenerating nerves could contain other neurohumors involved in trophic functions. This problem is discussed by Korr et al. (1967), who have observed the passage of proteins from the hypoglossus nerve to the muscle of the tongue and is reviewed in the Neuroscience Research Program No. 1, 1969.

Origin of the Vesicular Material in Regenerating Axons

As it was mentioned in the introduction two main lines of thought have tried to interprete the intense ultrastructural changes occurring at the proximal stump of a regenerating axon. Estable et al. (1957), Schlote and Hager (1960), Wechler and Hager (1962), Wettstein and Sotelo (1963), Melamed and Trujillo

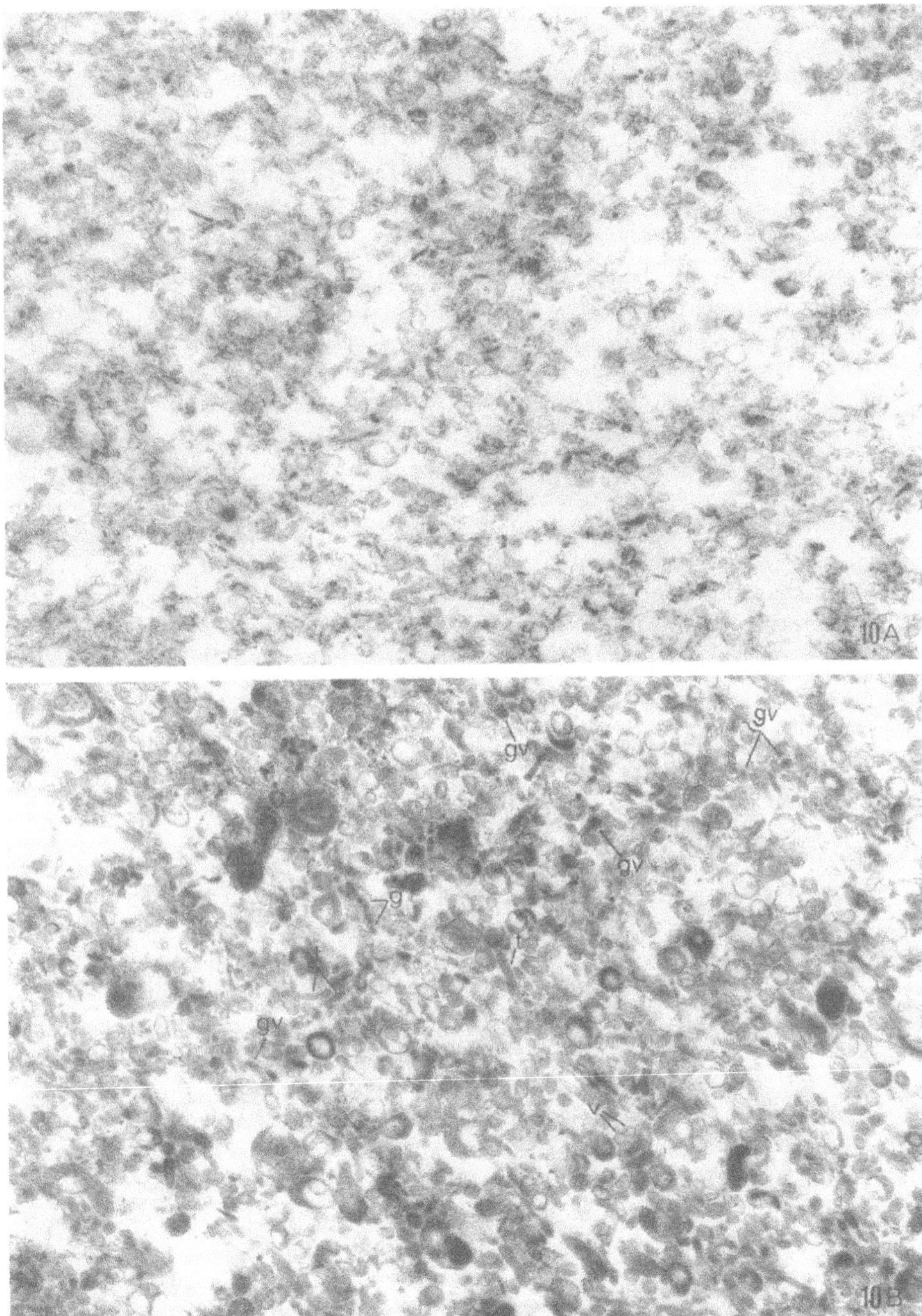

Fig. 10. A. Electronmicrograph of the G fraction in the control nerve (see Fig. 8). × 30.000. B. Similar to A. but from the compressed nerve. Observe the presence of numerous granulated and clear vesicles as well as tubular structures similar to those seen in sections of the axons (i.e. Fig. 2 and 3). × 30.000. (A. Pellegrino de Iraldi and Rodriguez de Lores Arnaiz, unpublished)

(1963) and Blümcke and Niedorf (1965) have interpreted their findings as due to a local axoplasmic reaction. Van Breemen et al. (1958) believed that the vesicles originate in the perikaryon and Weiss et al. (1962) considered the changes at the regenerating end as a manifestation of a dammed axonal flow. Similar interpretations have been advanced by Lubinska (1964), who postulated a both ways streaming in the axon, and by Dahlström (1965) and Dahlström and Häggendal (1966) and Kapeller and Mayor (1967) in their studies of catecholamines in compressed nerves.

Our findings confirm the general observation that in a few hours after ligation there are dramatic changes in the axoplasm of the regenerating fibres. The paucity of formed elements found in the normal axon is replaced by the extraordinary amount of the various components. One of the most notable is the large increase in neurotubular material which apparently undergoes rapid and intense morphological changes with dilatations, pinching off of clear vesicles and even the accumulation of dense material and the formation of granulated vesicles. The activity of this component seems greatly stimulated and the changes observed are difficult to explain simply on the grounds of hydrodynamics i.e. based on a passive flow of material synthesized at the perikaryon and conveyed by axon growth or cytoplasmic streaming. Particularly impressive is the observation of regions in which the tubular elements are disposed in a fashion that reminds the morphological organization and the secretory activity of the Golgi complex.

Similar observations were previously reported by Andres (1964, 1965) in synapses of the cerebral cortex and the olfatory bulb and by During (1967) in the motor end-plate. In all these synapses neurotubules were found to reach the presynaptic ending and to form a kind of loose reticulum, similar to the Golgi complex, from which the synaptic vesicles were pinched off.

Mori (1966) has pointed out the possible relationship between neurotubules and the Golgi region and Pellegrino de Iraldi and Jaim Etcheverry (1967), in ganglion cells of the retina, have observed continuities between the cisternae of the Golgi complex and the neurotubules irradiating from them. In their description of the synaptic vesicles de Robertis and Bennett (1954, 1955) suggested that the vacuolar system of the axon, mainly represented by the neurotubules, could have a role in the synthesis of this vesicular component and similar concepts were expressed by Palay (1956, 1958) referring to the endoplasmic reticulum. Blümcke and Niedorf (1965) also suggested that the neurotubules were directly involved in the formation of the vesicles at the regenerating ends. The results presented suggest that neurotubules are actively engaged in the formation of vesicular material both of the agranular and granular types. The idea that the neurotubules integrate a vast system, connected with the Golgi complex and that they are involved in secretory activity, is very appealing since it could explain not only the formation of vesicles at a regenerating and but also at a normal nerve terminal. It may be postulated that the neurotubules carry down the axon the essential building stones which may be then used for the formation of the vesicular elements at the endings.

The idea that parts of the Golgi complex could be present in the axon and then greatly increase at the regenerating stump is supported by the biochemical findings described here of an increase in thiamine pyrophosphatase, a marker enzyme for the Golgi complex.

Furthermore this enzyme is especially concentrated in a fraction of the axoplasm which has a similar specific gravity, as the Golgi complex (SCHNEIDER and KUFF, 1954) and which is composed of tubular and vesicular structures.

The transformation of neurotubules, into wider structures poses interesting problems from the point of view of the possible macromolecular mechanism involved. TILNEY and PORTER (1967) have observed that, by a cold treatment, the microtubules of 220 Å present in the axopodia of heliozoa become enlarged to 340 Å and tend to disintegrate. A macromolecular model for this transformation has been proposed which involved the twisting and sliding of macromolecules. Some of the enlarged neurotubules observed at the regenerating end of the axon have a remarkable similarity with the microtubules of cold-treated heliozoa. One point that is difficult to explain is the apparent lack of a "unit membrane" structure in the neurotubules and its appearance in the vesicles derived from them. Such an explanation could be easier if one assumes that the biological membranes may be constituted by macromolecular globular units in a two-dimentional array (KORN, 1966) and that at certain points lipid molecules could be attached to the protein framework to constitute lipoprotein structures. The possibility that the axon and nerve endings may have some local protein synthesis in addition to that produced in the perikaryon and transported by axonal flow should also be taken into consideration in the complex reaction taking place at the site of the ligature. Recently KOENIG and DROZ (1969) have brought radioautographic evidences that protein synthesis may occur locally in nerve endings.

Summary

In normal adrenergic and non-adrenergic axons a few clear and granulated vesicles of about 800 Å in diameter may be observed which within a few hours after compression increase in considerable number at the proximal stump. Since these vesicles are present in all kinds of axons its significance is still unsolved. However it is important to differentiate them from the small granular vesicles described by us in 1961 and which are characteristic of all peripheral adrenergic nerve endings.

During compression both clear and granular vesicles may be formed locally at the proximal stump by a process of dilatation and pinching off of neurotubules. Morphological and biochemical evidences suggest that a process of vesiculation similar to that taking place at the Golgi complex of the perikaryon may be occurring in compressed nerves. The neurotubules could integrate a vacuolar system going from the Golgi complex down to the nerve ending as previously postulated by DE ROBERTIS and BENNETT (1954 to 1955). This system could be involved in secretory activity at both ends and the neurotubules could carry down the axon the essential building stones used for the formation of vesicles at the regenerating end.

References

ANDRES, K. H.: Mikropinozytose im Zentralnervensystem. Z. Zellforsch. **64**, 63—73 (1964).
— Der Feinbau des Bulbus olfactorius der Ratte unter besonderer Berücksichtigung der synaptischen Verbindungen. Z. Zellforsch. **65**, 530—561 (1965).

Blümcke, S., Niedorf, H. R.: Elektronenoptische Untersuchungen an Wachstumsendkolben regenerierender peripherer Nervenfasern. Virchows Arch. path. Anat. **340**, 93—104 (1965).
Breemen, V. L., van Anderson, E., Reger, J. F.: An atempt to determine the origin of synaptic vesicles. Exp. Cell Res., Suppl. **5**, 153—167 (1958).
Dahlström, A.: Observations on the accumulation of noradrenaline in the proximal and distal parts of peripheral adrenergic nerves after compression. J. Anat. (Lond.) **99**, 677—689 (1965).
— Axonal flow of amine storage granules. Second. Int. Meeting of the Int. Soc. for Neurochem., Milan Sept 1—5 (1969), pag. 7—8. Paoletti, R., Fumagalli, R., Galli, C., Eds.
—, Haggendal, J.: Studies on the transport and life-span of amine storage granules in a peripheral adrenergic neuron system. Acta physiol. scand. **67**, 278—288 (1966).
De Robertis, E., Bennett, H. S.: Submicroscopic vesicular component in the synapse. Fed. Proc. **13**, 35 (1954).
— — Some features of the submicroscopic morphology of the synapses in frog and earthwork. J. biophys. biochem. Cytol. **1**, 47—56 (1955).
—, Franchi, C. M.: The submicroscopic organization of axon material isolated from myelin nerve fibres. J. exp. Med. **98**, 269 (1953).
—, Pellegrino de Iraldi, A.: A plurivesicular component in adrenergic nerve endings. Anat. Rec. **139**, 299 (1961).
—, Sotelo, J. R.: Electron microscopic study of cultured nervous tissue. Exp. Cell Res. **3**, 433—452 (1952).
Düring, M. v.: Über die Feinstruktur der motorischen Endplatte von höheren Wirbeltieren. Z. Zellforsch. **81**, 74—90 (1967).
Estable, C., Acosta-Ferreira, W., Sotelo, J. R.: An electron microscope study of the regenerating nerve fibres. Z. Zellforsch. **46**, 387—399 (1957).
Jaim Etcheverry, G., Zieher, L. M.: Cytochemistry of 5-hydroxytryptamine at the electron microscopic level. II. Localization in the autonomic nerves of the rat pineal gland. Z. Zellforsch. **86**, 393—400 (1968).
Kapeller, K., Mayor, D.: The accumulation of noradrenaline in constricted sympathetic nerves as studied by fluorescence and electron microscopy. Proc. roy. Soc. B **167**, 282—292 (1967).
Koenig, H. L., Droz, B.: Incorporation *in vitro* d'acides aminés marqués dans les terminaisons nervenses du ganglion ciliaire du Poulet. J. Microscopie 8, 63a (1969).
Korn, E. D.: Structure of biological membranes. The unit membrane theory is revaluated in light of the data now available. Science **153**, 1491—1498 (1966).
Korr, I. M., Wilkinson, P. N., Chornock, F. W.: Axonal delivery of neuroplasmic components to muscle cells. Science **155**, 343—345 (1967).
Lubinska, L.: Axoplasmic streaming in regenerating and in normal fibres. Prog. in Brain Res. **13**, 1—71 (1964).
Melamed, J., Trujillo-Cenóz, O.: Electron microscopic observations on reactional changes occurring in insect nerve fibres after transection. Z. Zellforsch. **59**, 851—856 (1963).
Mori, S.: Some observations on the fine structure of the corpus striatum of the rat brain. Z. Zellforsch. **70**, 461—488 (1966).
Novikoff, A. B., Goldfisher, S.: Nucleoside diphosphatase in the Golge apparatus and its usefulness for cytological studies. Proc. nat. Acad. Sci. (Wash.) **47**, 802—810 (1961).
Palay, S. L.: Synapses in the central nervous system. J. biophys. biochem. Cytol., Supp. **2**, 193—202 (1956).
— The morphology of synapses in the central nervous system. Exp. Cell Res., Suppl. **5**, 275—293 (1958).
Pellegrino de Iraldi, A., De Robertis, E.: Ultrastructure and function of catechol-amine containing systems. In: Proc. 2nd Internat. Congr. Endocr. London, Excerpta med. (Amst.) Internat. Congr. Ser. **83**, 355—363 (1965).
— — The neurotubular system of the axon and the origin of granulated and non-granulated vesicles in regenerating nerves. Z. Zellforsch. **87**, 330—344 (1968).
—, Gueudet, R.: Catecolamine and serotonin in granulated vesicles of nerve endings in the pineal gland of the rat. Int. J. Neuropharmacol. 8, 9—14 (1969).
—, Jaim Etcheverry, G.: Granulated vesicles in retinal synapses and neurons. Z. Zellforsch. **81**, 283—296 (1967).

SCHLOTE, W., HAGER, H.: Elektronenmikroskopische Befunde zur Feinstruktur von Axonveränderungen in peritranmatischen Bereich nach experimenteller Strangdurchtrennung am Rückenmark der weißen Ratte. Naturwissenschaften **47**, 448—451 (1960).
SCHNEIDER, W. C., KUFF, E. L.: On the isolation and some biochemical properties of the Golgi substance. Amer. J. Anat., **94**, 209—224 (1954).
THORNBURG, W., DE ROBERTIS, E.: Polarization and electron microscope study of frog nerve axoplasm. J. biophys. biochem. Cytol. **2**, 475—482 (1956).
TILNEY, L. G., PORTER, K. R.: Studies on the microtubules in heliozoa. II. The effect of low temperature on these structure in the formation and maintenance of the axopodia. J. Cell Biol. **34**, 327—343 (1967).
WECHSLER, W., HAGER, H.: Elektronenmikroskopische Befunde zur Feinstruktur von Axonveränderungen in regenerierenden Nervenfasern des Nervus ischiadicus der weißen Ratte. Acta neuropath. (Berl.) **1**, 489—506 (1962).
WEISS, P., TAYLOR, A. C., PILLAI, P. A.: The nerve fibre as a system in continuous flow; microcinematographic and electronmicroscopic demonstrations. Science **136**, 330 (1962).
WETTSTEIN, R., SOTELO, J. R.: Electron microscopic study on the regenerative process of peripheral nerves of mice. Z. Zellforsch. **59**, 708—730 (1963).

Dr. A. PELLEGRINO DE IRALDI
Instituto de Anatomia General y Embriologia
Facultad de Medicina-Univ. de Buenos Aires
Buenos Aires, Argentina

Discussion

GEFFEN: There are two things which I am not very clear about which perhaps could be cleared up. The first is the distinction made between neurotubules and smooth endoplasmic reticulum. As I understand it, the type of membrane in the neurotubules is very different to that in smooth endoplasmic reticulum. The smooth endoplasmic reticulum has a unit membrane—a double membrane—whereas the neurotubules are composed of radially arranged filaments. Therefore, one has in some way to get a transformation from one type of membrane to another since I think there is little doubt that the vesicles have a unit membrane rather than a membrane resembling the neurotubules. That's the first question.

DE ROBERTIS: This is one of the problems which is difficult to answer. We know that in the axon there are neurotubules and very little endoplasmic reticulum of the smooth type. Much evidence has been shown here that neurotubules are able to undergo dilatations at many points and pinching off of vesicular material from them. There are different types of tubules in different cells and even the microtubules of different cells have different properties. Neurotubules have no typical unit membrane but that does not mean that a unit membrane could not be obtained by modification of the structure of the neurotubules. There are other examples. In the case of the Golgi membranes, MORRÉ has observed that near the nucleus they show no unit membrane and then this structure appears in more distal membranes by some kind of transformation. We think in the dynamic structure of the membrane, i.e. the membranes may change in their composition and ultrastructure. Also the size of the tubules may change. TILNEY and PORTER have shown that microtubules of 250 Å may become larger by treatment with cold. We would like to emphasize that the tubular structure is probably connected with the Golgi on one side and that they are able to produce vesicles at the compressed end. In both sides, vesicles are being produced and it seems possible that at the nerve endings vesicles can be produced by the same type of mechanism.

GEFFEN: May I reply on the same point—I accept that one must take a dynamic view of membranes. However, it seems to me that one also ought to take a biochemical view of the problem. I understand that thiamine pyrophosphatase is a good marker of smooth endoplasmic reticulum whereas the best marker I know for neurotubules, is their capacity to bind colchicine. I wonder whether you have considered examining whether the proteins which accumulate above the ligature in this compressed nerve have the capacity to bind colchicine. That would be stronger evidence, in my opinion, that the membranes are formed from the neurotubules.

DE ROBERTIS: The best enzyme to mark the endoplasmic reticulum of the smooth type is glucose-6-phosphatase; thiamine pyrophosphatase is characteristic of the Golgi complex. The smooth endoplasmic reticular and the Golgi membranes have apparently a different enzymic make-up. The recent work of MORRÉ demonstrates that there are many enzymes associated with the Golgi which are not present in the smooth endoplasmic reticulum. Of course, it would be interesting to study the binding of colchicine to see if in the compressed nerve it increases or if it is present in the regenerating vesicles. This is a way to demonstrate that the protein making the neurotubule becomes incorporated into the vesicular wall.

DAHLSTRÖM: I wonder, Professor DE ROBERTIS, if you consider these dilatations of the microtubules above a nerve ligation as true signs of a formation of large or small dense core vesicles—or both? — It has been suggested by several investigators that some liquid substances possibly are transported distally inside the microtubules and that the flow rate may be rather high. Is it possible that the dilatations of the neurotubules can rather be due to the inhibition of this proposed flow by the ligations than to a pinching off of dense core vesicles?

DE ROBERTIS: There is evidence that within neurotubules granular material is present. The work done in Mendoza (Argentina) by RODRIGUEZ and PIEZZI has shown that within some neurotubules there are small granules. In the compressed nerve we find that above the ligature there are several zones. The one immediately above the ligature is amorphous and the axoplasm appears rather fluid, then we see the zone containing vesicles and then the zone with the neurotubules in which dilatations and pinching off are observed and a type of Golgi sometimes is formed. It may very well be that the fluid material moves much faster within these neurotubules and this may help in the process of vesicle formation. However, these changes may not only be due to mechanical forces but macromolecular mechanisms may be involved in the formation of these small structures. We adopt a dynamic view point instead of a purely mechanical one when we think about changes in structure. The fact that thiaminopyrophosphatase increases considerably at the regenerating end and in the so-called Golgi fraction, is a good indication that the process of formation of the regenerating vesicles is related to the Golgi complex.

Dahlström: A few more questions: Firstly, I wonder if this granular material, observed within the microtubules, does necessarily have to have the same molecular base as the granular material in the dense core vesicles? Could some other material appear dense in your preparations? (de Robertis—Yes) Secondly, in the adrenergic neurons rather few microtubules seem to be present. I once asked Dr. Sandy Palay to tell me how many microtubules he had observed in adrenergic axons and he said between 4 and 6. I think it is very difficult to imagine that the large accumulation of dense core vesicles developing so rapidly above a nerve ligation is due to a pinching off of vesicles from such few neurotubules. To make that reasonable, to my mind at least, you would have to suppose that the neurotubules themselves are pushed down—or grow down—at a tremendous speed. What is your opinion about this?

de Robertis: We found that the number of neurotubules above the ligation increased in comparison with a normal axon. In a normal axon there are very few neurotubules, four of five, in a section but in compressed nerves the number increases considerably. This may possibly be due to the rapid flow of neurotubules. There is evidence that proteins which are particulate, flow much faster than the proteins which are in a soluble form in the axoplasm. It may be that the neurotubules move faster in the axon or perhaps they can grow locally.

Bayer-Symposium II, 20—36 (1970)

The Effects of Drugs on Axonal Transport of Amine Storage Granules

ANNICA DAHLSTRÖM

With 8 Figures

During the last years evidence has been provided that the axoplasmic flow, described at first by WEISS and HISCOE (1948) in myelinated nerves, also occurs in unmyelinated adrenergic axons. Particular interest has been paid to the axoplasmic transport of amine storage granules, since these particles appear essential for a normal function of the adrenergic nerves (cf. IVERSEN, 1967; ANDÉN et al., 1969; HÄGGENDAL and MALMFORS, 1969).

I. Evidence for a Proximo-Distal Transport of Amine Storage Granules in Adrenergic Neurons

A. Accumulation of Storage Granules Above a Ligation

When a peripheral nerve containing adrenergic axons is ligated, noradrenaline (NA) accumulates proximal to the ligation [DAHLSTRÖM and FUXE, 1964; DAHLSTRÖM, 1965; ERÄNKÖ and HÄRKÖNEN, 1965; DAHLSTRÖM and HÄGGENDAL, 1966 (1), 1967; KAPELLER and MAYOR, 1965, 1967; GEFFEN and RUSH, 1968; BOYLE and GILLESPIE, 1968; LADURON and BELPAIRE, 1968]. The onset of this accumulation appears to be immediate, since as early as 5 to 10 min after the ligation, small amounts of accumulated material with a strong NA fluorescence can be observed within enlarged axons just above the ligation. This fluorescence (Fig. 1) is obtained after formaldehyde treatment according to the histochemical fluorescence method of HILLARP, FALCK and coworkers (for ref. and description see e.g. CORRODI and JONSSON, 1967). The amount of NA above the ligation, measured by quantitative chemical methods, then increases approximately linearly with time up to about 48 h [DAHLSTRÖM and HÄGGENDAL, 1966 (1), 1967].

Some observations indicate that the accumulated NA is stored within the so-called amine storage granules[1], for instance. a) Reserpine treatment, which probably causes a very long-lasting block of the amine storage mechanism of the storage granules (cf. CARLSSON, 1965), results in disappearance of the accumulated NA [DAHLSTRÖM, 1965, 1967 (1); KAPELLER and MAYOR, 1967]. This is in all probability due to enzymatic degradation, by e.g. monoamine oxidase (MAO), of the amine, after leaking from the granules after the storage blockade. Inhibition of the MAO (by

[1] The term amine storage granules refers to the subcellular particles, which have certain physiological and pharmacological properties, e.g. to take up, store, synthesize and release NA (in the case of peripheral adrenergic neurons in mammals). The dense cored vesicles, observed by the electron microscopists, probably correspond morphologically to the amine storage granules (e.g. HÖKFELT, 1968).

administration of e.g. nialamide) prior to the reserpine treatment prevents the disappearance of NA caused by reserpine alone [DAHLSTRÖM, 1967 (1)]. b) Electron microscopical studies of adrenergic axons just above a ligation have revealed that there is an accumulation of dense cored vesicles, the relative number of which appears to increase approximately parallel to the increase in NA with the time after ligation (KAPELLER and MAYOR, 1966, 1967; DAHLSTRÖM and HÖKFELT,

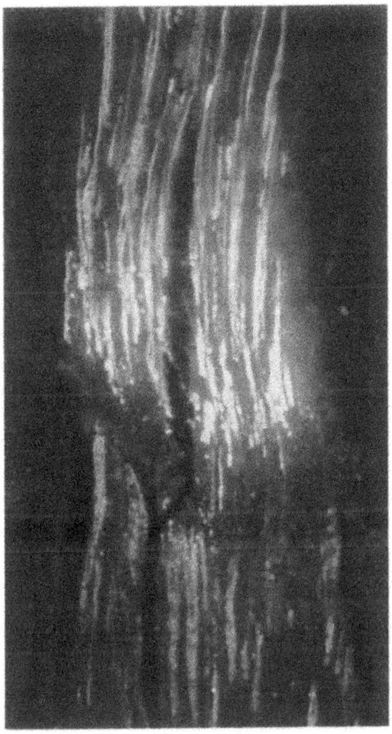

Fig. 1. Longitudinal section of rat sciatic nerve ligated 24 h before death. Above the ligation large amounts of strongly green fluorescent material (the reaction product of NA and formaldehyde) can be seen accumulated within adrenergic axons. The axons are enlarged and distended with a bloated appereance. Below the ligation small amounts of green fluorescent material is observed. This retrograde accumulation of NA is very small compared to the accumulation above the ligation. It is doubtful whether it represents signs of a retrograde transport of NA storage granules, or merely an effect of the type of ligation used. Fluorescence microphotograph, × 80

unpubl. obs.). The dense cored vesicles, observed both with glutaraldehyde-osmium tetroxide and potassium permanganate fixation, varied in diameter between 500 to 1,000 Å. At least the small type (about 500 Å) of the dense cored vesicles may correspond to the amine storage granules (for discussion and references see e.g. HÖKFELT, 1968). Evidence that also the large dense cored vesicles are involved in the storage of amines has been presented by TRANZER and THOENEN (1968). Thus, these observations also indicate that the NA accumulating above a ligation of an adrenergic nerve is bound in amine storage granules.

B. Transport of Storage Granules

The accumulation of NA occurring above a ligation has been interpreted to be due to the arrest of amine storage granules, containing NA, during their continuous proximo-distal transport through the axon from the site of manufacture, probably the cell body. Some evidence supporting this view will be given: (1) The very rapid onset of NA accumulation observed within some minutes after ligation indicates that the accumulation of NA was not due to a changed neuron reaction due to the trauma, but rather to the effect of the ligation on some continuous process in the axon, e.g. a proximo-distal flow of amine storage granules (cf. Dahlström, 1966). (2) The increase in NA above the ligation was approximately linear, also indicating a continuous and steady shift of amine storage granules towards the ligation [Dahlström and Häggendal, 1966 (1), 1967].

It may be argued that the observed accumulation of NA above a ligation may not necessarily reflect a proximo-distal transport of granules, but could be due to the accumulation or transport of one or more enzymes, which may influence the amount of NA found above a ligation. Of the known enzymes, taking part in NA synthesis, only one has so far been found to accumulate above a ligation at about the same rate as NA, namely dopamine (DA)-β-hydroxylase [Laduron and Belpaire, 1968 (1)], which is believed to be localized to the storage granules themselves (e.g. Kaufmann and Friedmann, 1965). Tyrosine-hydroxylase has not been found to accumulate above a ligation within 24 h [Laduron and Belpaire, 1968 (2)]. DOPA-decarboxylase likewise did not appear to accumulate above the ligation during the first 24 h, but during the following days an increase in enzyme activity was noticed. The time course curve for this enzyme accumulation, however, followed an entirely different pattern than that of NA (Dahlström and Jonason, 1968). Therefore, it appears unlikely that the NA accumulation above a ligature is largely due to local synthesis, induced by rapidly down-transported NA synthetising enzymes.

In double ligation experiments, where two ligations about 2 cm apart were tied simultaneously on the sciatic nerve, it was found that a shift of NA had occurred towards the distal ligation. The net NA content of the nerve appeared in these experiments to be essentially unchanged. These results indicate that a proximo-distal transport of NA, probably bound to amine storage granules (as indicated by reserpine depletion) took part in the separated nerve. Thus, the transport of amine storage granules appeared to be independent of the cell body, since the rate of proximo-distal shift in the separated nerve was about the same as that calculated in single ligated nerves [several mm/h, Dahlström and Häggendal, 1966 (1); Dahlström, 1967 (2); Banks et al., 1969]. The results also indicate that local NA synthesis, resulting in a net increase of NA per average granulum, plays a minor role in the increase of NA above a ligation. This increase may, in view of our present knowledge, rather be attributed to a piling up above the ligation of down-transported granules, probably formed in the perikarya.

C. Origin of Storage Granules

It may be argued that the amine storage granules above a ligation need not necessarily be formed in the cell body, but may have been formed locally above the ligation by the fusion of smaller molecules that were produced in the cell body and transported down the axon. Recent results from the laboratory of Ochs may support evidence *against* this view. Labelled amino acids were injected in the lumbar ganglia of cat, and the transport and accumulation of labelled proteins in the sciatic nerve were studied. The first peak of radioactive proteins moving down the sensory axons in the sciatic nerve, was found to travel very fast. A rate of 17 mm/h was calculated (410 mm/day) (Ochs et al., 1969). This first peak consisted in the main of high molecular weight proteins. A later, more slowly moving peak was found to contain for the most part small protein molecules (Kidway and Ochs, 1969). If a ligation was put around the nerve, mainly large protein molecules accumulated within the first 24 h. In order to rule out the possibility that the large molecules had been formed from small molecules in the axon above the ligation, experiments with two simultaneous ligations were performed. A proximo-distal shift of labelled proteins was found also in the separated nerve part. Analysis of the molecular weight of the proteins in the nerve between the two ligations showed that the ratio between high and low molecular weight proteins was essentially the same as that in the unligated nerve (Ochs and Sabri, 1969; Ochs, personal communication). Thus, a transformation in the axon of small molecules into large molecules appeared unlikely in the neuron system studied by these authors.

Another possibility may be that the storage granules are formed above the ligation by pinching off from some tubular structures, as suggested by Pellegrino de Iraldi and de Robertis (1968). These authors discuss the neurotubules as the "mother-tubule" of dense cored vesicles. This idea may well fit in with the results obtained in adrenergic neurons, provided that it is supposed that the tubules are formed in the cell body, transported down the axons, and contain an arrangement of macromolecules essentially similar to that of the amine storage granules. However, in order to be able to pinch off such large amounts of granules (or vesicles in cholinergic neurons) as indeed observed above ligations (see above) within a rather short time interval, it must be assumed that the neurotubules themselves are formed and grow out from the cell body, at high rates. Otherwise it is difficult to imagine how the neurotubules can provide enough material for the formation of vesicle membranes. Recent experiments (Sjöstrand and Karlsson, personal communication; Karlsson, 1970; Weiss, personal communication) appear to demonstrate that neurotubules grow very slowly and move down the axon in the slow phase, i.e. 1 to 2 mm/day (see also Droz, 1967). Thus, it appears unlikely that the storage granules are formed from neurotubules.

Some reasons indicated in the literature for not accepting the perikaryon as the site of origin of the amine storage granules appear to be: a) dense cored vesicles, probably corresponding to amine storage granules (for references and discussion see Bloom and Aghajanian, 1968; Hökfelt, 1968) are rarely seen in the unligated axon or in the perikarya, b) the dense cored vesicles, as observed in glutaraldehyde-osmium fixation, may be seen also in myelinated nerves, and may thus not represent amine storage granules (e.g. Pellegrino de Iraldi and de Robertis, 1968), c) dense cored vesicles are often seen attached to—or even continuous with—neurotubular-like elements above a ligation (Pellegrino de Iraldi and de Robertis, 1968). However, it may be argued that: a) with potassium permanganate ($KMnO_4$) fixation according to Richardson (1966)—which apparently has not been used by the above authors in their described findings (Pellegrino de Iraldi and de Robertis, 1968)—a large number of dense cored vesicles has been observed in adrenergic ganglion cell bodies. Also in the unligated axons comparatively many dense cored vesicles are seen (Hökfelt, 1969). Furthermore, amine storage granules have been isolated from sympathetic ganglia by Philippu et al. (1967), and from adrenergic axons of the splenic nerve by von Euler et al. (e.g. 1963). b) As judged from pharmacological studies, and studies on electrically stimulated tissues, the small dense cored vesicles (500 Å) in $KMnO_4$-fixed adrenergic nerve terminals almost certainly correspond to amine storage granules (Hökfelt, 1968). Also the larger type of dense cored vesicles in adrenergic neurons have been shown to store amines according to Tranzer and Thoenen (1968). Of course, the mere presence of electron dense material in various types of vesicles or tubes cannot be accepted as evidence that the particle is involved in catecholamine storage. Conversely, the presence of vesicles with electron dense cores in tissues other than adrenergic neurons cannot be used as an argument against the view that dense cored vesicles in adrenergic neurons morphologically correspond to amine storage granules. Particularly, since only the dense cored vesicles in adrenergic neurons have been shown to react like amine storage granules to a wide variety of experimental treatments (Hökfelt, 1968). Preferably, electron microscopic studies should perhaps include at least two different fixation techniques (to compare the pictures obtained). Identification of amine storage particles must be based on (1) the electron dense core *and* (2) the reaction to experimental (e.g. pharmacological) treatments. c) If the amine storage granules are transported distally by some coupling mechanism with active sites on the outer walls of the neurotubules, as suggested by Schmitt (1968) or by a mechanism as proposed by Weiss (1969), one may expect to find them attached to the neurotubules in the axon, at least occasionally. In fact, quite recently Järlfors and Smith (1969) in the lamprey larvae observed synaptic vesicles grouped around and attached to neurotubules in regular patterns in the axons, indicating some coupling between vesicles and tubules rather than budding off of vesicles.

By histochemical immunofluorescence methods Livett et al. (1968) has demonstrated the presence of chromogranin and DA-β-hydroxylase in adrenergic cell bodies. These two proteins were found to accumulate to about the same degree above a nerve ligation. These findings may indicate that amine storage granules (which contain the two proteins), accumulating proximal to a ligation, are formed in the cell bodies of the adrenergic neurons.

Experiments with reserpine have also given results that clearly suggest the perikaryon as the site of synthesis of amine storage granules. A large dose of reserpine blocks the storage mechanism of the granules for a long time (cf. Carlsson et al., 1963; Carlsson, 1965). It may even be that a large dose of reserpine (10 mg/kg i.p.) to rats irreversibly blocks the granules [cf. Dahlström and Häggendal, 1966 (2)]. After reserpine the amines in the central and peripheral nervous systems are depleted, probably due to leakage of the amines from the granules and enzymatic degradation of the extragranular amines. After the initial period of depletion, the recovery sets in. However, the recovery of NA in adrenergic neurons does not occur simultaneously in all parts of the neuron, but starts in the nerve cell bodies long before any recovery can be seen in the nerve terminals

[DAHLSTRÖM et al., 1965; DAHLSTRÖM and FUXE, 1965; NORBERG, 1965; DAHLSTRÖM, 1967 (1)]. In the fluorescence microscope the first reappearing NA fluorescence is seen in a perinuclear zone (Fig. 2a), which gradually widens, finally filling the cytoplasm. This reappeared fluorescence (as well as the reappeared fluorescence in all parts of the neuron) can be abolished by a second small dose of reserpine, or by tetrabenazine which is a short-lasting blocker of the granular storage mechanism. This indicates that the reappeared NA is bound to unblocked, functioning granules, or to some macromolecules having the same properties, regarding reserpine-sensitive amine storage, as the amine storage granules have.

The first reappeared NA fluorescence in the cell bodies could be seen at about 12 h after the administration of 10 mg/kg i.p. of reserpine to a rat. At 15 h clear fluorescence were seen in almost all cell bodies (Fig. 2). In the axons small amounts of NA could be detected at 15 h above a 1 h ligation of the sciatic nerve

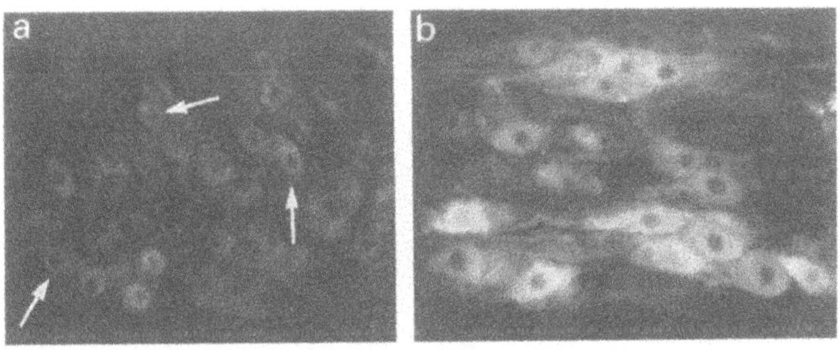

Fig. 2 a. A lumbar sympathetic ganglion from a rat given reserpine (10 mg/kg i.p.) 15 h before death. Very weak green NA-fluorescence can be seen in a narrow perinuclear zone in most cell bodies (→). b. A lumbar sympathetic ganglion from a normal rat. The whole perikaryal cytoplasm in the cells excert various intensities of NA-fluorescence. Only the nuclea are dark. Fluorescence microphotograph × 150

performed at a high level (Fig. 3). If the ligation was placed at a level 1.5 cm more distally, no NA fluorescence was seen at 15 h, but at 18 h small amounts were observed. At 18 h the amounts of NA fluorescence above the high level ligature were clearly increased as compared to the amounts seen at 15 h. As mentioned above, this reappeared NA could be abolished by a second injection of reserpine, indicating storage binding to some reserpine-sensitive sites, probably the storage granules [DAHLSTRÖM, 1967 (1)]. If the new granules were formed locally above the ligation, one would have expected the NA accumulation to be observed above any ligation at a certain time, regardless of the level. In this case a 3 h delay was observed with a level difference of 1.5 cm, indicating a transport rate of about 5 mm/h [DAHLSTRÖM, 1967 (1)], the same rate as was calculated in single ligated nerves without drug treatment [DAHLSTRÖM and HÄGGENDAL 1966 (1)].

Thus, after reserpine depletion due to a long-lasting blockade of the granular storage mechanism, the reappearance of NA, bound to a storage and protection protein, can be detected first in the cell bodies, some hours later above a high level, short term ligation, and some hours later again above a low level ligation. In the most distal parts of the neuron, the nerve terminals, no NA fluorescence has been detected before 24 h after reserpine (e.g. Norberg, 1965). These observations clearly indicate a proximo-distal migration from the cell body and distally, of proteins capable of storing NA, possibly amine storage granules. More electron microscopical data from reserpine treated animals appears necessary to confirm this view. For further description of experiments with reserpine and tetrabenazine on single and double ligated nerves see Dahlström [1967 (1)].

The above mentioned data indicate that new storage granules are formed in the adrenergic perikarya, transported down the axon at a rate of several mm/h, and reach the nerve terminal varicosities where they take part in the synthesis, storage, reuptake, and release of the transmitter. It must be pointed out, that the

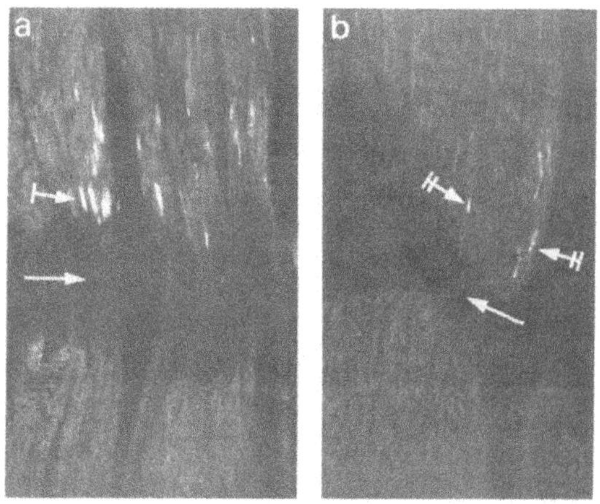

Fig. 3. Ligated sciatic nerves of a rat given reserpine (10 mg/kg i.p.) 18 h before death. The ligations (→), performed 1 h before death were placed at a high level on the left nerve (a), and 1—5 cm lower at the right nerve (b). The amount of NA accumulated above the high level ligation (a) is larger (|→) than above the low level ligation (b), where the NA fluorescence is very weak (||→). Fluorescence microphotographs × 68. [From Dahlström, 1967 (1)]

local synthesis of NA in the nerve terminals is by far the most important mechanism for maintenance and economy of the transmitter stores. However, *the factories* (i.e. the storage granules) for this synthesis (by the β-hydroxylation of DA) are probably formed in the perikarya and transported down the axons.

Newly formed granules are able to store NA soon after their formation in the cell body, this property enabling us to approximately calculate the flow and life-span of these granules by estimations of one of their contents—namely NA [Dahlström and Häggendal, 1966 (1), 1967]. The life-span of the storage granules in the nerve terminals has in recent recalculations, considering the degree of NA loading in the granules in different parts of the neuron, been found to be about 3 weeks in the rat [see e.g. Dahlström and Häggendal, 1970 (1, 2)]. This figure may represent the turnover of NA factories (granules), while the turnover of their product (NA) appears to be in the order of hours (e.g. Iversen, 1967).

II. Effect of Reserpine on the Synthesis and Transport of Amine Storage Granules

As mentioned above, the reappearance of NA fluorescence in axons above a ligation after reserpine can be seen at 15 to 18 h after the reserpine injection. This may indicate that at this time new functioning granules are being transported down the axon, although their number, as judged from fluorescence histochemical observations, is very low compared to that in the normal animal.

Quantitative chemical estimations of the amount of NA accumulating above a 6 h ligation of the rat sciatic nerve after reserpine reveal the time course curve seen in Fig. 4 (Dahlström and Häggendal, 1969). As mentioned above, it was

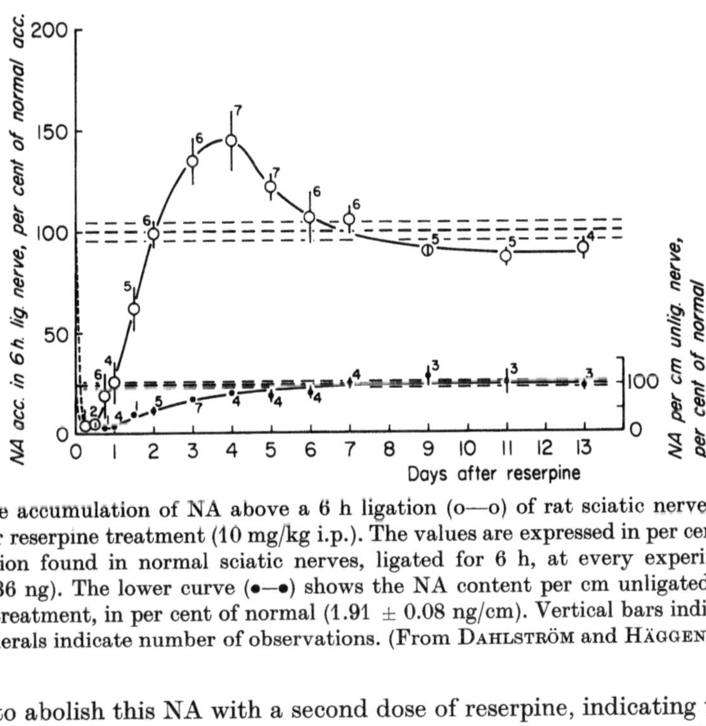

Fig. 4. The accumulation of NA above a 6 h ligation (o—o) of rat sciatic nerve at different times after reserpine treatment (10 mg/kg i.p.). The values are expressed in per cent of the NA accumulation found in normal sciatic nerves, ligated for 6 h, at every experimental serie (7,99 ± 0,36 ng). The lower curve (•—•) shows the NA content per cm unligated nerve after reserpine treatment, in per cent of normal (1.91 ± 0.08 ng/cm). Vertical bars indicate S.E.M., and numerals indicate number of observations. (From Dahlström and Häggendal, 1969)

possible to abolish this NA with a second dose of reserpine, indicating the storage to be in particular macromolecules, possibly amine storage granules. The amount of NA accumulated was thus considered to reflect the number of functioning granules transported to the nerve part above the ligature during 6 h. As seen from the figure, supranormal levels of NA was found to accumulate on the 3rd to 6th day after reserpine, reaching a maximum of about 145% of normal at day 4 (Fig. 4). The increase probably reflected an increase above normal in the amount of NA granules accumulated, and not an increased local NA synthesis above the ligation. In experiments with *delayed* double ligations (a low ligature was followed 6 h later by a high ligature tied 1 cm proximally) it was found that the NA content within the separated nerve part increased somewhat during the 9 h after the second ligation, and that this rise (probably due to an increased degree of NA loading of the granules) was about similar both in normal animals and in animals given reserpine 4 d beforehand (Dahlström et al., unpubl). Thus, the supranormal accumulations are probably not due to an increased local synthesis of NA, but may rather depend on an increased number of granules.

If an increased *number* of granules are transported down the axon for several days, a marked increase of production of granules must exist in the perikarya. An increased *rate* of transport may also occur during this period after reserpine as indicated by the subnormal NA content in the unligated nerve at day 4 when the accumulation of NA was about 145% of normal. An increased rate of transport alone can not explain the increased granule accumulation, since the distance between the cell bodies and the ligation is about 3 to 4 cm. With a transport rate of more than 5 mm/h the increased accumulation of granules would be declining already 6 to 8 h after the onset of the speculated increased flow rate. Therefore, an increased granule synthesis is likely to take place in the perikaryon.

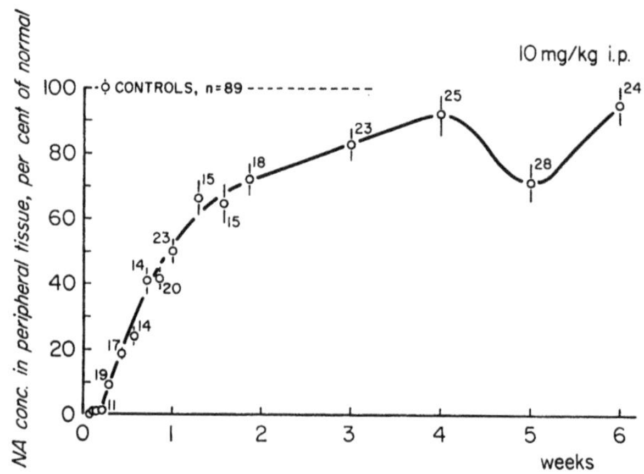

Fig. 5. The recovery of endogenous NA levels in rat peripheral tissues (heart, skeletal muscle and submandibular gland taken together) after one single dose of reserpine (10 mg/kg i.p.). The values are expressed in per cent of the NA content found in normal tissues at every experimental serie. Vertical bars indicate the S.E.M., and numerals indicate number of estimations. The drop in NA levels observed at week 5 is statistically significant from the values at weeks 4 and 6. During the first 7 to 10 days the rise in the NA levels is faster than during the following time, up to 4 weeks. (From Häggendal and Dahlström, 1970)

The basis for such an increase in production of the amine storage granules, essential for the function of the adrenergic neuron, may be a feed-back mechanism which starts to operate in a situation where the adrenergic transmission is impaired, as it is after reserpine treatment. The feed-back mechanism may for instance operate via the decrease in blood pressure, occurring after reserpine. This decreased blood pressure may induce an increased impulse activity in the preganglionic neurons. Such an increased impulse activity has been reported to occur in cat after reserpine treatment (Iggo and Vogt, 1960). Increased impulse activity has been shown to induce an increased protein synthesis in neurons (e.g. Hydén, 1960). An increase in the formation of the protein-containing amine storage granules may thus be expected to occur in the adrenergic neuron in this situation. Another protein in adrenergic neurons, tyrosine hydroxylase, has been shown to increase in sympathetic ganglia after reserpine treatment (Thoenen, et al., 1968, see also

AXELROD this volume). This increase in tyrosine hydroxylase could be prevented if the impulse flow to the ganglia was abolished by preganglionic denervation or administration of ganglion blocking agents.

The increased formation and transport of amine storage granules observed in the axons during the 3rd to 6th day after reserpine was reflected in the nerve terminals by a temporary increase in the rate of NA recovery in the nerve terminals (Fig. 5) (HÄGGENDAL and DAHLSTRÖM, 1970). The normalization or even slight decrease of the amount of granules transported down the axons, observed between the 7th to 13th day in the axons, was also followed by a decrease in the rate of NA recovery in the nerve terminals. These observations support the evidence for an increased formation and transport of granules during a certain time period after reserpine. For further discussion on NA recovery in reserpine treated animals see DAHLSTRÖM and HÄGGENDAL, (1970), HÄGGENDAL and DAHLSTRÖM (1970).

III. Effect of Colchicine and Vinblastine on the Transport of Amine Storage Granules

Colchicine

Colchicine has been found to depolymerize the microtubules of the mitotic spindle and thereby prevent mitosis. It also has the property of combining with the protein subunits of the neurotubules (possibly the same as the protein subunits in the mitotic spindle tubules) and causing a disruption of their organization (for references and discussion see SCHMITT, 1968).

It has been suggested that the neurotubules may take part in fast axoplasmic transport by different theoretical mechanisms (SCHMITT, 1968; WEISS, 1969). Since colchicine disrupts the neurotubules, the fast transport of e.g. amine storage granules should be inhibited if the neurotubules are essential for the transport. This was in fact found to be the case, provided that the concentration of the colchicine used was not less than 0.05 molar in saline (DAHLSTRÖM, 1968) for local applications.

Local application of the colchicine solution on the lumbar sympathetic ganglia of rat caused a highly increased NA fluorescence particularly in the intraganglionic axons and also in the cell bodies (Fig. 6a). These changes were already apparent some hours after the colchicine administration (Fig. 6c) (DAHLSTRÖM, unpublished observations). Often it could be seen that the increase in fluorescence intensity in the cell bodies was most prominent around the periphery. The cell bodies showed different intensities of fluorescence, some having extremely high intensities of fluorescence, and some having almost normal intensities. These differences may possibly reflect different stages of activities. In the sciatic nerve of rats treated this way 24 h beforehand, no or very little fluorescent material (NA) was observed to accumulate above a ligation performed 1 h before sacrifice (Fig. 7b). In control animals, the lumbar ganglia of which were treated with saline (Fig. 6b), normal amounts of NA accumulations were observed (Fig. 7a, DAHLSTRÖM, 1968). These observations indicate that colchicine when applied to the ganglia inhibits the flow of granules from the cell bodies and intraganglionic axons, and that no or very few granules could enter the adrenergic axons of the sciatic nerve, originating from the colchicine treated lumbar ganglia. Electronmicroscopic studies on

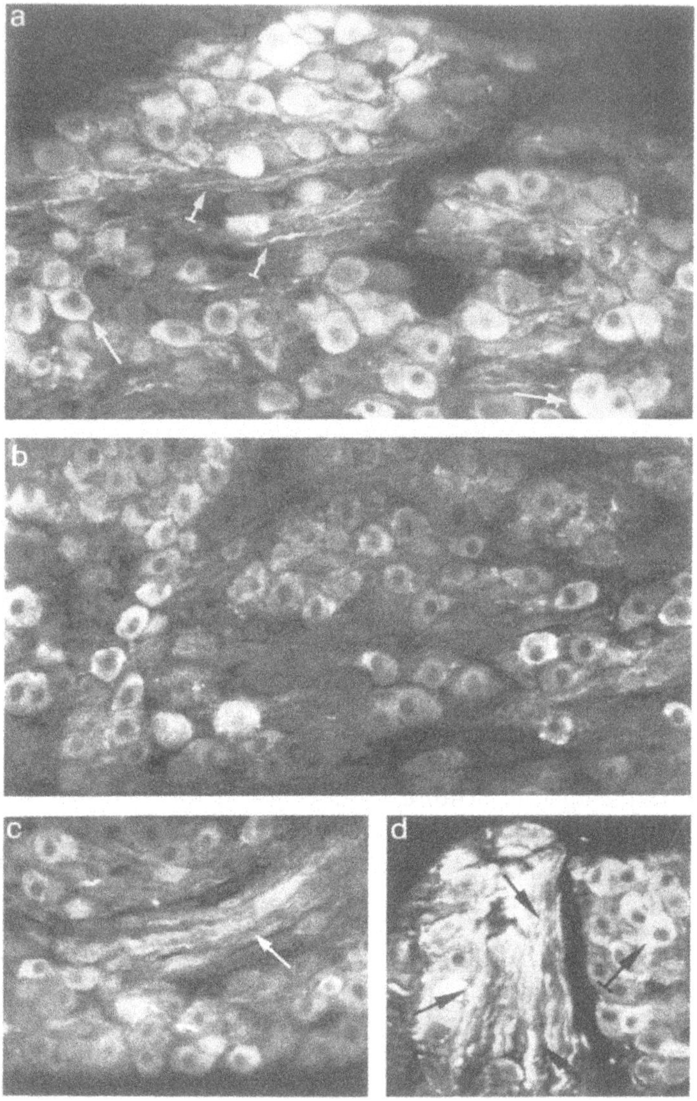

Fig. 6. Set of microphotographs demonstrating the effect of colchicine and vinblastine, locally applied to the lumbar sympathetic ganglia of rat. a) Ganglion treated with a 0.5 molar solution of colchicine for 10 min 24 h prior to sacrifice. Most nerve cell bodies show an increased fluorescence intensity as compared to that observed in saline treated ganglia (b). In many cells the NA fluorescence is concentrated to the periphery (→). The intraganglionic axons also excert an increased fluorescence intensity (|→). b) Ganglion treated with saline for 10 min 24 h prior to death. Weak to medium fluorescence intensities are seen in the nerve cell bodies. The intraganglionic axons have a weak to very weak fluorescence. c) Ganglion treated with a 0.5 molar solution of colchicine for 10 min 3 h before death. The intraganglionic axons show a clear increase in fluorescence intensity (→) while the nerve cell bodies have about normal intensities. d) Ganglion treated with a 0.001 molar solution of vinblastine for 10 min 24 h prior to sacrifice. The intraganglionic axons have very strong fluorescence intensities (→). Many nerve cell bodies with very strong fluorescence are observed (→). Fluorescence microphotographs × 160

colchicine treated axons in the sympathetic chain have demonstrated the accumulation of a large number of dense cored vesicles (together with other organelles) (HÖKFELT and DAHLSTRÖM, unpublished observations).

Local injection of colchicine solution under the epineurium of the sciatic nerve caused a tremendous increase in the fluorescence intensity of the adrenergic axons above and also within the area of the injection. The axons appeared enlarged and bloated, and had about the same appearance as the dilated axons above a ligation.

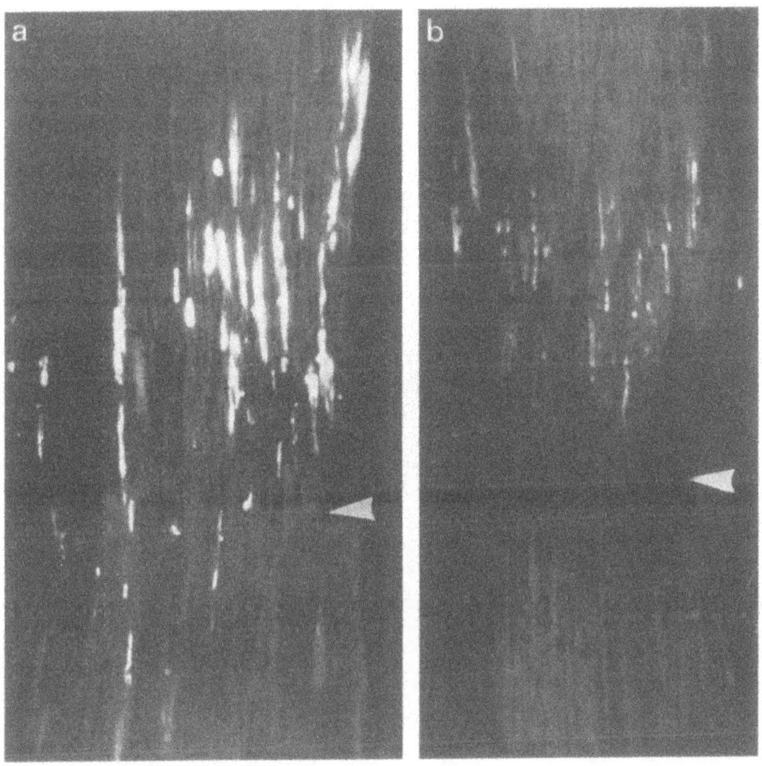

Fig. 7. The accumulation of NA above a 3 h ligation (→) in saline (a) and colchicine treated (b) rat. The colchicine solution (0.5 molar) was applied locally to the lumbar sympathetic ganglia (L2-L4) for 10 min 21 h before the ligation of the sciatic nerve. The NA accumulations seen in (b) are thin and of very weak fluorescence intensity as compared to the wide, strongly fluorescent NA accumulations observed in (a). Fluorescence microphotographs × 160

This was probably not due to mechanical pressure of the injection *per se*, since the perineural injection of a similar volume of saline produced no, or in some cases, a few weakly fluorescent accumulations of NA. When a ligation was performed 1 cm below the site of colchicine injection no or very little accumulation of NA was observed above ligation (DAHLSTRÖM, 1968). Thus, colchicine appeared to inhibit the transport of granules also if applied to the axons in the sciatic nerve.

The above mentioned results may support the view that the microtubules are essential for the fast proximo-distal transport of amine storage granules in adrenergic neurons. KREUTZBERG (1969) and KARLSSON and SJÖSTRAND (1969) have reported inhibition by colchicine of fast transport also in myelinated axons and optic tract. However, in our studies the concentrations of colchicine needed were rather high. A 0.005 molar solution caused no observable change, while the 0.05 molar solution (2%) induced moderate increases in fluorescence intensities in cell bodies and axons. The full effect appeared to be reached with a 0.5 molar solution. This high molar concentration is somewhat unsatisfactory, since it has been shown that colchicine antagonizes ATP, which takes part in any energy-dependent process, in different tissues (e.g. LANG et al., 1951). It has also been found that DNA synthesis may be effected by colchicine. However, this effect was seen only during a certain stage of mitosis (GUSTAVSSON, 1968), and may thus not interfere with the transport in adrenergic neurons, since these cells rarely divide.

Vinblastine

Another alkaloid, having pronounced effects on mitotic spindle tubules (MALAWISTA et al., 1968) and neurotubules in spinal cord motor neurons (WISNIEWSKI et al., 1968) is vinblastine. So far, no reports on ATP influence of this

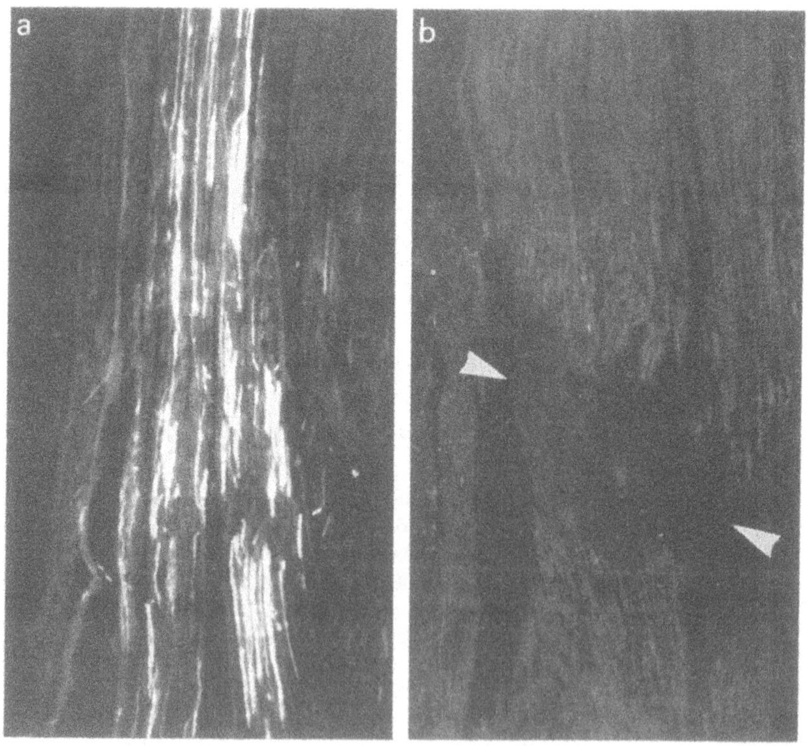

Fig. 8. The effect of an epineural injection of 0.001 molar vinblastine in the sciatic nerve of rat. The injection was made 24 h prior to death. a) The level of injection. Enlarged adrenergic axons with large amounts of strongly fluorescent material can be observed. b) The site of ligation, tied on the same nerve 1 h prior to death at a level 1 cm distally to the injection area. No accumulation of fluorescent material can be observed above the ligation (→). Fluorescence micrographs × 160

alkaloid seems to have been published. This alkaloid (Velbe, Lilly) was used for the same types of experiments as those performed with colchicine.

The same results in general were obtained with vinblastine as with colchicine. Local application of the solution caused a markedly increased fluorescence in the lumbar ganglia (Fig. 6d), and no or very small accumulations of NA was observed in the sciatic nerve of these animals. Perineural injection of about 1 μl solution in the sciatic nerve caused marked accumulations of NA in enlarged axons (Fig. 8a), and no NA fluorescence was seen proximal to a ligation of the same nerve performed 1 cm below the injection area 24 h later (Fig. 8b). Thus, also vinblastine appeared to inhibit the transport of storage granules (Dahlström, unpublished observations). However, in contrast to colchicine, vinblastine caused pronounced effects in very low concentrations. The above mentioned results were obtained with a 0.001 molar solution of vinblastine.

The similar effects of both colchicine and vinblastine on the transport of amine storage granules in adrenergic neurons may probably be related to the capacity of these alkaloids to disrupt neurotubules. This would thus support the view that the neurotubules are the essential structures for the fast proximo-distal transport of amine storage granules.

Summary

The fast proximo-distal transport of amine storage granules in adrenergic nerves is discussed. Evidence to indicate the perikaryon as the site of formation of these granules, is presented. Reserpine appears to induce a supranormal synthesis and down-transport of granules during a certain time period after the administration. This is discussed in relation to a possible feed-back mechanism. Colchicine and vinblastine, both potent mitosis inhibitors, are able to inhibit the transport of granules, presumably by the disruption of neurotubules, which may be the structural base for fast axo-plasmic transport of e.g. amine storage granules.

Acknowledgements

Support has been obtained by grants from the Swedish Medical Research Council (B69-14X-2207-03) from the Faculty of Medicine, University of Göteborg (G. & M. Lindgren Foundation and W. & M. Lundgren Foundation) and from Magnus Bergwalls Foundation.

For generous supplies of drugs I thank the Swedish CIBA, Stockholm (reserpine-Serpasil) and Lilly, Stockholm (vinblastine-Velbe).

For stimulating discussions I am indebted to Dr. J. Häggendal.

References

Andén, N.-E., Carlsson, A., Häggendal, J.: Adrenergic mechanisms. Ann. Rev. Pharmacol. **9**, 119—134 (1969).

Banks, P., Mangnall, D., Mayor, D.: The re-distribution of cytochrome oxidase, noradrenaline and adenosine triphosphate in adrenergic nerves constricted at two points. J. Physiol. (Lond.) **200**, 745—762 (1969).

Bloom, F. E., Aghajanian, G. K.: An electron microscopic analysis of large granular synaptic vesicles of the brain in relation to monoamine content. J. Pharmacol. exp. Ther. **159**, 261—273 (1968).

Boyle, F. C., Gillespie, J. S.: Relationship between the accumulation of noradrenaline and the development of fluorescence above a constriction in cat splenic nerves. J. Physiol. (Lond.) **195**, 27—28 P (1968).

Carlsson, A.: Drugs which block the storage of 5-hydroxytryptamine and related amines. In: Handbuch der exp. Pharmacol., pp. 529—592, (V. Erspamer, Ed.), Berlin-Heidelberg-New York: Springer 1965.

—, Hillarp, N.-Å., Waldeck, B.: Analysis of the Mg^{++}-ATP dependent storage mechanism in the amine granules of the adrenal medulla. Acta physiol. scand. **59**, Suppl. **215**, 1—38 (1963).

Corrodi, H., Jonsson, G.: The formaldehyde fluorescence method for the histochemical demonstration of biogenic amines. A review on the methodology. J. Histochem. Cytochem. **15**, 65—78 (1967).

Dahlström, A.: Observations on the accumulation of noradrenaline in the proximal and distal parts of peripheral adrenergic nerves after compression. J. Anat. (Lond.) **99**, 677—689 (1965).

— The intraneuronal distribution of noradrenaline and the transport and life-span of amine storage granules in the sympathetic adrenergic neuron. M. D. Thesis, Stockholm 1966.

— (1) The effect of reserpine and tetrabenazine on the accumulation of noradrenaline in the rat sciatic nerve after ligation. Acta physiol. scand. **69**, 167—179 (1967).

— (2) The transport of noradrenaline between two simultaneously performed ligations of the sciatic nerves of rat and cat. Acta physiol. scand. **69**, 158—166 (1967).

— Effect of colchicine on transport of amine storage granules in sympathetic nerves of rat. Europ. J. Pharmacol. **5**, 111—113 (1968).

—, Fuxe, K.: A method for the demonstration of adrenergic nerve fibres in peripheral nerves. Z. Zellforsch. **62**, 602—607 (1964).

— — Evidence for the existence of monoamine containing neurons in the central nervous system. II. Experimentally induced changes in the intraneuronal amine levels of bulbo-spinal neuron systems. Acta physiol. scand. **64**, Suppl. **247**, 1—36 (1965).

— —, Hillarp, N.-Å.: Site of action of reserpine. Acta pharmacol. (Kbh.) **22**, 277—292 (1965).

—, Häggendal, J.: (1) Studies on the transport and life-span of amine storage granules in a peripheral adrenergic neuron system. Acta physiol. scand. **67**, 278—288 (1966).

— — (2) Recovery of noradrenaline levels after reserpine compared with the life-span of amine storage granules in rat and rabbit. J. Pharm. Pharmacol. **18**, 750—753 (1966).

— — Studies on the transport and life-span of amine storage granules in the adrenergic neuron system of the rabbit sciatic nerve. Acta physiol. scand. **69**, 153—157 (1967).

— — Recovery of noradrenaline in adrenergic axons of rat sciatic nerves after reserpine treatment. J. Pharm. Pharmacol. **21**, 633—638 (1969).

— — (1) Axonal transport of amine storage granules in sympathetic adrenergic neurons. In: Biochemistry of simple neuronal models. (Costa, E., Giacobini, E., Eds.) New York: Raven Press 1970 (in press).

— — (2) A reevaluation of the life-span of amine storage granules with regard to their capacity to store endogenous noradrenaline in adrenergic nerve terminals of the rat. (1970) (In preparation).

—, Jonason, J.: DOPA-decarboxylase activity in sciatic nerves of the rat after constriction. Europ. J. Pharmacol. **4**, 377—383 (1968).

Droz, B.: Synthèse et transfert des protéines cellulaire dans les neurones ganglionnaires: étude radioautographique quantitative en microscopie électronique. J. Micr. **6**, 201—228 (1967).

Eränkö, O., Härkönen, M.: Effect of axon division on the distribution of noradrenaline and acetylcholinesterase in sympathetic neurons of the rat. Acta physiol. scand. **63**, 411—412 (1965).

Euler, U. S., Lishajko, F., Stjärne, L.: Cathecolamines and adenosine triphosphate in isolated adrenergic nerve granules. Acta physiol. scand. **59**, 495—496 (1963).

Geffen, L. B., Rush, R. A.: Transport of noradrenaline in sympathetic nerves and the effect of nerve impulses on its contribution to transmitter stores. J. Neurochem. **15**, 925—930 (1968).

Gustavsson, M.: Effect of colchicine on DNA synthesis in pleroceroids of *diphyllobothrium dendriticum* (Cestoda). Exp. Cell Res. **50**, 1—8 (1968).

Häggendal, J., Dahlström, A.: (1) Recovery of noradrenaline in adrenergic nerve terminals of the rat after reserpine. J. Pharm. Pharmacol. 1970 (in press).

—, Malmfors, T.: The effect of nerve stimulation on catecholamines taken up in adrenergic nerves after reserpine pretreatment. Acta physiol. scand. **75**, 33—38 (1969).

HÖKFELT, T.: In vitro studies on central and peripheral monoamine neurons at the ultrastructural level. Z. Zellforsch. **91**, 1—74 (1968).
— Distribution of noradrenaline storage particles in peripheral adrenergic neurons as revealed by electron microscopy. Acta physiol. scand. **76**, 427—440 (1969).
HYDÉN, H.: The neuron. In: The Cell, IV, 215—323 (BRACHET, J., MIRSKY, A., Eds.) New York-London: Academic press 1960.
IGGO, A., VOGT, M.: Preganglionic sympathetic activity in normal and in reserpine-treated cats. J. Physiol. (Lond.) **150**, 114—133 (1960).
IVERSEN, L. L.: The uptake and storage of noradrenaline in sympathetic adrenergic nerves. London: Cambridge University Press 1967.
KAPELLER, K., MAYOR, D.: Accumulation of noradrenaline proximal to the site of constriction of sympathetic nerves. J. Physiol. (Lond.) **182**, 44—45 P (1965).
— — Ultrastructural changes proximal to a constriction in sympathetic axons during first 24 hours after operation. J. Anat. (Lond.) **100**, 439—441 (1966).
— — The accumulation of noradrenaline in constricted sympathetic nerves as studied by fluorescence and electron microscopy. Proc. roy. Soc. B. **167**, 282—292 (1967).
KARLSSON, J.-O.: Transport of microtubule protein in the axon. (1970) (in preparation).
—, SJÖSTRAND, J.: The effect of colchicine on the axonal transport of protein in the optic nerve and tract of the rabbit. Brain Res. **13**, 617—619 (1969).
KAUFMANN, S., FRIEDMANN, S.: Dopamine-β-hydroxylase. Pharmacol. Rev. **17**, 71—100 (1965).
KIDWAI, A. M., OCHS, S.: Components of fast and slow phases of axoplasmic flow. J. Neurochem. **16**, 1105—1112 (1969).
KREUTZBERG, G.: Neuronal dynamics and axonal flow. IV. Blockage of intra-axonal enzyme transport by colchicine. Proc. nat. Acad. Sci. (Wash.) **62**, 722—728 (1969).
LADURON, P., BELPAIRE, F.: (1) Transport of noradrenaline and dopamine-β-hydroxylase in sympathetic nerves. Life Sci. **7**, 1—7 (1968).
— — (2) Evidence for an extragranular localization of tyrosine hydroxylase. Nature (Lond.) **217**, 1155—1156 (1968).
LANG, K., SIEBERT, G., ESTELMANN, T.: Hemmung der Adenosintriphosphatase durch Colchicin. Experientia (Basel) **7**, 379—380 (1951).
LIVETT, B. G., GEFFEN, L. B., RUSH, R. A.: Immunohistochemical evidence for the transport of dopamine-β-hydroxylase and a catecholamine binding protein in sympathetic nerves. Biochem. Pharmacol. **17**, 327—328 (1968).
MALAWISTA, S. E., BENSCH, K. G., SATO, H.: Vinblastine and griseofulvine reversibly disrupt the living mitotic spinde. Science **160**, 770—772 (1968).
NORBERG, K.-A.: Drug-induced changes in catecholamine levels in the sympathetic adrenergic ganglion cells and terminals. A histochemical study. Acta physiol. scand. **65**, 221—234 (1965).
OCHS, S., SABRI, M. I.: Somal synthesis and fast transport of labelled materials in nerve fibres. Proc. of the 2nd Meeting of the Internat. Soc. for Neurochem. 305—306 (1969).
— —, JOHNSON, J.: Fast transport system of materials in mammalian nerve fibres. Science **163**, 686—687 (1969).
PELLEGRINO DE IRALDI, A., DE ROBERTIS, E.: The neurotubular system of the axon and the origin of granulated and non-granulated vesicles in regenerating nerves. Z. Zellforsch. **87**, 330—344 (1968).
PHILIPPU, A., PFEIFFER, A., SCHÜMANN, H. J.: Amin-Aufnahme in Noradrenalin-speichernde Granula sympathischer Ganglien. Naunyn-Schmiedebergs Arch. Pharmak. exp. Path. **257**, 321 (1967).
RICHARDSON, K. G.: Electron microscopic identification of autonomic nerve endings. Nature (Lond.) **210**, 756 (1966).
SCHMITT, F. O.: The molecular biology of neural fibrous proteins. Neuroscl. Res. Progr. Bull. **6**, 119—144 (1968).
THOENEN, H., MUELLER, R. A., AXELROD, J.: Increased tyrosinehydroxylase activity after drug-induced alteration of sympathetic transmission. Nature (Lond.) **221**, 1246 (1969).
TRANZER, J. P., THOENEN, H.: Various types of amine-storing vesicles in peripheral adrenergic nerve terminals. Experientia (Basel) **24**, 484—486 (1968).
WEISS, P.: In: Cellular dynamics of the neuron. Cell Biol. (BARONDES, S., Ed.) 3—34 (1969).

WEISS, P, HISCOE, H.: Experimentals on the mechanism of nerve growth. J. exp. Zool. **107**, 315—396 (1948).
WISNIEWSKI, H., SHELANSKI, M. L., TERRY, R. D.: Effects of mitotic spindle inhibitors on neurotubules and neurofilaments in anterior horn cells. J. Cell Biol. **38**, 224—229 (1968).

Dr. A. DAHLSTRÖM
Institute of Neurobiology
University of Göteborg
Faculty of Medicine
Medicinargatan 5
Göteborg/Sweden

Discussion

GEFFEN: You quote increased impulse activity is an explanation for the overshoot in the amount of noradrenaline which accumulates above a ligation after reserpine. We have done the experiments which are actually the converse, that is, we have decentralized the ganglion. When we cut the preganglionic nerves to the spleen we found that decentralization for 24 h has no effect upon the accumulation of noradrenaline above a ligature on the splenic nerves, if anything, there is slightly more noradrenaline in the decentralized nerves than in the normal nerves. Do you think there is any explanation for this discrepancy ?

DAHLSTRÖM: I think it is possible that an acute decrease in the impulse flow cannot perhaps influence the synthesis or transport of amine granules—at least not to any pronounced degree. It may, however, be that a strongly *increased* impulse flow can have an effect on the synthesis of granules. The possible explanation for the effects of reserpine on the synthesis and transport of amine granules is based partially on the observations by IGGO and VOGT of an increased nerve activity in cervical sympathetic trunk of cat after reserpine treatment. I think that the explanation may be a reasonable one also in view of the findings by THOENEN, AXELROD and MUELLER, which agree with our findings.

GEFFEN: The fact that there is an increased production of tyrosine hydroxylase after nerve stimulation does not necessarily mean that the amine storage vesicles are also produced in increased amounts. After all, it is not certain that tyrosine hydroxylase is associated with the amine storage vesicles.

DAHLSTRÖM: Of course, tyrosine hydroxylase—as has been expected for a long time—is probably not associated with the amine storage granules. I think evidence has been given by LADURON and BELPAIRE that this enzyme has no connection with the storage granules. What I intended to say was the following: An increased production of proteins has been demonstrated in nerve cell bodies at increased stimulation. In adrenergic neurons increased amounts of (1) tyrosine hydroxylase, which is a protein and of (2) amine storage granules that are protein containing structures have been found in the nerve cell bodies after reserpine. In the case of tyrosine hydroxylase the increase was prevented by preganglionic denervation. I think that the same will be found to be the case with the storage granules when the experiment is done.

AXELROD: I would like to elaborate on this discussion. Dr. MOLINOFF, in our laboratory, has found that increase nerve impulse flow can elevate dopamine-β-oxidase in both the cell body and nerve terminals. This is of particular interest since this enzyme can serve as a marker for the noradrenaline storage vesicle.

THOENEN: The finding that dopamine-β-hydroxylase is present in a soluble form in the ganglion fits well into the general picture, in that the properties of the granules in the ganglion are quite different from those in the nerve endings. SNYDER and AXELROD have found that in the ganglion—in contrast to the periphery—the main part of the norepinephrine is not located in the microsomal fraction but in the supernatant and it has to be assumed that some alterations in the properties of the granules take place during transport to the periphery. I have a question to Dr. DAHLSTRÖM: Do you assume that a recovery from the effect of reserpine takes place in the following way: Reserpine produces irreversible damage to the storage vesicles and the recovery occurs by replacing the damaged vesicles by new ones transported down the axon to the periphery? You agree, this is correct? TRANZER and I have performed experiments to check this possibility. The experiments were based on earlier observations that 5-hydroxydopamine is a useful electronmicroscopic marker for the storage vesicles of the adrenergic nerve endings. After administration of this amine—in vivo or in vitro—every single vesicle is completely filled with a dense osmiophilic material. Pretreatment with reserpine prevents the accumulation of 5-hydroxydopamine; all the vesicles remain empty. If the assumption of a recovery by replacing the damaged vesicles by new ones is correct, one would expect in the time course of recovery two different vesicle populations: the new ones, which can be completely filled with 5-hydroxydopamine, and the irreversibly damaged ones, which remain empty. Our findings do not support this hypothesis since during the recovery we have found every kind of transitory stages, empty vesicles, vesicles with a few osmiophilic granules, half and completely filled vesicles. Our findings are more in favour of a recovery of the single vesicles than of irreversible damage and replacement by new ones.

DAHLSTRÖM: May I first ask what is the difference between the picture obtained with 5-hydroxydopamine in normal animals and in reserpine-treated animals?

THOENEN: Unfortunately I have no slides with me. In animals not pretreated with reserpine you can completely fill every single vesicle of the adrenergic nerve terminals with 5-hydroxydopamine. This is quite different from the pictures you obtain if the nerve endings contain the physiological transmitter norepinephrine, where you always have a few empty or half empty vesicles even under optimal fixation conditions. The unique properties of 5-hydroxydopamine should enable us to distinguish the two vesicle populations occurring in the recovery phase from treatment with reserpine.

AXELROD: I don't think anybody has asked the question—what is the population of the dense core vesicles in the cell body as compared with the nerve terminals?

DAHLSTRÖM: There appear to be many dense core vesicles in the nerve cell bodies, in fact surprisingly quite a lot, as observed by Dr. HÖKFELT in permanganate fixed tissues.

Axelrod: Are there dense core vesicles in the cell bodies of 500 Å?

Dahlström: Well, I do not exactly remember the figures in Dr. Hökfelt's study but I think that the percentage of large dense core vesicles in the nerve terminals was about 2% and the rest mainly small 500 Å in size. In the perikarya the number of large dense core vesicles was proportionally somewhat larger, say about 5%, while the rest were intermediate and mainly of the small type. The observation that the relative number of large dense core vesicles is larger in the perikarya may indicate that larger vesicles perhaps correspond to younger granules and thus, that the morphological appearance of the amine granules may vary with age. It is also possible that the function of the granules may vary with age. Some of the results obtained, together with Dr. Häggendal, seem to indicate that especially the capacity to take up amines may vary with the age of the granules.

Bayer-Symposium II, 39—41 (1970)

Invited Comment

L. B. GEFFEN

We have heard evidence that synaptic vesicles are formed in the soma of the neurone as well as the opposing view that they are formed locally in the axon terminals. In the discussion that follows, it is clearly necessary for us to distinguish synthesis and transport of components of organelles from assembly of their structure.

Our experiments on this problem were at first similar in principle to those described by DAHLSTRÖM. They were designed to examine both the contribution of noradrenaline transported along axons to transmitter stores in the axon terminals, and its dependence upon nerve impulses (GEFFEN and RUSH, 1968). We ligated one branch of the postganglionic sympathetic nerves to the cat spleen and left the other branch intact. The amount of noradrenaline which accumulated in the constricted axons in 24 h was only 1% of the total transmitter stores remaining in the portion of spleen supplied by these nerves. Abolition of reflex activity for a day by cutting the preganglionic nerves did not significantly alter the accumulation of noradrenaline in the ligated nerves and the noradrenaline contents of the two decentralized portions of spleen were the same. This suggests that transport of noradrenaline per se is quantitatively unimportant for the maintenance of transmitter stores and is not dependent on nerve impulses. The rate of axonal transport of noradrenaline calculated by the method of DAHLSTRÖM and HÄGGENDAL (1966) was between 1.4 and 3.3 mm/h, or about 1/3 the rate they obtained in the sympathetic fibres of the cat sciatic nerve. We do not attribute much significance to this disparity since a measured difference of 10^{-8} g/cm in the normal noradrenaline content of the nerves doubled the calculated flow rate. Moreover, it is not known whether the rate of transport is dependent on the length of the axons which vary greatly between, for example the sympathetic fibres of the sciatic, the splenic and the very short pelvic nerves.

Certain assumptions are necessary for the direct extrapolation from measurements of noradrenaline to calculations of the transport rate and the lifespan of the organelles that synthesize, take up, store and release the transmitter in the synaptic regions of the axon. They include:

1. that the proportion of vesicular to extra-vesicular noradrenaline remains unchanged throughout the neurone and at sites of injury;

2. that the amount of noradrenaline in each vesicle remains constant throughout its lifespan, increments due to uptake and synthesis being balanced by losses due to release and deamination;

3. that ligation does not alter either the production or the transport rates of the vesicles;

4. that noradrenaline is only transported in a somatofugal direction in the axon;

5. that other components of vesicles such as proteins, lipids and nucleotides have the same transport characteristics as noradrenaline;

6. that there is morphological evidence in the electron microscope of characteristic granular synaptic vesicles distributed throughout the neurone and accumulated proximal to the axonal ligature.

We therefore injected ^{14}C-noradrenaline directly into the coeliac ganglion to obviate some of these difficulties (LIVETT, GEFFEN and AUSTIN, 1968). The preganglionic nerves were cut to prevent release of noradrenaline by nerve impulses, and the cat was pretreated with iproniazid, a mono-amine oxidase inhibitor, to block intraneuronal deamination. A second tie was placed more distally on the ligated branch of the splenic nerves to control for uptake of ^{14}C-noradrenaline from the circulation at the site of injury. Local synthesis of ^{14}C-noradrenaline, was, of course, not possible in this experiment.

There was an accumulation of ^{14}C-noradrenaline within hours above the proximal but not the distal ligature. Since the method was essentially a "pulse label", the size of the accumulation was as dependent on the uncontrollable pulse duration as on the transport rate. However, in a few experiments of shorter duration, we detected a peak of radioactivity in the spenic nerves before it reached the constriction, travelling at about 5 mm/h. We repeated the experiments using ^{14}C-leucine and found labelled proteins transported in an identical way. In experiments of 2 to 3 weeks duration, where the nerves had regenerated past the ligature, it was still possible to detect these rapidly transported proteins persisting in the outgrowing axon sprouts while in addition, higher up the axon, there was a second component of ^{14}C-proteins which advanced at 1/100 the fast rate.

We then introduced ^{14}C-noradrenaline directly into the axon terminals of the spleen by close intra arterial infusion to determine whether it would also be transported somatopetally. Although the spleen was heavily labelled radioactivity did not accumulate on the distal side of a ligature on the splenic nerves (GEFFEN, HUNTER and RUSH, 1969). Since this could have been due to some degeneration of the distal portion of the axons, we also injected ^{14}C-noradrenaline into one eye of the sheep where it was taken up by the sympathetic nerve terminals but again there was no difference in the labelling of the cell-bodies in the contra and ipsilateral superior cervical ganglia.

Finally, I wish to refer briefly to some electron microscope studies on material from these experiments (GEFFEN and OSTBERG, 1969). We compared the sizes of granulated vesicles in the axon terminals of the cat splenic nerves with those found proximal to a ligation. In osmium fixed material, there were two populations in the terminals (diameters measured in Angström units $\pm$ S.D. of mean)—numerous small vesicles (443 $\pm$ 113) with small electron dense cores (192 $\pm$ 87) and occasional larger, more completely filled vesicles. Granulated vesicles were very sparse in the soma and in unligated axons, and were of the larger variety only. On the other hand, granular vesicles accumulated in very large numbers above an axonal constriction and were almost exclusively the large type (688 $\pm$ 136) with large dense cores (478 $\pm$ 102). Numerous large electron lucid vesicles were also present above the ligature. Like the granulated vesicles they possessed unit membranes and hence appeared to be derived from endoplasmic reticulum rather than neurotubules. Occasionally two granules joined by a dumbell shaped membrane were seen suggesting they were either budding or fusing.

Reserpine greatly reduced the noradrenaline content of the ligated axon as well as its terminals and depleted the granular cores of the vesicles, often leaving a central clear droplet.

These findings make it necessary to define the relationship of the two populations of vesicles before conclusions can be drawn about the origin of synaptic vesicles, let alone their fate. It seems probable, however, that noradrenaline in some situations may be located in large as well as small vesicles with electron dense cores. If the rapidly transported noradrenaline and proteins were located in these large vesicles, as seems likely from their identical transport rates and the distribution of both radioactivity and vesicles in ligated nerves, then the large variety may be the precursors of the smaller synaptic vesicles. However, it cannot be excluded that the large and small vesicles have separate origins and liberate different substances.

References

DAHLSTRÖM, A., HÄGGENDAL, J.: Studies on the transport and lifespan of amine storage granules in a peripheral adrenergic neuron system. Acta physiol. scand. **67**, 278—288 (1966).

GEFFEN, L. B., HUNTER, C., RUSH, R. A.: Is there bidirectional transport of noradrenaline in sympathetic nerves? J. Neurochem. **16**, 469—474 (1969).

—, OSTBERG, A.: Distribution of granular vesicles in normal and constricted sympathetic neurones. J. Physiol. (Lond.) **204**, 583—592 (1969).

—, RUSH, R. A.: Transport of noradrenaline in sympathetic nerves and the effect of nerve impulses on its contribution to transmitter stores. J. Neurochem. **15**, 925—930 (1968).

LIVETT, B. G., GEFFEN, L. B., AUSTIN, L.: Proximodistal transport of ^{14}C-noradrenaline and protein in sympathetic nerves. J. Neurochem. **15**, 931—938 (1968).

PELLEGRINO DE IRALDI, A., DE ROBERTIS, E.: The neurotubular system of the axon and the origin of granulated and non-granulated vesicles in regenerating nerves. Z. Zellforsch. **87**, 330—344 (1968).

Dr. L. B. GEFFEN
Department of Physiology
Monash University
Clayton, 3168, Victoria/Australia

Discussion

THOENEN: Dr. GEFFEN, what is the evidence that the large dense core vesicles above the ligature contain norepinephrine? Have you, for instance, used the bichromate method or 5-hydroxydopamine as a marker?

GEFFEN: When we gave reserpine to these animals, the vesicles lost their dense cores and the noradrenaline which had accumulated above the ligature also disappeared. This is at least one good indication that the large dense core vesicles contain noradrenaline—after all it is the primary evidence that the small dense-core vesicles contain noradrenaline.

THOENEN: It is very surprising that you have been able to deplete the dense core of the large vesicles by reserpine in such a convincing way, since this is not the case in the adrenergic nerve endings where reserpine does not produce a distinct change in the electron density of the osmiophilic material in the large vesicles, and it has been assumed that the dense core of the large vesicles represents a "reserpine-resistant" pool of norepinephrine. However, TRANZER and I have shown, by help of the bichromate method and with 5-hydroxydopamine as a marker, that the large vesicles contain an electron-dense material in addition to the adrenergic transmitter and that the norepinephrine contained in the large vesicles can also be depleted by reserpine.

GEFFEN: We are not the only ones who have shown this, KAPELLAR and MAYOR have done similar experiments. The dense core vesicles, which one finds above a ligature, are in such large numbers that it is quite easy to tell when their dense cores have been depleted.

DE ROBERTIS: I agree that the vesicles above the constriction are not of the type which are found in adrenergic nerve endings. However, a few medium size (i.e. 800 Å) dense core and also clear vesicles are found in all types of nerve endings. In compression all the nerve types present in the sciatic nerve react in the same way. The cholinergic, the adrenergic, the motor and the sensory, every axon has a similar reaction. I think that this is an important point which should be taken into consideration.

GEFFEN: I fully accept the proposition that there is more than one species of dense core vesicle and that large dense-core vesicles occur in *non* adrenergic nerves. The only point that I wish to make is that throughout nerves which are unequivocally adrenergic, only the large species of dense core vesicle has the same ubiquitous distribution as noradrenaline. I do not think the dense core is peculiar to the catecholamines but in this particular instance, the vesicles which we have in abundance are depleted by reserpine in the same way as the small granular vesicles. I take Dr. THOENEN's point that there are now more specific histochemical methods for doing this and it ought to be repeated, using those methods.

PELLEGRINO DE IRALDI: I would agree that similar vesicles are found in the neurosecretory systems which are neither cholinergic nor adrenergic, in the hypothalamus, where they are called elementary neurosecretory vesicles or granules. I would also like to say that vesicles of about 800 Å in diameter present in adrenergic and cholinergic nerves react in a different way to some histochemical technique, like that of WOOD. Such vesicles are positive in the adrenergic nerves but negative in the nonadrenergic ones. This does not mean, of course, that noradrenaline is not present in these vesicles. The amine may be in a different storage and could be lost during the histochemical technique.

KIRSHNER: I would like to ask Professor DE ROBERTIS about the origin of the vesicle components that he sees,—Dr. GEFFEN mentioned this in his first statements—and about the possibilities that the vesicles one sees might be due to formation from existing soluble compounds after injury or after dissolution of the existing membranes—recombination of the components to form new vesicles rather than just a vesicle budding off of pre-existing membranes.

DE ROBERTIS: I would like to raise a point which may answer this question. Together with Dr. LUNT and LAPETINA, we have measured the half-life of synaptic vesicles in brain by injecting C-14-choline and determining the radioactive phosphatidylcholine. It was found that the synaptic vesicles have a half-life of about

30 days, while the nerve ending membranes have a much longer half-life. At the same time all the measurements which have been done on the half-life of the different transmitters may be expressed in minutes. This is a very important point which I think will come out during further discussions about this problem, because it shows that the vesicle, as such, should be re-used many times before it disintegrates.

CARLSSON: When I came here I had what I thought was a relatively clear picture of these electronmicroscopical phenomena but now I must say that I am a bit confused and I should like to ask the electronmicroscopists about the role of the fixation technique. Dr. HÖKFELT told me that this problem of dense-core versus non-dense-core vesicles is very much dependant on the technique of fixation. If he uses potassium permanganate then he gets results that are specific from the point of view of monoamines, but if you use other techniques you will end up with these rather unspecific phenomena, dense core vesicles in all kinds of nerves. I would like to ask whether you agree on these points or whether it is a point of disagreement?

GEFFEN: It seems to me that one way in which we can resolve this problem of the fixative is by electronmicroscope-autoradiography. By this method, DE SCARRIES and DROZ have shown that in the central nervous system, fixation with potassium permanganate does not retain radioactivity due to noradrenaline—whereas glutaraldehyde/osmium or the glutaraldehyde/dichromate methods both produced a core which is associated with radioactivity due to noradrenaline.

CARLSSON: I do not think this is a decisive point really because it could very well be that the phenomenon of dense core is indeed related to the presence or absence of catecholamines, and yet that your radioactivity experiment will turn negative because it is not necessarily so that the dense core phenomenon is due to the presence of catecholamines at the time when you look at the phenomenon in the electronmicroscope. It may have been there at the time of the histochemical reaction but when you do your radioactivity measurement the amine has left. Another possibility, I would like to ask you about is, whether it could not possibly be so that this dense core material is an aggregate in which you have the radioactive material, you then get absorption of the radiation from the surrounding material, so that your reaction will not become positive in the autoradiograph.

DE ROBERTIS: One of the best histochemical reaction for norepinephrine which can be used is the one shown in this slide which was made in our laboratory with the dichromate technique of WOOD. In this adrenergic nerve from rat pineal all dense core vesicles shown a positive reaction with the chromate. Since this chromate reaction is specific for norepinephrine the depleting action of reserpine may be demonstrated. In fact, the reaction becomes negative in the vesicles. We have information that in the compressed nerve above the ligature in the sciatic nerve, most of the large vesicles do not react with the chromate reaction.

KIRSHNER: I should like to mention the report by LADURON and BELPAIRE where they find that dopamine hydroxylase also accumulates above ligatures. This observation is important for the transport of specific vesicle proteins down the axon.

Bayer-Symposium II, 44 (1970)

Session 2a

Mechanism of Transmitter Release with Special Reference to the Problem of Exocytosis (I)

Chairman: E. de Robertis

Opening Remarks

E. DE ROBERTIS

Because I think that electronmicroscopists are in a very small minority in this audience, I would like to convey to you the idea of what exocytosis means from the point of view of ultrastructure. This process was first observed by us in the adrenal gland. We found, with VAZ FERREIRA in 1957, that when the gland is stimulated through the splanchnic nerve the first changes that occur in the catecholamine containing vesicles (or granules) are the following: the granule attaches to the plasma membrane and it swells; at this moment there is no distinction between the clear zone on the outside and the dense granule inside; then there is a decrease in the amount of material contained in the granule. The material flows out but the membrane of the granule remains within the adrenal cell. It has been very interesting to see throughout the years that, thanks to the work done mainly in Professor BLASCHKO's laboratory by several workers and also by Professor DOUGLAS and Dr. KIRSHNER, this mechanism of exocytosis has been confirmed biochemically. In fact, it has been shown that not only the norepinephrine content is released but also that ATP and the chromogranins are released together when the gland is stimulated. Today's session will deal with some of these problems and the first paper will refer to the chromogranins.

Dr. HELLE will present her paper on the Immunological properties of chromogranin.

Bayer-Symposium II, 45—56 (1970)

Immunological Properties of Chromogranin

Karen B. Helle

With 5 Figures

In the search for a specific macromolecule involved in the retention of catecholamines and ATP within the chromaffin granule attention has been directed to the study of the water-soluble proteins of these cellular structures.

When chromaffin granules are subjected to hypotonic conditions they lyse and release about 70% of their total protein in a water-soluble form (Eade, 1957). The first observation on the chemical properties of the water-soluble protein fraction was made by Hillarp in 1958 when he showed that the bulk of the lysate protein has an isoelectric point close to pH 4.0. By isoelectric precipitation in 50% ethanol Hillarp isolated a protein fraction which appeared homogenous by paper electrophoresis. Making use of protein fractionation on Sephadex columns, Blaschko and Helle (1963) were able to separate the ethanol-precipitated protein into two main fractions, a major one with isoelectric point near pH 4.1 and a minor one with pH 6.0. Surprisingly the protein material in the two peaks did not differ in their electrophoretic patterns [Helle, 1966 (1)]. This phenomenom was explained by assuming that the fraction with isoelectric point at pH 6.0 contained firmly bound cations reducing the number of accessible anionic charges on the protein, but that the latter in the fully dissociated state had an isoelectric point at pH 4.1.

The amino acid analysis of the ethanol-precipitated protein is given in Table 1, together with that of other preparations of the major lysate protein component. It may be seen that the amino acid composition of the different preparations is closely similar, and that the protein is rather unique in its high content of glutamyl and prolyl residues and its low values of cysteine and cystine.

Although Hillarp (1958) had reported an adenylate kinase activity for the ethanol-precipitated protein such an activity could not be detected in our preparations.

More gentle procedures for the purification of the lysate protein were used by Kirshner and his group (1966, 1967) and by Smith and Winkler (1965, 1967), in which the ethanol-precipitation step was omitted. In these studies a major component of the water-soluble protein fraction could be isolated by means of ion-exchange and molecular sieve chromatography which appeared very similar to the ethanol-precipitated protein in amino acid composition (see Table 1) but differed in sedimentation properties and molecular weight.

Chromogranin

The term "chromogranins" was suggested by Blaschko et al. (1967) as a common name for all the protein components of the granule lysate and the major component characterized by its unique amino acid composition (see Table 1) was lettered "A". However, in view of recent reports on the detection of the enzyme dopamine-β-hydroxylase as one of the water-soluble protein components (Laduron and Belpaire, 1968; Viveros et al., 1969) it seems advisable to restrict the term

Table 1. *Amino acid composition of the major lysate protein*

Amino acid	g amino acid/100 g protein			
	C^{+}	A^{++}	S_1	VI-I
1/2Cys	0.7	0.6	0.6	0
Asp	8.5	8.4	8.5	8.4
Thr	2.5	2.5	2.5	2.5
Ser	6.2	6.2	6.1	5.8
Pro	9.3	8.7	8.0	8.2
Glu	26.8	26.0	27.3	25.4
Gly	4.0	3.9	3.5	4.3
Ala	5.4	5.0	5.3	5.4
Val	3.7	3.3	3.5	3.4
Met	2.1	2.2	2.3	1.6
Leu	7.4	7.3	6.5	7.2
Ile	1.7	1.1	1.4	1.3
Tyr	2.1	1.7	1.2	2.1
Phe	2.5	2.1	1.8	2.3
Lys	9.2	9.4	10.2	8.3
His	2.5	2.4	2.0	2.4
Arg	9.0	8.5	7.0	9.8

C^{+} The preparation of chromogranin used as antigen in the preparation of antisera AC3 and AC2 in the present work.
A^{++} Chromogranin A, as prepared by Smith and Winkler (1967).
S_1 The vesicle protein S_1 prepared by Kirshner et al. (1965).
VI-I The major lysate protein VI-I, isolated by ethanolprecipitation at pH 4.0 [Helle, 1966 (4)].

"chromogranins" to the major lysate protein component and to subunits and aggregates thereof. In the following the major protein component of the lysate accordingly will be referred to by the term "chromogranin".

Immunological Assay of Chromogranin

In order to overcome the difficulties inherent in assaying a protein devoid of enzymatic activity an immunological assay was introduced [Banks and Helle, 1965; Helle, 1966 (2)]. Antisera against a protein may serve two purposes: first they provide a test for the purity of the antigen, and secondly they lend themselves as means of a convenient, sensitive and specific assay for a non-enzymatic protein.

Three different antigen preparations were used for immunisation in order to test the purity of the antigen. In Table 2 are listed the antisera made against the three preparations of chromogranin. In accordance with immunological practice each antigen was used to prepare two antisera.

In the first preparation of antigen [Helle, 1966 (2)] the phospholipids appearing in the void volume of the Sephadex G-200 column (peak I, Fig. 1) were not separated from the peak containing chromogranin (peak II, Fig. 1). Amino acid analyses of the small amounts of protein in the phospholipid-rich peak have indicated that the protein material in this fraction was identical in composition to that of chromogranin (Banks et al., 1969). The antigen preparations differed in the temperature at which they were stored. Since electrophoresis experiments using polyacrylamide gels had shown that repeated freezing and thawing caused a change in the electrophoretic mobility pattern of the antigen, antisera were made

Table 2. *Preparations of chromogranin used as antigen and the corresponding immune sera*

Antigen preparation	Temp. of storage	Immune sera
1. Protein peak I + II [Helle, 1966 (1)]	—20°	AC 1 and AC 2
2. Protein peak II	—20°	AC 3 and AC 4
3. Protein peak II	+ 4°	AC 5 and AC 6

Each antigen was used to produce antisera from two rabbits. The peak numbers refer to the peaks of the chromatogram obtained when the granule lysate was passed through columns of Sephadex G—200, as described in Fig. 1.

not only against an antigen preparation stored at —20° but also against one which had not been stored in the frozen state.

Only one antibody reactive against the antigen solution could be detected since only one precipitin arch was obtained by immunelectrophoresis with each of the six antisera. In Fig. 2 (A) such a pattern is given. Immunelectrophoretic patterns obtained with crude granule lysate as the source of chromogranin did not reveal any precipitin line near the well, even at high protein concentration. Such a line has been described by Sage et al. (1967) to be due to an antibody against dopamine-β-hydroxylase present in trace amounts in their preparations of chromogranin. The pattern of precipitin lines obtained with the six antisera on an Ouchterlony plate against crude granule lysate is given in Fig. 2 (B) and it may be seen that a pattern of complete identity was obtained for the six antisera. A small amount of contaminating protein that might have been present when the unresolved peaks I and II were used for immunisation (antisera AC 1 and AC 2) would have made itself evident as a spur in the diffusion pattern obtained against the highly concentrated crude lysate.

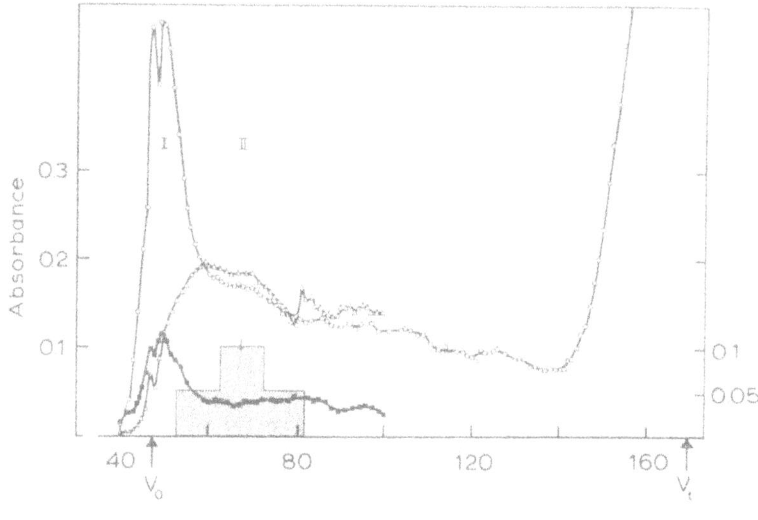

Fig. 1. *Molecular sieve chromatography of the granule lysate protein.* Concentrated granule lysate was applied to a column of Sephadex G-200. The column was eluted with sodium succinate buffer of pH 5.95 and fractions containing 4.5 ml eluate were collected at a rate of 11 ml/h. The void volume (V_0) was 216 ml and the total volume was 720 ml. Eluate fractions were assayed for material absorbing at 280 nm (o—o), for protein by the Folin method (Δ—Δ), for organic phosphate by the Fiske-SubbaRow method after digestion of the sample in 10N H_2SO_4 (■—■). Chromogranin reactive protein was assayed immunologically using the antiserum AC 1 (▩)

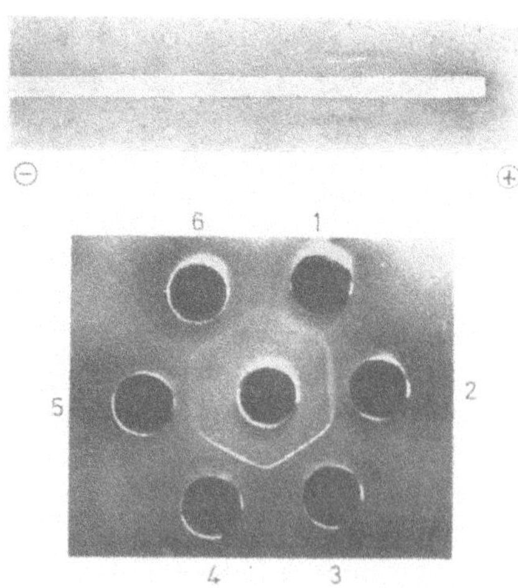

Fig. 2. *Precipitin patterns obtained by immunelectrophoresis and double diffusion of chromogranin with the six antisera.* A) Immunelectrophoresis pattern obtained with purified chromogranin (antigen preparation 2, Table 2) at 0.7 mg protein/ml in the wells and immune serum AC 3 in the trough. B) The double diffusion pattern obtained when the six antisera were allowed to diffuse against crude granule lysate at 12.5 mg protein/ml in the center well. The wells numbered 1 to 6 contained the immune sera AC 1 to 6

In view of these observations it can be concluded that the antisera produced were reactive specifically against chromogranin.

In our work we have used the immunological assay for both qualitative and quantitative purposes. For quantitative estimations of chromogranin the protein solutions have been titrated by means of the double diffusion technique using 1% agarose for the Ouchterlony plates. This method does allow a detection of 2.5 ± 1 μg chromogranin in a total volume of 50 μl. A more sensitive assay employing the microcomplement fixation technique has been used by SAGE et al. (1967) and by SCHNEIDER et al. (1967). This method allows the detection of as little as 0.025 μg chromogranin.

Chromogranin in the Granule Membrane

During the lysis procedure conventionally used to collect the water-soluble fraction, the chromaffin granules obtained as a pellet in 1.6 M sucrose were resuspended in 10 volumes of 170 mM saline. The sediment of whole granules and membranes of lysed particles was washed twice in 10 volumes of the same solvent. The last wash was usually free of immunologically reactive protein. When the pink sediment obtained from lysed granules was solubilised in detergents such as Triton-X-100 or sodium deoxycholate, we were astonished by the high titer of chromogranin that we found in the sedimented fraction.

In a number of experiments we have been able to verify that the water-soluble protein hitherto known to be a specific marker of the interior of the chromaffin granule is also a conspicuous component of the water-insoluble granule membrane (HELLE and SERCK-HANSSEN, 1969; SERCK-HANSSEN, 1969).

With two different methods of lysis, by sonication and by osmotic shock, the solubilisation of catecholamines was 98 and 77% respectively. The amount of chromogranin detected in the lysates was 37 to 40% of the total lysate protein, a figure which is in good agreement with that previously reported (see review by SMITH, 1968).

After lysis, either by sonication or by osmotic shock, the pellets were solubilised in 1% sodium deoxycholate. Of the protein present in these fractions 73 to 77% could be accounted for as immunologically reactive chromogranin.

In studies of other biological membranes the importance of divalent cations such as Mg^{++} and Ca^{++} for the stability of membrane structure has been pointed out (MUNOZ et al., 1968). Removal of divalent cations leads to a dissociation of the membrane structure. The authors have shown that this effect, when accompanied by a reduction in ionic strength, results in a release of membrane-associated enzyme activities. Somewhat analogous conditions were chosen for the lysis of chromaffin granules: 170 mM saline containing 18 mM $MgCl_2$ was used to lyse chromaffin granules obtained as a pellet in 1.6 M sucrose. The lysis procedure was in other respects similar to that already described (see page 5). As shown in Table 3 the first supernatant (SN 1) and the two supernatants obtained when the pellet was washed twice with the same solvent (SN 2 and SN 3) contained 95% of the catecholamines and 70% of the protein of the whole granule. While fractions SN 1 and SN 2 contained 24 and 21% respectively of the protein as immunologically reactive chromogranin, SN 3 was found to be free of detectable amounts of chromogranin. The washed pellet was then treated in a hypotonic medium containing 0.02% EDTA in 50 mM TES buffer of pH 6.5. After centrifugation more protein and chromogranin was discovered in the supernatant fluid (SN 4) in a ratio similar to that found in the earlier washings. After repeating the treatment with the same hypotonic medium and removal of a supernatant fluid (SN 5), sodium deoxycholate was added to the remaining pellet. Table 3 shows that the solubilized pellet (S 6) also obtained chromogranin. Fig. 3 shows the electrophoretic patterns obtained. It can be seen that the sample SN 3 which gave no titer for chromogranin by immunological technique did not show the electrophoretic pattern indicative of the presence of this protein. However, in agreement with

Table 3. *Solubilisation of granule constituents by stepwise reduction in tonicity*

Fraction	Medium	10^{-6} moles CA / mg protein	mg chromogranin / mg protein	Recovery in % of total: Ca	protein	chromogranin
SN 1[a]	170 mM NaCl + 18 mM $MgCl_2$	5.5	0.24	52	38	40
SN 2	170 mM NaCl + 18 mM $MgCl_2$	4.9	0.21	27	22	20
SN 3	170 mM NaCl + 18 mM $MgCl_2$	5.8	0	16	10	0
SN 4	50 mM TES[a] + 0.02% EDTA pH 6.5	1.5	0.26	5	14	16
SN 5	50 mM TES[a] + 0.02% EDTA pH 6.5	0	0.77	0	2	8
S 6	2% DOC +170 mM NaCl	0	0.26	0	14	16

[a] The supernatants were obtained after centrifugation at 130000 × g for 30 min.
CA = catecholamines determined fluorimetrically.
TES = N-tris (hydroxymetyl)metyl-2-aminoethane sulfonicacid.
DOC = sodium deoxycholate.

immunological titrations, subsequent treatment with the hypotonic medium (SN 4 and SN 5) yielded the electrophoretic pattern characteristic of the granule lysate.

Fig. 4 shows an experiment in which samples of SN 1, SN 4 and S 6 were tested on an Ouchterlony plate. The precipitin lines indicate immunological identity in these three fractions; there is no sign of the presence of any contaminants.

These observations taken together lead to the conclusion that chromogranin constitutes a considerable amount of the water-insoluble protein of the granule membrane. Another protein constituent of the chromaffin granule dopamine-β-hydroxylase was at first believed to be confined solely to the granule membrane (LEVIN et al., 1960). However, recently this enzyme has been shown to be present in a soluble state as part of the water-soluble protein of the granule lysate (LADU-

RON and BELPAIRE, 1968; VIVEROS et al., 1969). The distribution of chromogranin over both the water-soluble and membrane protein fraction is therefore analogous to that of another protein component characteristic of the chromaffin granule.

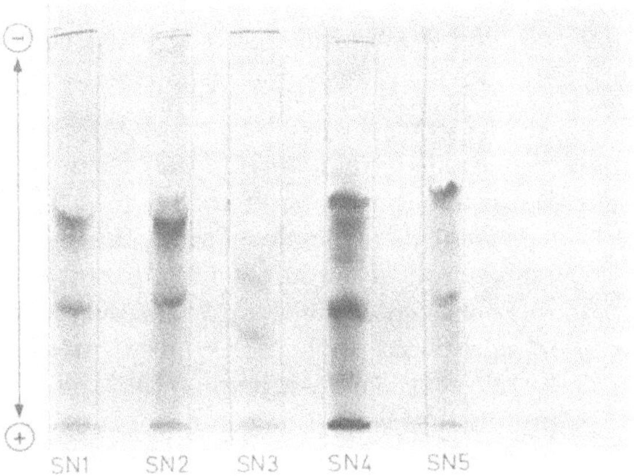

Fig. 3

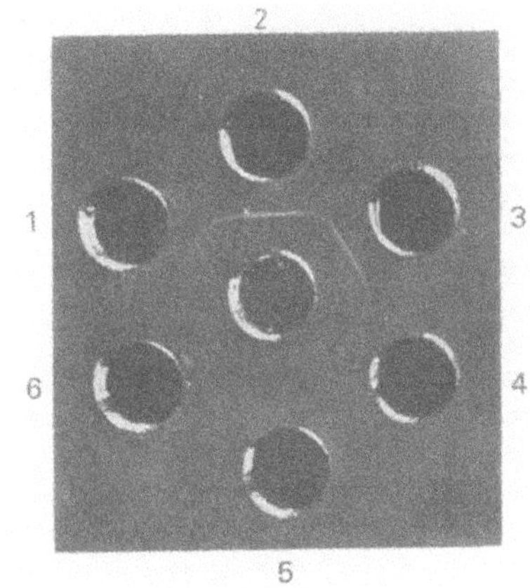

Fig. 4

Fig. 3. *Disc electrophoresis patterns of granule fractions.* Water-soluble granule fractions obtained by stepwise reduction in tonicity of the suspension medium were subjected to disc electrophoresis on 7.5% polyacrylamide gels, at pH 8.9. The samples were applied to the top and moved towards the anode. SN 1, SN 2, SN 3, SN 4, and SN 5 represent the supernatants obtained when chromaffin granules were subjected to lysis as described in the text. The content of chromogranin in these fractions are given in Table 3

Fig. 4. *Immunodiffusion patterns of granule fractions.* The water-soluble granule fractions SN 1 and SN 4 and the water-insoluble pellet after solubilisation in 2% deoxycholate (S 6) were allowed to diffuse against antiserum to chromogranin (AC 3) in the center well. Wells 1, 2 and 3 contained S 6, SN 1 and SN 4 at protein concentration 2 mg/ml respectively, while wells 4, 5 and 6 contained SN 1, S 6 and SN 4 at 0.25 mg/ml

Chromogranin in Nervous Tissue

In the search for chromogranin-like protein in tissues other than the adrenal medulla, HOPWOOD (1967) was able to detect cross-reacting protein in the brainstem and splenic nerve. The cross-reacting protein was observed by immunohistochemical methods using an antiserum against whole chromaffin granules of the ox.

Chromaffin granules of the adrenal medulla and the amine-storing particles of sympathetic nerve fibres show a number of biochemical similarities. For instance, the molar ratios of catecholamines to ATP in these two types of organelles are similar. Secondly the immunological method has revealed structural similarity in the protein obtained from these tissues. These considerations led us to a search for chromogranin-like protein associated with isolated amine-storing particles of the splenic nerve.

In the particulate fraction prepared from press juice of bovine splenic nerves by differential centrifugation we obtained evidence for the presence of a protein with immunological properties identical to that of purified chromogranin (BANKS et al., 1969). Our observation thus reinforces and supplements the finding of HOPWOOD (1967, 1968). The concentration of chromogranin-like material in the nerve granule lysate was estimated to be 6% of the total lysate protein. Assuming that all of the noradrenaline detected in the particulate fraction, 5 nmoles/mg total protein, was associated with the chromogranin-like protein in the particles these granules would be responsible for the accomodation of 80 nmoles of noradrenaline/mg of chromogranin. This is a figure much lower than that obtained for the chromaffin granules of the adrenal, and it was therefore suggested that only part of the chromogranin-containing nerve granules was associated with noradrenaline.

More recently the particulate fraction obtained from press juice of splenic nerves has been sedimented on a sucrose gradient ranging from 0.6 to 1.2 M sucrose. Under these conditions the highest amount of noradrenaline/mg protein was obtained at the level of 1.0 M sucrose, in agreement with earlier observations by Potter and Axelrod (1963). The protein in this fraction contained 2.3% as immunologically reactive chromogranin after treatment with Triton-X-100. The amount of chromogranin associated with the peak of noradrenaline in the gradient accounted for 67% of the total chromogranin in the particulate fraction (Helle and Serck-Hanssen, unpublished observations). The chromogranin-like protein detected in the noradrenaline peak of the gradient was tested immunochemically on an Ouchterlony plate and gave a pattern of complete identity with that of chromogranin; this is shown in Fig. 5. In this fraction the ratio of noradrenaline to protein was 13 nmoles/mg and the ratio of noradrenaline to chromogranin was 660 nmoles/mg.

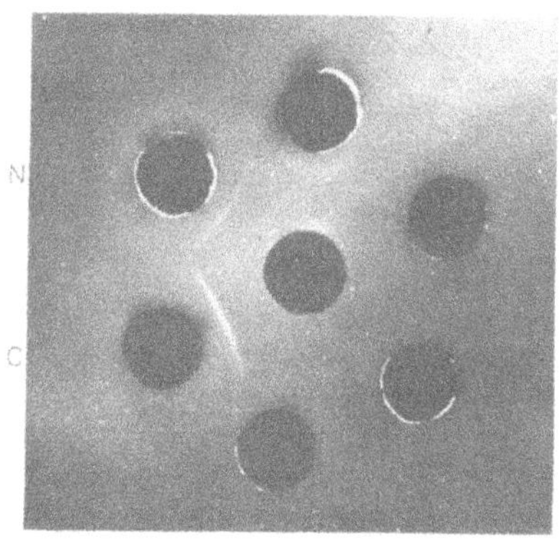

Fig. 5. *Cross-reaction of nerve granule protein with anti-chromogranin serum.* Immundiffusion was carried out on an Ouchterlony plate in which the center well contained antiserum against chromogranin (AC 3). Well C contained purified chromogranin and well N contained Triton-solubilized protein obtained from the noradrenaline rich particles isolated on the sucrose gradient at the level of 1.0 M sucrose

These observations support the results obtained by Banks et al. (1969) in that chromogranin-containing granules apparently are not fully loaded with transmitter while on their way down to the nerve terminals.

Recent reports by Livett et al. (1969) have confirmed our observations on the presence of chromogranin-like protein in the splenic nerve. The Australian group used rabbit antisera against both bovine chromogranin and dopamine-β-hydroxylase from the adrenal medullary chromaffin granules and an immunohistochemical method for the detection of these two proteins. They found both proteins in high concentrations in the superior cervical and stellate ganglia of the sheep and in lower concentrations in the sympathetic fibres emanating from these ganglia. In constricted nerves these two proteins were shown to accumulate proximal to the constriction, an observation indicative of a proximo-distal transport of both chromogranin and dopamine-β-hydroxylase.

Chromogranin-like Protein in Species other than the Ox

The water-soluble protein fraction of the chromaffin granules from the adrenal medulla has been studied in a number of species.

Immunochemically it has been shown that a structural homology exists between chromogranin of the ox and that of the sheep, pig and horse [HELLE, 1966 (3); HOPWOOD, 1967]. A pattern of complete identity was obtained between chromogranin of the ox and the sheep, while a pattern of partial identity was obtained between the water-soluble proteins of either horse or pig and chromogranin of the ox. A cross-reaction between bovine chromogranin and water-soluble protein from the adrenal medulla of the deer (*Cervus elaphus* L.) and elk (*Alces alces* L.) has since been detected (HELLE, unpublished results).

The immunological homology detected between chromogranin of these different species of ungulates is also supported by electrophoretic studies of water-soluble protein from chromaffin granules of the sheep, horse and pig (WINKLER et al., 1966). In these four species the electrophoretic mobility of the major lysate protein appeared to be very similar while the minor components differed in their electrophoretic distribution patterns.

The bovine chromogranin, however, is distinguishable from that of the dog, cat, rabbit, rat, mouse and hamster since no cross-reaction can be detected between the adrenal medullary proteins of these species and that of the ox chromogranin (HOPWOOD, 1967; HELLE, unpublished observations).

Chromogranin Released from the Adrenal Medulla

The physiological importance of the immunochemical assay developed for chromogranin became evident when in 1965 BANKS and HELLE detected the presence of immunochemically active protein identical to chromogranin in the perfusates of stimulated, but not of the unstimulated, bovine adrenal medulla. In this study we used a method of perfusion similar to that described by SCHÜMANN and PHILIPPU (1962). In our work a quantitative relationship between amounts of catecholamines secreted and protein secreted was found in a ratio that was of the same order as that in the lysate of isolated chromaffin granules. In view of the fact that macromolecules of protein nature do not usually pass through intact cell membranes the detection of the specific protein characteristic of the chromaffin granule in the perfusate of the stimulated gland provided the first biochemical support for the secretion mechanism by exocytosis suggested by DE ROBERTIS and VAZ FERREIRA in 1957. These authors presented their hypothesis on basis of electron microscopic evidence; they reported a fusion of chromaffin granules with the cell membrane and they detected emptied membrane sacs near the cell membrane.

The biochemical evidence just quoted was soon confirmed and extended by SAGE et al. (1967) who found by microcomplement fixation technique a quantitative relationship between catecholamines and chromogranin in the perfusates of the stimulated adrenal with a ratio of the same order as that found in the isolated chromaffin granules. SCHNEIDER et al. (1967) also used the microcomplement technique; they found that the secretion of chromogranin was not followed by a parallel secretion of cytoplasmic constituents from the chromaffin cell, nor by a release of phospholipids of an order comparable to that known to be characteristic of the granule membrane fraction. The secretion of chromogranin from the stimulated gland was abolished when Ca^{++} was omitted from the perfusion medium (Schneider et al., 1967), parallel to the Ca^{++}-dependent release of catecholamines from perfused adrenals upon stimulation with carbachol (Banks, 1965).

However, it could still be argued that the secretion of immunologically active protein identical to the specific protein enclosed in the chromaffin granule from glands stimulated *in vitro* and perfused in a retrograde fashion did not necessarily prove that this also happened under more physiological conditions. A final proof for secretion of chromogranin *in vivo* was provided by Blaschko et al. (1967) when they detected immunologically the secretion of this protein in the venous outflow from calf adrenal medulla stimulated *in situ*.

Once more the enzyme dopamine-β-hydroxylase has been shown to resemble chromogranin in its pattern of distribution, since this enzyme was detected in the perfusates of stimulated bovine adrenal medulla (Viveros et al., 1968).

Chromogranin Released from Sympathetic Nerve Endings

In the search for a parallelism between release of transmitter, noradrenaline, from the endings of adrenergic nerves and the exocytotic mechanism of secretion established for the catecholamine-containing granules of the adrenal medulla, the presence of chromogranin and dopamine-β-hydroxylase were looked for in the perfused and stimulated spleen. de Potter et al. (1969) obtained evidence for the secretion of both these proteins in the perfusates from bovine spleen after stimulation. Chromogranin detected immunologically and dopamine-β-hydroxylase assayed enzymatically were both found to be secreted, but the ratio of either protein to that of noradrenaline secreted was lower than that found in the intact nerve granules. Using immunohistochemical methods for both proteins Geffen et al. (1969) were able to detect a secretion of both chromogranin and dopamine-β-hydroxylase from the perfused spleen of the sheep.

These observations strongly suggest that a structural similarity in protein composition exists between the catecholamine-storing particles of the adrenal medulla and those of sympathetic nerves. The discovery that the two proteins, chromogranin and dopamine-β-hydroxylase, are secreted when the spenic nerve is stimulated, lends support to the idea of an exocytotic release mechanism for the adrenergic transmitter similar to that established for the secretory response of the chromaffin cell of the adrenal medulla.

Summary

1. Chromogranin, the major water-soluble protein of the chromaffin granule, has been purified by molecular sieve chromatography of granule lysates from bovine adrenal medulla.

2. With the use of specific rabbit antisera against the purified chromogranin this protein has been found to be a major constituent also of the granule water-insoluble fraction.

3. Immunological assay reveals the presence of chromogranin in noradrenaline-containing particulate fractions of bovine splenic nerve.

4. The immunological technique has led to the detection of chromogranin in perfusates from stimulated adrenal medulla.

5. A similar release of chromogranin has been discovered in perfusates from spleen after nervous stimulation.

6. Immunological assay of chromogranin thus has provided evidence for an exocytotic mechanism not only of the secretion of the water-soluble contents of the adrenal medulla, but also of the release of the adrenergic transmitter in sympathetic nerves.

Acknowledgements

I wish to thank my collegue Dr. GULDBORG SERCK-HANSSEN in cooperation with whom part of this work was carried out. Thanks are directed to Prof. Dr. H. BLASCHKO, F.R.S., for his valuable suggestions during the preparation of this manuscript. Thanks are also due to Dr. philos. A. GROV for preparing the rabbit immune sera and to Dr. med. T. FLATMARK for carrying out the amino acid analyses.

References

BANKS, P.: Effects of stimulation by carbachol on the metabolism of the bovine adrenal medulla. Biochem. J. **97**, 550—560 (1965).

—, HELLE, K. B.: The release of protein from the stimulated adrenal medulla. Biochem. J. **97**, 40C—41C (1965).

— —, MAJOR, D.: Evidence for the presence of a chromogranin-like protein in bovine splenic nerve granules. Molec. Pharmacol. **5**, 210—212 (1969).

BLASCHKO, H., COMLINE, R. S., SCHNEIDER, F., SILVER, M., SMITH, A. D.: Secretion of a chromaffin granule protein, chromogranin, from the adrenal gland after splanchnic stimulation. Nature (Lond.) **215**, 58—59 (1967).

DEPOTTER, W. P., DESCHAEPDRYVER, A. F., MOERMAN, E. J., SMITH, A. D.: Evidence for the release of vesicle-protein together with noradrenaline upon stimulation of the splenic nerve. J. Physiol. (Lond.) **204**, 102P—104P (1969).

DEROBERTIS, E. D. P., VAZ FERREIRA, A.: Electron microscopic study of the excretion of catechol-containing droplets in the adrenal medulla. Exp. Cell Res. **12**, 568—574 (1957).

EADE, N. R.: D. Ph. Thesis, Oxford University 1957.

GEFFEN, L. B., LIVETT, B. G., RUSH, R. A.: Immunological localization of chromogranins in sheep sympathetic neurons, and their release by nerve impulses. J. Physiol. (Lond.) **204**, 58P—59P (1969).

HILLARP, N.-Å.: Enzymatic systems involving adenosinephosphates in the adrenaline and noradrenaline containing granules of the adrenal medulla. Acta physiol. scand. **42**, 144—165 (1969).

HELLE, K. B.: (1) Some chemical and physical properties of the soluble protein fraction of bovine adrenal chromaffin granules. Molec. Pharmacol. **2**, 298—310 (1966).

— (2) Antibody formation against soluble protein from bovine adrenal chromaffin granules. Biochim. biophys. Acta (Amst.) **117**, 107—110 (1966).

— (3) Comparative studies on the soluble protein fractions of the bovine, equine, porcine and ovine adrenal chromaffin granules. Biochem. J. **100**, 6C—7C (1966).

—, SERCK-HANSSEN, G.: Chromogranin: the soluble and membrane-bound lipoprotein of the chromaffin granule. Pharmacol. Res. Comm. **1**, 25—29 (1969).

HOPWOOD, D.: An immunohistochemical method for chromaffin granule protein. J. Anat. (Lond.) **101**, 619 (1967).

— An immunohistochemical study of the adrenal medulla of the ox. A comparison of antibodies against whole ox chromaffin granules and ox chromogranin A. Histochemie **13**, 323—330 (1968).

KIRSHNER, N., HOLLOWAY, C., SMITH, W. J., KIRSHNER, A. G.: Uptake and storage of catecholamines. In: Mechanisms of release of biogenic amines, pp. 109—123. (v. EULER, U. S., ROSELL, S., UVNÄS, B., Eds.) Oxford: Pergamon Press 1966.

LADURON, P., BELPAIRE, F.: Tissue fractionation and catecholamines II. Intracellular distribution patterns of tyrosine hydroxylase, dopadecarboxylase, dopamine-β-hydroxylase, phenyletylamine-N-metyltransferase and monoamine oxydase in adrenal medulla. Biochem. Pharmacol. **17**, 1127—1140 (1968).

LEVIN, E. Y., LEVENBERG, B., KAUFMAN, S.: The enzymatic conversion of 3,4-Dihydroxyphenylethylamine to norepinephrine. J. biol. Chem. **235**, 2080—2086 (1960).

LIVETT, B. G., GEFFEN, L. B., RUSH, R. A.: Immunohistochemical evidence for the transport of dopamine-β-hydroxylase and a catecholamine binding protein in sympathetic nerves. Biochem. Pharmacol. **18**, 923—924 (1969).

MUNOZ, E., NACHBAR, M. S., SCHOR, M. T., SALTON, M. R. J.: Adenosinetriphosphatase of *Micrococcus Lysodeikticus*: Selective release and relationship to membrane structure. Biochem. biophys. Res. Commun. **32**, 539—546 (1968).

POTTER, L. T., AXELROD, J.: Subcellular localization of catecholamines in tissues of the rat. J. Pharmacol. **142**, 291—298 (1963).

SAGE, H. J., SMITH, W. J., KIRSHNER, N.: Mechanism of secretion from the adrenal medulla. I. A microquantitative immunological assay for bovine catecholamine storage vesicle protein and its application to studies of the secretory process. Molec. Pharmacol. **3**, 81—89 (1967).

SCHNEIDER, F., SMITH, A. D., WINKLER, H.: Secretion from the adrenal medulla: biochemical evidence for exocytosis. Brit. J. Pharmacol. **31**, 94—104 (1967).

SCHÜHMANN, H. J., PHILIPPU, A.: Der Einfluß von Calsium auf die Brenzcatechinaminfreisetzung. Experientia (Basel) **18**, 138—140 (1962).

SERCK-HANSSEN, G.: Chromogranin, a constituent of the chromaffin granule membrane. Acta physiol. scand. Suppl. **330**, 63 (1969).

SMITH, A. D.: Biochemistry of adrenal chromaffin granules. In: The interaction of drugs and subcellular components on animal cells. (CAMPBELL, P. N., Ed.) London: Churchill 1968.

—, WINKLER, H.: Studies of soluble protein from adrenal chromaffin granules. Biochem. J. **95**, 42P (1965).

— — Purification and properties of an acidic protein from chromaffin granules of bovine adrenal medulla. Biochem. J. **103**, 483—492 (1967).

SMITH, W. J., KIRSHNER, N.: A specific soluble protein from the catecholamine storage vesicles of bovine adrenal medulla. I. Purification and chemical characterization. Molec. Pharmacol. **3**, 52—62 (1967).

VIVEROS, O. H., ARQUEROS, L., CONNETT, R. J., KIRSHNER, N.: Mechanism of secretion from the adrenal medulla. III. Studies of dopamine-β-hydroxylase as a marker for catecholamine storage vesicle membranes in rabbit adrenal glands. Molec. Pharmacol. **5**, 60—68 (1969).

WINKLER, H., ZIEGLER, E., STRIEDER, N.: Studies on the proteins from chromaffin granules of ox, horse and pig. Nature (Lond.) **211**, 982—983 (1966).

Dr. K. B. HELLE
Institute of Physiology
University of Bergen
Årstadiven 19
Bergen/Norway

Discussion

KIRSHNER: I am a little confused about the lysis experiments and the conclusions you have drawn from them. Are you saying that the chromogranin is a part of the vesicle membrane?

HELLE: Yes.

KIRSHNER: I think this would largely depend on how well you wash the membranes and how well the vesicles were lysed.

Helle: Yes, this is an important problem. Subjecting granules to lysis in distilled water is not a very physiological method.

Kirshner: Well, that is the question—I mean, just because by your procedure of lysing the granules you find chromogranin associated with the membranes, after subjecting them to detergent treatment, does not mean that the chromogranin is a constituent of the membrane itself. It is a very highly charged molecule and may be associated with some other charged particles in the membrane. One should really have lysed the vesicles and washed them with, perhaps, a salt solution to show that one is washing things away which are ionically bound and trying as hard as one can to really wash things away. Then break down the membrane to see whether the component is still a real constituent of the membrane or associated by some sort of loose binding to the membrane.

Helle: I am sure you will remember the slide that was shown of the electropherograms obtained in the release experiment which I have just discussed. We could not detect chromogranin in the second washing (SN 3). The lack of chromogranin in the last washing is quite important because we obtained a release of chromogranin on further treatment of the pellet in hypotonic medium. This must imply that chromogranin is bound to the membrane in a way different from what we have believed up to now. The studies have been done on isolated granules. I think it is quite important when we want to discuss what is in the membrane and what is in the water-soluble fraction, that we do not try to solubilize the membrane before we study it. This is the reason why we tried to use more physiological conditions for the lysis experiments, with medium compositions similar to those used in biochemical work for the study of membrane structures of other organisms.

Kirshner: I think we should stress a point that Dr. Helle said—we do not know what constitutes a membrane, but one can, by some device of standard methods of washing and preparing membranes, define what we can call a membrane. One has to be careful what one says is in the membrane and what is merely bound to a membrane. This could largely depend upon the procedures used in preparing the particular fraction of the membrane.

de Robertis: I understand that the bulk of chromogranin comes first, if that is true, which would be the proportion bound to the membrane in comparison with the soluble?

Helle: Forty per cent of the total chromogranin is still bound to the membrane, when 95% of the catecholamines have been released into the water-soluble fraction.

Kirshner: I should like to make another comment. In the earlier experiments by Hillarp after lysing the vesicle in distilled water, he recovered about 80% of the total protein of the vesicle as the soluble protein, and from later studies in our laboratory and by others, about 50% of the total soluble protein is chromogranin A. Again this depends on how you lyse the vesicles and what you use to wash them. I think that 40% bound to the membrane is a rather high figure.

Helle: I agree that until we have done lysis experiments under conditions which do not destroy lipid-protein interactions in the membrane, I think it is very difficult to say precisely what is the water-soluble part and what is in the membrane.

Bayer-Symposium II, 58—72 (1970)

Immunohistochemical Localization of Chromogranins in Sheep Sympathetic Neurones and their Release by Nerve Impulses

L. B. GEFFEN, B. G. LIVETT*, and R. A. RUSH

With 10 Figures

The theory that chemical transmission of nerve impulses at synapses occurs by the release of a number of discrete packages or "quanta" of transmitter, is based primarily on electrophysiological evidence. Since the discovery of synaptic vesicles 15 years ago it has often been suggested that they constitute the morphological basis of the quanta but direct evidence is still lacking that they contain the required amount of transmitter or indeed that they release it during transmission. In the adrenal medulla, however, considerable progress on the mechanism of secretion of catecholamines has been made possible by the recognition that the catecholamine storage vesicles also contain a large proportion of specific soluble proteins called chromogranins, and that these proteins are secreted together with the catecholamines and the other major constituent of the vesicles, adenine nucleotides (SMITH, 1968). The vesicles appear to release their entire content and it has been calculated that each nerve impulse to the adrenal medulla releases the equivalent of the contents of one vesicle per cell (VIVEROS et al., 1969).

The extrapolation from adrenal medulla to adrenergic neurones is a long one, which, in spite of obvious phylogenetic, embryonic and functional affinities, has to take into account differences in the rate and quantity of secretion, the presence of recapture mechanisms, of postjunctional effector cells with trophic interactions, and perhaps most pertinently, the considerable distances between the synaptic apparatus of the nerve terminals and the soma of the neurone. We have used the opportunity presented by the adrenal medulla as a homogeneous and concentrated source of the protein components of catecholamine containing vesicles to study by immunological methods whether these proteins are also present in sympathetic neurones, and to examine their distribution, transport and participation in the transmission process [GEFFEN et al., 1969 (1, 2)].

Apart from dopamine-β-hydroxylase, which controls the final step in noradrenaline synthesis, the function of the chromogranins in the adrenal medullary vesicles is not known. Nevertheless, it was clear that they might be a most useful marker for studies on vesicles from noradrenergic nerves. In a previous session our work on radioactive labelling of a rapidly transported component of axonal protein associated with noradrenaline transport has been described (LIVETT et al., 1968). The identification of these proteins might go some way to elucidating the

* Present address: Department of Pharmacology, University of Oxford, Oxford OX1 3QT, England.

site of synthesis of components of the synaptic vesicles and also help resolve the relationship between the small granular vesicles, characteristic of the synaptic regions of sympathetic nerves and the bigger ones, more sparsely distributed throughout the neurone, which accumulate in such large numbers above an axonal ligature (KAPELLER and MAYOR, 1967; GEFFEN and OSTBERG, 1969).

Preparation of Antibodies to Chromogranins

Antibodies to two chromogranins, dopamine-β-hydroxylase and chromogranin A, were prepared from pooled sheep adrenal medullae as follows (Fig. 1). The medullae were homogenized in 0.3 *M* sucrose and a large particle fraction was prepared by differential centrifugation. The catecholamine vesicles were separated from less dense material by discontinuous density gradient centrifugation with

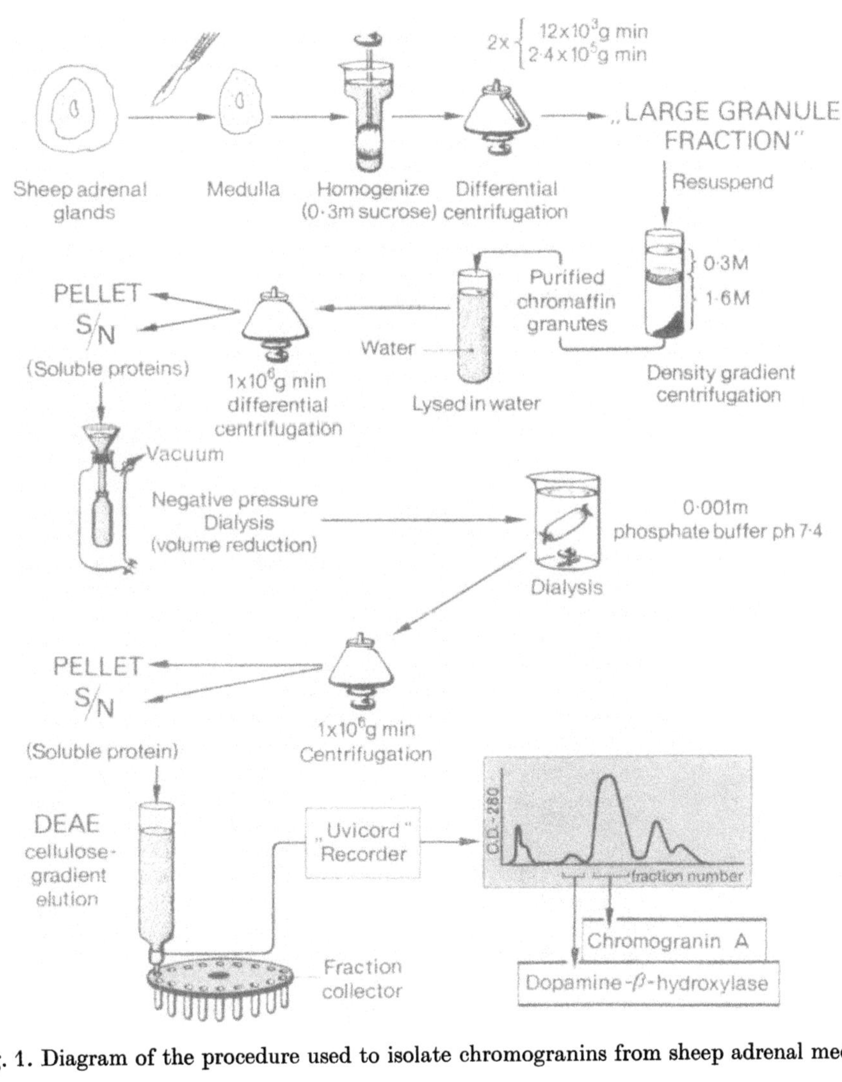

Fig. 1. Diagram of the procedure used to isolate chromogranins from sheep adrenal medullae

0.3 and 1.6 *M* sucrose (Smith and Winkler, 1967). An osmium fixed pellet examined in the electron microscope contained mostly dense cored vesicles. Some of these were partially disrupted and contained dispersed clusters of aggregated

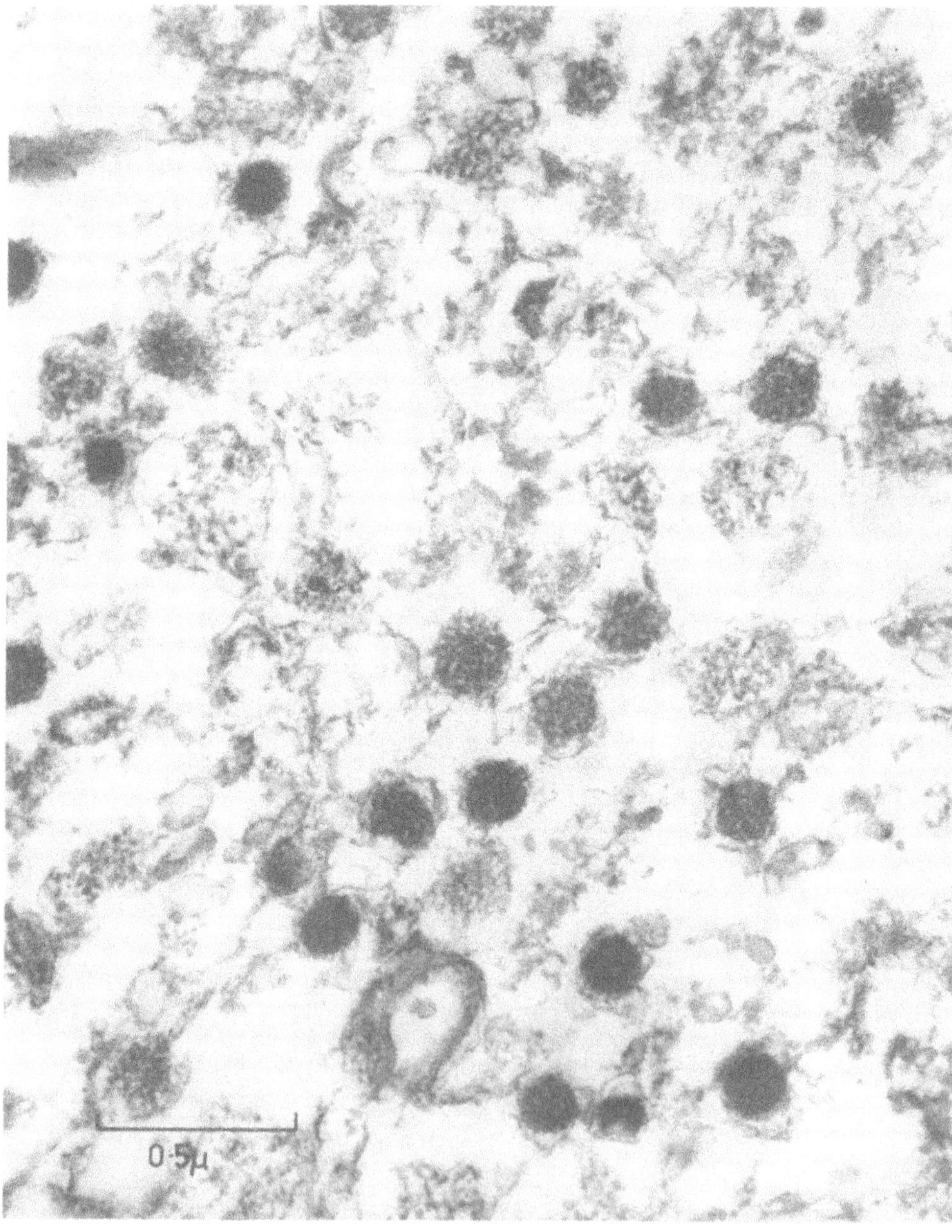

Fig. 2. Electron micrograph of osmium fixed catecholamine storage vesicles of the adrenal medulla isolated by differential and density gradient centrifugation to show the aggregates of electron dense particles comprising the granular core of the vesicles

particles which appeared to form the basis of the single very dense core of the intact vesicles (Fig. 2).

The vesicles were osmotically lysed, centrifuged to remove insoluble material, and the soluble proteins were concentrated by negative pressure dialysis and separated by ion exchange chromatography on DEAE cellulose columns. The eluted proteins were followed by UV absorption at 280 mμ. A small peak containing all the dopamine-β-hydroxylase activity preceded a series of three protein peaks, the first of which contained the bulk of the eluted proteins. When the proteins of the lysate were prepared by molecular sieving, the corresponding major fraction was called chromogranin A (SCHNEIDER et al., 1967), and when this was preceded by a preliminary separation of dopamine-β-hydroxylase on DEAE-cellulose, the peak was called S 1 (SMITH and KIRSHNER, 1967). S 1 and chromogranin A appear to be identical (KIRSHNER and KIRSHNER, 1969). The homogeneity of the eluted protein fractions was examined by polyacrylamide disc gel electrophoresis. The dopamine-β-hydroxylase fraction produced only a single band widely separated from the more mobile chromogranin A fraction which had one main and two minor components. Antibodies were then produced in rabbits to the two antigen fractions and when tested by immunodiffusion in agar, precipitations only occurred with the specific antigen. We have been able to show only slight binding of ^{14}C-noradrenaline to chromogranin A by equilibrium dialysis, achieving a concentration ratio of 1.5 without the addition of ATP or calcium and have not been able to detect significant binding by gel filtration.

However, in the preparation of chromogranin A from lysed vesicles the proportions of the diffusable constituents of the vesicles would be greatly altered. Recent evidence that catecholamines, ATP and calcium in critical proportions form an insoluble miscellar complex in the absence of chromogranins (BERNEIS, PLETSCHER and DA PRADA, 1969) raises the question of the role of chromogranin A in binding catecholamines within vesicles. Chromogranin A in high concentrations can form a gel (SMITH, 1968) and it is possible that this highly acidic protein creates within its interstices, the ionic conditions necessary for the formation of the miscellar complex. This could permit the presence of high concentrations of these otherwise diffusable substances in the vesicle without osmotic effects.

Chromogranin A may also play a role in release when the vesicle interior is exposed to the high sodium concentration of the extracellular fluid. Neutralization of the acidic charges on the protein reduces its hydrodynamic volume (vide SMITH, 1968) and this may assist in the release of the vesicle constituents.

Immunofluorescent Localization in Neurones

Immunofluorescence histology combines the sensitivity of fluorescence microscopy with the specificity of the immune reaction. We have used this technique to study the distribution of chromogranins in various sheep tissues including adrenal medulla, sympathetic ganglia, postganglionic nerve trunks, both normal and ligated, spleen, vas deferens and blood vessels.

Tissues were sectioned either unfixed and frozen, or after alcohol fixation, and then treated with the rabbit antisera. At first we conjugated the rabbit antibodies directly to a fluorescent dye, either fluorescein isothiocyanate or rhodamine. In subsequent work we have used the indirect "sandwich" technique (vide NAIRN, 1969) in which the unlabelled rabbit antiserum adsorbed on to the section constitutes an antigen which is used to bind goat anti rabbit γ-globulin conjugated to fluorescein (Fig. 3). Apart from its convenience, this method confers added sensitivity with little loss of specificity. At present we do not have equipment for

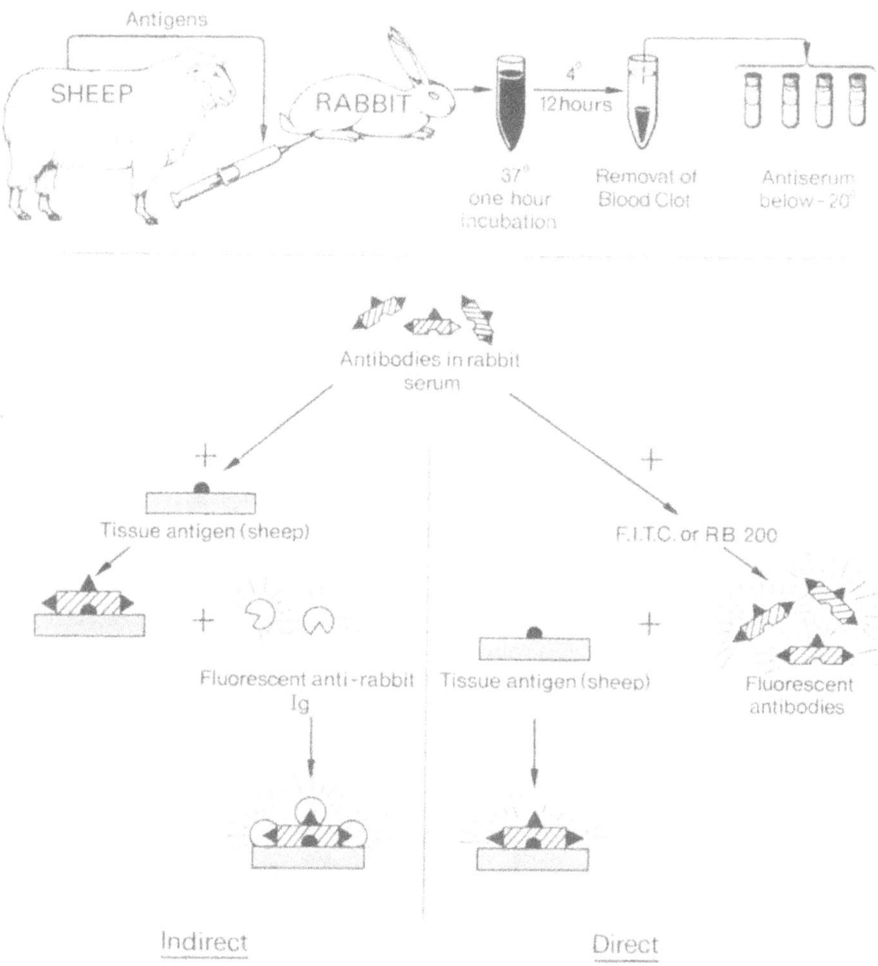

Fig. 3. Diagram of the procedures used for immunofluorescence histology

quantitating the distribution and intensity of the fluorescence other than by colour photography.

The results were clear-cut. Both antibodies localized in the adrenal medulla and the postganglionic sympathetic neurones but not in the adrenal cortex or preganglionic nerves (Fig. 4). Extra-adrenal chromaffin tissue and the central nervous system are at present being studied. The cell bodies of sympathetic neurones showed bright homogeneous staining apart from their nuclei (Fig. 5). The axons and their terminals were less intensely stained, and it was often difficult to dis-

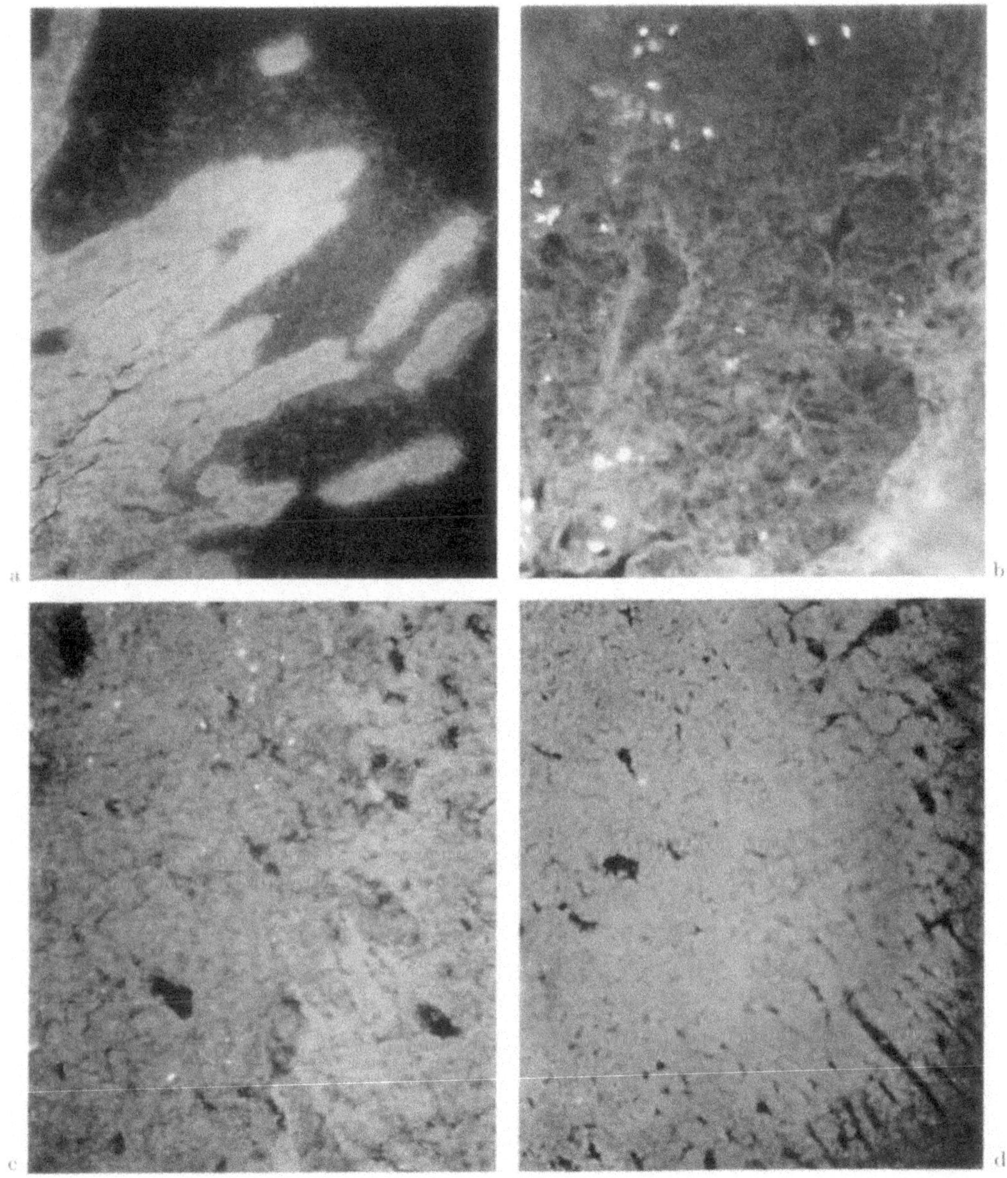

Fig. 4. Comparison of immunofluorescence with chromogranin antibodies, and catecholamine fluorescence induced by formaldehyde in sections of the cortico-medullary junction of sheep adrenals. (a) Catecholamine fluorescence of adrenal medulla which was bright yellow-green but appears more yellow in colour photographs. Some diffusion has occurred. (b) Direct immunofluorescence using antibodies to chromogranin conjugated with Rhodamine B200. Note the blue autofluorescence of the adrenal cortex and vein and the specific red fluorescence of the medulla. (c) Indirect immunofluorescence using pre-immune rabbit serum and fluorescein labelled goat anti-rabbit globulin. Both cortex and medulla show only blue autofluorescence. (d) Indirect immunofluorescence with anti-chromogranin rabbit serum and fluorescein labelled anti-rabbit globulin showing specific green fluorescence of the medullary cells

criminate the fine varicose terminals from the staining of the surrounding tissue especially smooth muscle. In ligated axons, both antigens accumulated on the proximal side of the constriction.

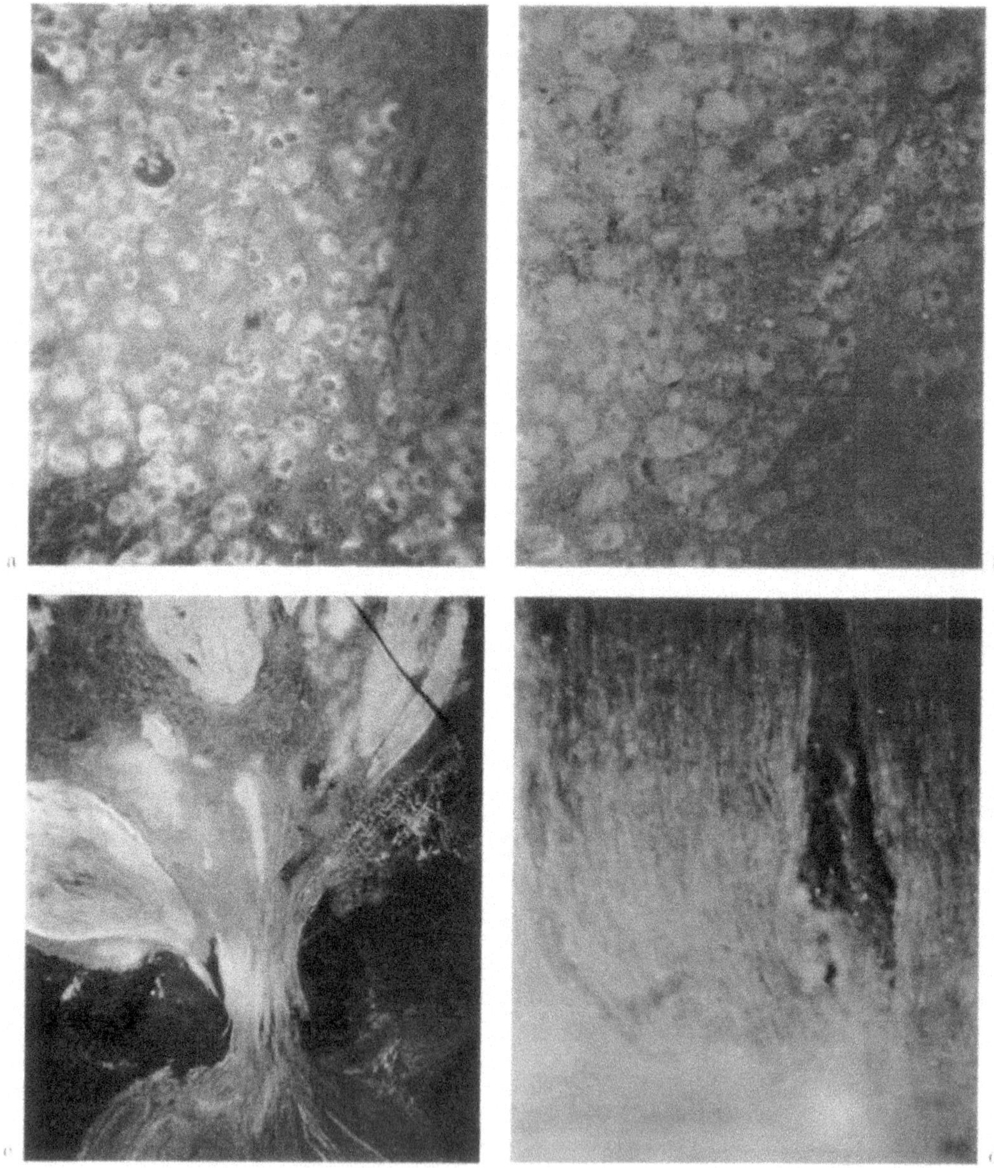

Fig. 5. Comparison of chromogranin immunofluorescence and noradrenaline fluorescence in sheep sympathetic neurones. (a) Noradrenaline fluorescence (greener in the microscope) of coeliac ganglion cells. (b) Indirect immunofluorescence of same cells using fluorescein conjugated antibodies. (c) Ligated splenic nerves showing a proximal accumulation of noradrenaline fluorescence after 24 h. (d) Immunofluorescence of ligated nerves showing increased specific green fluorescence immediately proximal to the ligated area which has only blue autofluorescence

On parallel sections, we also studied the distribution of catecholamines by the histochemical technique of FALK and OWMAN (1965). The distribution of noradrenaline fluorescence corresponded well to that of the immunofluorescence apart

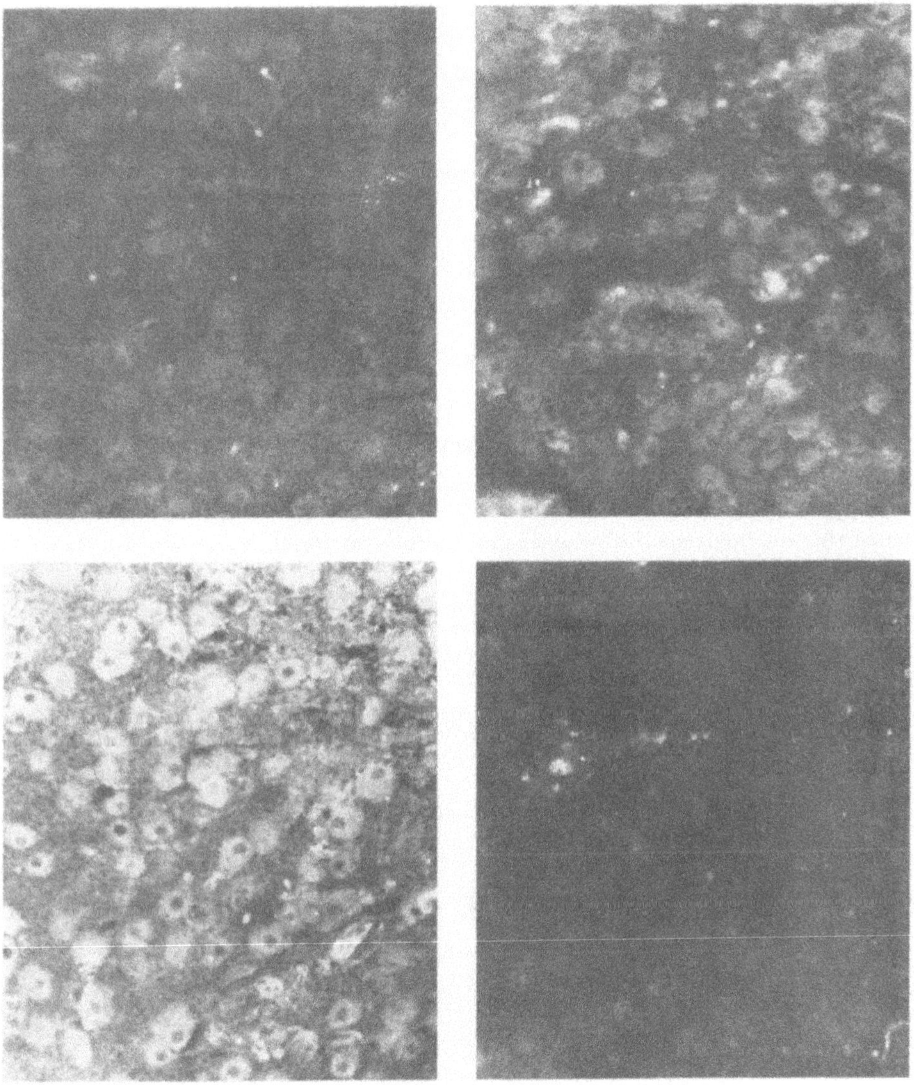

Fig. 6. Indirect immunofluorescent staining of sympathetic ganglion cells with antibodies to dopamine-β-hydroxylase showing. (a) Autofluorescence in the absence of antiserum. (b) Nonspecific, mainly extracellular, fluorescence with pre-immune serum. (c) Specific fluorescence of the cytoplasm but not the nuclei of the neurones in the presence of immune serum. (d) Elimination of specific fluorescence by pre-incubation of immune serum with antigen

from the more discrete and intense localization in the axon terminals of catecholamine compared to immuno-fluorescence. Some of these differences in localization may be due to diffusion. The staining and washing with antisera is of necessity carried out in aqueous solutions, whereas the induction of catecholamine fluorescence is carried out on freezedried sections in humidity conditions carefully controlled to prevent diffusion. Formaldehyde penetrates easily whereas the reaction of the fluorescein labelled antibody is mainly restricted to the cut surface of the tissue section. Since the axon terminals are only 1 μ in diameter many would remain uncut in a 5 μ section.

The following controls were used to establish the specificity of the staining: (1) unstained sections show blue autofluorescence which is easily distinguished with appropriate filters from the apple-green fluorescence of the antibodies. (2) sections treated with pre-immune serum from the same rabbit to check for binding of conjugated antibodies other than those induced in the specific immune serum. (3) omission of the rabbit serum altogether to examine direct binding of the goat serum to the sheep tissue sections. (4) pre-incubation of the immune serum with the specific antigen to absorb and hence eliminate the specific antibody (Fig. 6).

Subcellular Fractionation

The next step was to investigate the subcellular localization of the chromogranins in different parts of the sympathetic neurones. We prepared microsomal fractions by differential centrifugation of homogenates from sheep sympathetic ganglia, post-ganglionic nerve trunks and from adrenergically innervated organs such as the vas deferens, heart and spleen. Both the final pellet and supernatant

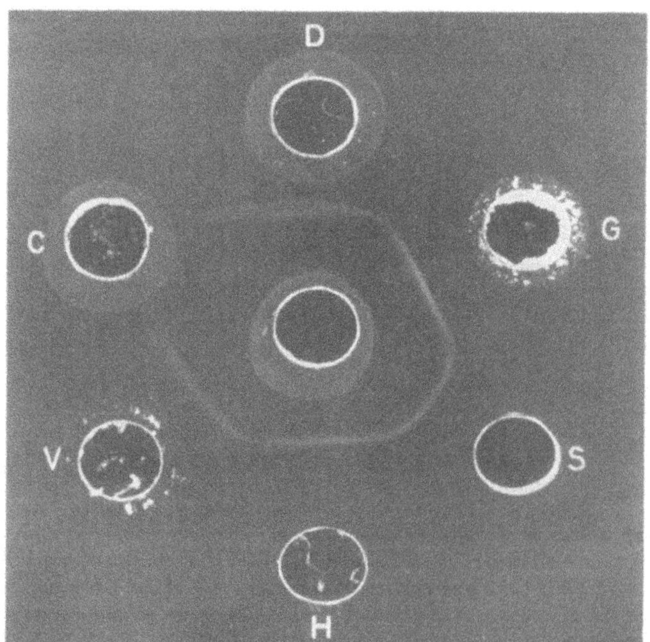

Fig. 7. Immunodiffusion in agar showing precipitin lines of identity between antiserum to dopamine-β-hydroxylase (centre well), the purified antigen (A) and lysed microsomal fractions from sympathetic ganglia (G), spleen (S), heart (H), vas deferens (V). There is no cross reaction with purified chromogranin A (C)

were examined for the presence of chromogranins by two immunological techniques, immunodiffusion and complement fixation. The preliminary results (as yet unpublished) indicate that chromogranins in ganglia, nerves and peripheral tissues are present in both the microsomal pellet and in the supernatant in approximately the same proportions as noradrenaline (Fig. 7). Other evidence that chromogranins are present in sympathetic nerves comes from immunofluorescence histology (Hopwood, 1968), and from subcellular fractionation of splenic nerve tissue using immunoprecipitation (Banks, Helle and Mayor, 1969) and complement fixation (de Potter, de Schaepdryver, Moerman and Smith, 1969).

Release of Vesicle Contents During Stimulation

We have also studied whether sympathetic nerves released any protein components of the synaptic vesicles together with noradrenaline during electrical stimulation. We have not yet measured adenine nucleotides because of the

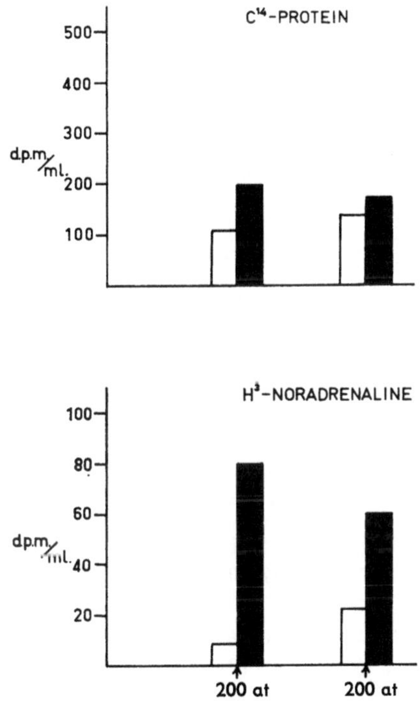

Fig. 8. Release of radioactivity from the cat spleen by nerve stimulation 24 h after injecting ^{3}H-noradrenaline and ^{14}C-leucine into the coeliac ganglion. The concentrations of radioactivity in the venous blood due to ^{14}C proteins (above) and ^{3}H-noradrenaline and metabolites (below) are shown before (white bars) and after stimulation

difficulties introduced by extra neuronal sources during effector cell activity. We chose the spleen because it has a high noradrenaline content produced by a relatively homogeneous and accessible sympathetic nerve supply, because it is easy to perfuse and because considerable data is available on the venous overflow of noradrenaline during stimulation. An additional advantage of the spleen is that it has a discontinuous vasculature with blood lakes not bounded by capillary walls which might favour the appearance of proteins in the venous blood.

We had some preliminary evidence on the cat spleen that proteins labelled by injection of radioactive amino acids into the coeliac ganglion were transported to the axon terminals and released by nerve stimulation (Fig. 8), (GEFFEN and LIVETT, 1968). This latter evidence was not adequate, however, to state that the released proteins were of neural origin and not other proteins synthesized in the spleen and labelled by circulating ^{14}C-leucine. Such proteins could have been

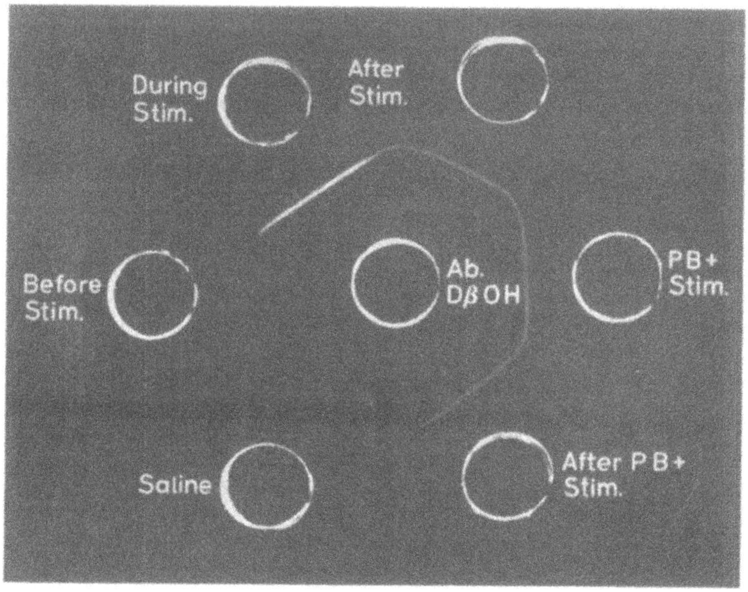

Fig. 9. Immunodiffusion in agar showing the presence of dopamine-β-hydroxylase (DβOH) in saline perfusates of sheep spleen during nerve stimulation in the presence and absence of phenoxybenzamine (PB)

expelled along with blood elements by the contraction of the spleen in response to stimulation.

Since it was apparent that if proteins were released from nerves together with transmitter the amounts of protein involved would be very small, we felt that an immunological approach would be best suited to their identification. In the sheep spleen we have been able to identify both dopamine-β-hydroxylase and chromogranin A in the perfusate during and immediately after a brief period of stimulation (10 sec for 60 sec) at a time when noradrenaline also overflows. We have used both immunoprecipitation and complement fixation analysis (Figs. 9 and 10). We do not yet have adequate quantitative data, but DE POTTER, DE SCHAEPDRYVER, MOERMAN and SMITH (1969) have independently reported preliminary, more extensive evidence on the perfused calf spleen. Chromogranin A and dopamine-β-hydroxylase were released during nerve stimulation by a calcium dependent mechanism whereas the enzyme dopa decarboxylase which is soluble in the axoplasm was not liberated. The ratio of chromogranin A and of dopamine-β-hydroxylase to noradrenaline in the perfusate was very much lower than those they obtained in subcellular fractions of splenic nerve vesicles. However, the composition of the perfusate does not necessarily reflect the proportions in which the vesicle contents were released into the synaptic gap, since the amounts of the different components would be affected by their respective uptake, binding and diffusion properties.

Until these factors are delineated, it is not possible to decide whether the whole content of each vesicle is released or only a part. It should be pointed out that

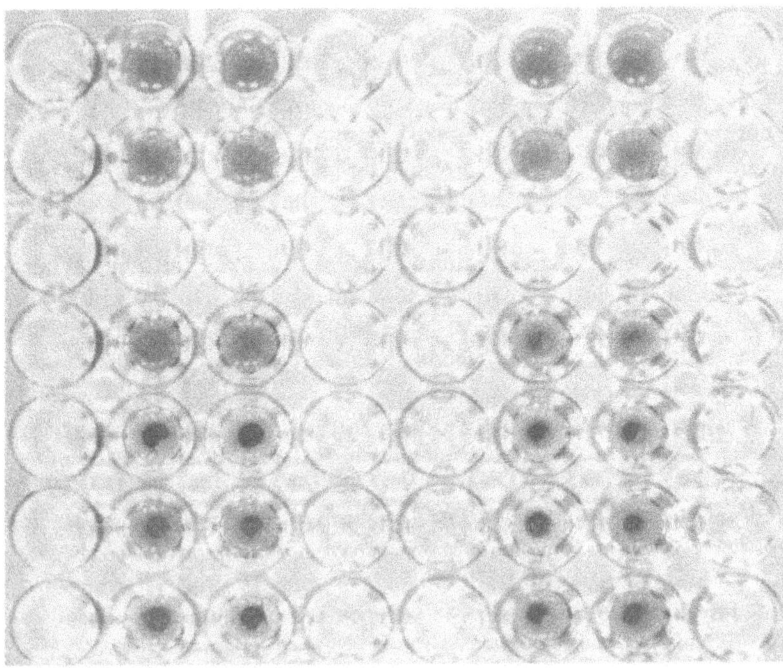

Fig. 10. Complement fixation analysis of sheep spleen perfusates for the presence of chromogranin A and dopamine-β-hydroxylase. The analysis is in duplicate and the lack of haemolysis during stimulation of the splenic nerves indicates the presence of the specific antigen

quantal release is not incompatible with release of portion of the vesicle's contents. The conjunction of vesicle and axon membranes, if it occurs, may be transitory but of fixed duration, the quantum being produced by a temporal gate. Alternatively, the vesicle may contain a more readily diffusable compartment in equilibrium with a firmly complexed store.

Origin and Fate of Synaptic Vesicles

Certain tentative conclusions may be drawn from these results for discussion.

1. Noradrenergic neurones contain the same proteins as those in the catecholamine storage vesicles of the adrenal medulla.

2. The histological distribution of these proteins corresponds to that of noradrenaline being present throughout the neurone, although the relative intensity of immunofluorescence was greater in the cell body than the axon terminals, whereas the converse applied to noradrenaline. This could be ascribed to differences in histochemical technique.

3. In tissue extracts it is not at present possible to assess the relative concentrations of chromogranins in ganglia, nerve trunks and axon terminals since the latter are diluted by the considerable mass of the tissue they innervate. Chromogranins are found both in microsomal and soluble fractions as is noradrenaline. The presence of both in the supernatant may be due to vesicle disruption during homogenization.

4. The accumulation of chromogranins above an axonal ligature corresponds to that of noradrenaline and of large granular vesicles. These vesicles are also found sparsely in the cell body and in axon terminals and they may be the form in which chromogranins are transported peripherally.

5. Chromogranins are released from sympathetic nerves together with noradrenaline by electrical stimulation, suggesting that release from vesicles occurs by exocytosis.

Noradrenaline is synthesized predominantly in the axon terminals whereas the protein components of the synaptic vesicles appear to be synthesized in the soma of the neurone and then transported down the axon. Yet large granular vesicles are distributed throughout the neurone whereas the small granular variety are confined to the axon terminals.

Whether the large granular vesicles are transformed into the smaller synaptic variety, and if so what the mechanism is, budding or gradual depletion of contents during impulse transmissions, or whether the vesicle components are separately transported and locally assembled are largely open questions. Since dopamine-β-hydroxylase is mainly membrane bound whereas chromogranin A is contained within the medullary vesicles, it should be possible to establish whether "packaging" is a function of soma or axon. It is possible that membranes and soluble contents are both synthesized in the soma and assemble into the organelle in the axon terminals or in the artificial cul-de-sacs in axons produced by ligation.

The life span of the vesicles is thought to be of the order of weeks (Dählstrom, 1967). If release of part or whole of the vesicle contents occurs by exocytosis, the vesicles may be recharged. Whether the proteins that are released are subsequently recaptured and reused by the nerves as is most of the transmitter (Blakeley, Brown and Geffen, 1969), whether they have a role to play in the synaptic gap modifying the concentration of noradrenaline at receptor sites, or by regulating the receptors themselves through a trophic mechanism, or whether protein loss is an inevitable consequence of the transmission mechanism, are problems for future work. The most pressing evidence required is the fate of the proteins after release from the nerve terminal and to this end we are infusing labelled chromogranins into sympathetically innervated organs.

Summary

The possibility that sympathetic nerves release noradrenaline from vesicles by a process of exocytosis, as in the adrenal medulla, was examined in sheep. Antibodies were prepared against two specific protein components of adrenal chromaffin vesicles, chromogranin A and dopamine-β-hydroxylase. Their distribution throughout sympathetic neurones was shown by immunofluorescence histology. Both proteins accumulated in axons proximal to a ligature in a corresponding manner to that shown for noradrenaline, for ^{14}C-protein from the ganglion, and for large granular vesicles, suggesting a rapid somatofugal transport of the precusors of synaptic vesicles. In immunoprecipitation studies of subcellular fractions of adrenal medullae, ganglia and nerves, dopamine-β-hydroxylase and chromogranin A were present with noradrenaline in both the microsomal pellet, and the high speed supernatant presumably due to disruption of some vesicles. Both dopamine-β-hydroxylase and chromogranin A could be detected by immunoprecipitation and complement fixation analysis in perfusates of isolated spleens during nerve stimulation. The results favour the hypothesis that the transmitter in liberated by partial exocytosis of vesicles.

References

Banks, P., Helle, K. B., Mayor, D.: Evidence for the presence of a chromogranin-like protein in bovine splenic nerve granules. Molec. Pharmacol. **5**, 210—212 (1969).

Berneis, K. H., Pletscher, A., Da Prada, M.: Metal dependent aggregation of biogenic amines: A possible explanation for their storage and release. Nature (Lond.) **224**, 281—283 (1969).

Blakeley, A. G. H., Brown, G. L., Geffen, L. B.: Uptake and re-use of sympathetic transmitter in the cat spleen. Proc. roy. Soc. B **174**, 51—68 (1969).

Dahlström, A.: The intraneuronal distribution of noradrenaline and the transport and life-span of amine storage granules in the sympathetic adrenergic neuron. Naunyn-Schmiedeberg's Arch. Pharmak. exp. Path. **257**, 93—115 (1967).

De Potter, W. P., Schaepdryver, A. F., Moerman, E. J., Smith, A. D.: Evidence for the release of vesicle-proteins together with noradrenaline upon stimulation of the splenic nerve. J. Physiol. (Lond.) **204**, 102—104 (1969).

Falck, B., Owman, C.: A detailed methodological description of the fluorescence method for cellular localization of biogenic monoamines. Archiv. Univ. Lund. **II**, **7**, 1—14 (1965).

Geffen, L. B., Livett, B. G.: Axoplasmic transport of ^{14}C-noradrenaline and protein and their release by nerve impulses. Proc. Int. Union. Physiol. Soc. **7**, 152 (1968).

— — Rush, R. A.: (1) Immunohistochemical localization of protein components of catecholamine storage vesicles. J. Physiol. (Lond.) **204**, 593—605 (1969).

— — — (2) Immunological localization of chromogranins in sheep sympathetic neurones and their release by nerve impulses. J. Physiol. (Lond.) **204**, 58—59 P (1969).

— Ostberg, A.: Distribution of granular vesicles in normal and constricted sympathetic neurones. J. Physiol. (Lond.) **204**, 583—292 (1969).

Hopwood, D.: An immunohistochemical study of the adrenal medulla of the ox. Histochemie **13**, 323—330 (1968).

Kapeller, K., Mayor, D.: The accumulation of noradrenaline in constricted sympathetic nerves as studied by fluorescence and electron microscopy. Proc. roy. Soc. B. **167**, 282—292 (1967).

Kirshner, A. G., Kirshner, N.: A specific soluble protein from the catecholamine storage vesicles of bovine adrenal medulla. Biochim. biophys. Acta (Amst.) **181**, 219—225 (1969).

Livett, B. G., Geffen, L. B., Austin, L.: Proximo-distal transport of ^{14}C-noradrenaline and protein in sympathetic nerves. J. Neurochem. **15**, 931—939 (1968).

Nairn, R. C.: Fluorescent protein tracing. 3rd ed. London: Livingstone 1969.

Schneider, F. H. Smith, A. D., Winkler, H.: Secretion from the adrenal medulla: biochemical evidence for exocytosis. Brit. J. Pharmacol. **31**, 94—104 (1967).

Smith, A. D.: Biochemistry of adrenal chromaffin granules. In: The interaction of drugs and subcellular components on animal cells. (Campbell, P. N., Ed.), London: Churchill, 1968.

— Winkler, H.: A simple method for the isolation of adrenal chromaffin granules on a large scale. Biochem. J. **103**, 480—482 (1967).

SMITH, W. J., KIRSHNER, N.: A specific soluble protein from the catecholamine storage vesicles of bovine adrenal medulla. Molec. Pharmacol. **3**, 52—62 (1967).
VIVEROS, O. H., ARQUEROS, L., KIRSHNER, N.: Quantal secretion from adrenal medulla: all or none release of storage vesicle content. Science **165**, 911—913 (1969).

Dr. L. B. GEFFEN
Department of Physiology
Monash University
Clayton, 3168, Victoria/Australia

Discussion

DE ROBERTIS: I am very pleased that it has been said that the synaptic vesicle may be reused many times in the adrenergic nerve. I think that the evidence for that on the cholinergic nerves is now very abundant and convincing.

AXELROD: Is it possible to determine the concentration of dopamine-β-oxidase and chromogranin by measuring the intensity of fluorescence of the antibody? If the vesicle is reused again it has to be very rapidly released again. This would mean that there is a special mechanism for rapid protein synthesis or a unique type of sealing mechanism that works very fast.

GEFFEN: The answer to the second question is—I do not know—the answer to the first question is that one must be careful because of the problem of quenching not to equate intensity of fluorescence with the concentration of the fluorophor. Nor can the intensity of catecholamine fluorescence be equated with immuno-fluorescence. Whereas the formaldehyde vapour penetrates the entire section, immunofluorescence is due mainly to a surface reaction. While one after sections across cell bodies, the nerve terminals are very fine and in 5 μ sections, many of the axon terminals within the section, remain intact so that their membranes may well limit inward diffusion of labelled antibody. There are many other technical problems associated with the equation of fluorescence intensity with concentration of different chromogranins, such as antibody titre. I personally have no evidence, as yet, on whether there is a higher concentration of chromogranins in the nerve terminals compared with the cell body. I am sorry both answers are so negative but these are really early days.

VON EULER: Did I understand you right, Dr. GEFFEN, that according to your opinion the release should be directly from the vesicles in some way through the axon membrane? This was one question and the second—have you any data about the proportions of released noradrenaline and chromogranin in relation to the proportions in the resting tissue?

GEFFEN: Yes, you understand me correctly to say that I believe the release of the transmitter is directly from the vesicle to the outside of the axon.

VON EULER: Exocytosis.

GEFFEN: Yes. I must stress that I have no quantitative data but at the same time as we did these experiments on the release of chromogranin by nerve stimulation in the sheep spleen, DE POTTER, DE SCHAEPDRYVER, MOERMAN and SMITH independently showed this in the calf spleen. Their experiments were more quantitative and indicate that the proportion of noradrenaline to chromogranin in the perfusate is much greater than in vesicles isolated from axons. However, we do not know anything yet about the possible recapture of chromogranins or of their binding onto structures within the organ before they appear in the perfusate. I believe that the spleen will be one of the very few organs in which it will be possible to demonstrate a release of proteins into the perfusate because of the nature of the circulation in the spleen. It has a discontinuous vasculature so that there is not always a capillary barrier interposed between the synaptic gap and the circulation. Similarly, in the adrenal medulla there are a large fenestrae in the capillary membranes. I am not optimistic that it will be easy to show the release of these proteins in the perfusate from other organs.

CARLSSON: You had contraction of the spleen when you stimulated. I must say I am sorry that you left out the phenoxybenzamine experiment because that is, as I see it, a very important control experiment. Therefore, I would like to ask whether you did get release of the material when you blocked the alpha receptors? I have another question: Did you do an experiment where you induced contraction of the spleen without stimulating, for instance, by injecting noradrenaline to see what then came out?

GEFFEN: I thank you for the second suggestion—it certainly ought to be done. We have found release of chromogranin in the presence of phenoxybenzamine and one of the slides did show this. However, the result is inconsistent and all I can say is that phenoxybenzamine does not produce the same elevation in the overflow of chromogranins as it does for noradrenaline. Nevertheless, we always put in some phenoxybenzamine at the end of the experiment before we pack up and go home!

PALM: Have you any idea of the turnover rate of this protein? Is it possible to compare the turnover rate of this specific protein and the turnover rate of noradrenaline? I think one could find a correlation, a functional correlation between these two substances. There may be one indication—the same speed of migration in your experiments in the splenic nerves which you have found for noradrenaline and as well for your leucine labelled protein. Do you think that this could mean nearly identical turnover rate of this chromogranin?

GEFFEN: Noradrenaline turns over at a very much faster rate—of the order of hours—whereas the protein turnover may be of the order of days, or even weeks.

PALM: The total protein?

GEFFEN: Yes. If not all the chromogranins in the vesicle were released during nerve activity, the vesicle may still have the capacity to go on synthesizing and storing noradrenaline. However, it may gradually lose its content of chromogranins and eventually become a ghost, I presume.

SCHÜMANN: One could imagine that always some vesicles are destroyed by lysosomes and that therefore protein, chromogranin and dopamine-β-hydroxylase from these vesicles are free in the axoplasma. If now a stimulus changes the permeability of the axon-membrane chromogranin and β-hydroxylase could also be released, but in the case the release of noradrenaline had really nothing to do with that of chromogranin and β-hydroxylase. Would you exclude this possibility?

GEFFEN: No, I cannot exclude that or other possibilities such as that chromogranins are released from a different sort of vesicle to noradrenaline. One simple idea is that the small granular vesicles release noradrenaline and the large dense-core vesicles release chromogranins, in which case one would no longer have evidence for exocytosis. At present I do not think we can definitely exclude any of these suggestions. However, it seems to me that the finding of newly synthesized noradrenaline being preferentially released by nerve stimulation and the finding that dopamine-β-hydroxylase and chromogranin are found in the vesicle and are released together with noradrenaline, whereas other axoplasmic constituents are not, is at least circumstantial evidence favouring the idea of exocytosis. But I do not wish to leave the impression that these experiments prove exocytosis.

VON EULER: I think you have answered some of the questions that I was going to ask but still I must put this question: If you assume exocytosis—although I understand now you are a little more hesitant about it—how do you imagine that so much more noradrenaline should be poured out and only some chromogranin trickling out, as it were? In which state is noradrenaline pouring out—if I may ask for your opinion.

GEFFEN: The problem which Professor VON EULER raised is a very real one for which there are several possible explanations. First of all, the amine is a very much smaller molecule than the protein and so if the vesicle is open to the synaptic gap for only a limited period of time—if there is a time-dependent gate involved—there would be more rapid diffusion out of the smaller molecule in a very much higher concentration. However, we also have to consider the possibility that not all the noradrenaline within the vesicles is in a soluble form. PLETSCHER's Group has recently produced very important evidence to show that ATP and noradrenaline, in association with divalent ions, form an insoluble aggregate independent of chromogranin. I have already mentioned the evidence that chromogranin only binds catecholamines weakly. One of the functions of chromogranin may be that it is able to form a gel-like matrix in which noradrenaline and ATP, in the presence of the very high concentrations of calcium within the vesicle, form aggregates within the interstices of the protein matrix of the vesicle. In this way both a bound and a free pool of noradrenaline could exist within the vesicle. It may well be the soluble pool which preferentially is released and this would be the newly synthesized noradrenaline moreover.

CARLSSON: I am a bit confused about the conclusion, in so far as exocytosis is concerned. I quite agree that we are in a very preliminary stage but if we take the data as they stand as given by Dr. GEFFEN, is not the weight of evidence rather against the idea of exocytosis? If I understand you correctly, the proportion of the protein coming out is very much lower than that of the catecholamines and the turnover rates that you estimate on the basis of your idea also argue in the same direction. So why do you tentatively conclude that your data support exocytosis?

Geffen: When one talks of exocytosis, one does not necessarily imply that the entire soluble contents of the vesicle are released. Exocytosis, as I understand it, now means that something is released from the inside of the vesicle directly to the outside of the axon without passing through the axoplasm. But it does not have to be the total contents. There could be partial exocytosis and that is an important point to bear in mind.

Blaschko: I think that we should remember that the catecholamines as chemical messengers have two relationships to distance. Dr. Geffen, in the introduction to his talk, referred to distance in relation to the intracellular (intraneuronal) migration, but distances have also to be travelled between the point at which the messengers set free and its point of action on the effector cell. Although it is an interesting and a curious historical fact that Professor Paton first discussed the phenomenon of uptake in relationship to the adrenal medulla on the basis of his work with Marley, it is obvious that here it is the amine that has escaped from the site of liberation that carries the message. Now, when the catecholamine functions as a mediator, the extracellular distance that it has to travel is very small. Thus anatomical and other considerations make me very hesitant to consider the perfusate as perfect a vehicle to transmit other material that has been released as I would consider the adrenal venous blood in relation to the medullary hormone.

I think that the biochemists should agree with Professor Carlsson at this stage and say that the biochemical evidence in favour of exocytosis at the adrenal nerve endings is based on a few very interesting observations. For the adrenal medulla there have been reports by Schneider et al. in 1967 on the absence of release of chromogranin in the absence of calcium and also the fact which was quoted by Dr. P. Smith in his review about a year ago, that proteins present in the cytoplasmic sap do not appear in the perfusate after stimulation. I mention these two facts because I feel that here are ways in which the evidence for exocytosis in the adrenergic neuron might be supplemented.

Kirshner: Are cytoplasmic proteins, such as lactic dehydrogenase, coming out in these experiments?

Geffen: I can only answer this from the experiments of de Potter et al. who studied whether dopadecarboxylase, which is generally accepted as a soluble cytoplasmic constituent, is released from the spleen by nerve stimulation. It is not so that here is an enzyme located in the nerve outside the vesicles which does not appear in the perfusate. The other point I want to make is that the perfusate is not an accurate measure of what is released into the synaptic gap. These are differences in diffusion between amines and proteins and we do not yet know if the chromogranins are bound by other tissues.

Häggendal: I think it would be interesting to know if the proportion between the released noradrenaline and chromogranin is changed when the stimulation is continued for a very long time. In this connection I would like to ask—how long did you treat the tissues before you performed the stimulation? In other words, when did you kill the animal and how long did it take before you performed your experiments after having treated the tissue?

Geffen: It takes some 10 to 12 min to set up the perfusion after killing the animal and we then perfused the spleen for half an hour without any stimulation. We also gave one train of stimuli to produce a contraction and empty the spleen, which we discarded.

Dahlström: May I say that in my opinion the evidence that Dr. Geffen presented clearly contradicted the view on exocytosis of whole nerve granules. Can it perhaps be that the chromogranin molecules are released more or less by accident from the nerves? Together with Dr. Häggendal, I have preliminary data indicating that the capacity of new amine granules are probably different from that of older granules, for instance, in taking up exogenous noradrenaline. There are also these electronmicroscopical observations indicating that the new granules perhaps have a morphology different from that of old granules. Can it perhaps be that the small repeated losses of proteins from the granules—by accident or not—may be responsible for the ageing of the granules and change their capacity? In this way we would have the basis for the life-span of the granules: gradual losses of proteins.

de Robertis: I think this could be a very good possibility. We know that the vesicle is reused but that it has a definite life-span. Probably the loss of these proteins may be an important factor in determining the life-span of the vesicle.

Dahlström: This may also imply that you do not need to have any protein synthesis going on in the axon terminals to refill the storage granules—they are gradually depleted of proteins, die off and are replaced by new ones.

Glowinski: I would like to mention that it has been shown that dopamine could be released as well as, or instead of, norepinephrine from peripheral noradrenergic neurons during nerve stimulation in pharmacological situation, such as partial inhibition of dopamine-β-oxidase. In this case, dopamine taken up in the vesicle and not converted in norepinephrine was released. This is not a direct argument in favour of partial exocytosis, but is in agreement with this hypothesis. Similarly, we have observed that dopamine synthesized from tyrosine in slices obtained from a noradrenergic containing structure of the central nervous system was spontaneously released as well as norepinephrine in the incubating medium. This again does not prove definitely that dopamin and norepinephrine contained in vesicles of central noradrenergic neurons are released by partial or complete exocytosis but, nevertheless, it supports this concept.

Thoenen: I think that it is a general feature observed with all false transmitters studied so far—in this case dopamine can be considered as a false transmitter in the periphery since usually it is undetectable or present only in very small amounts—that they are liberated together with norepinephrine in the same proportion as they are stored in the adrenergic nerve endings. This was first shown for α-methylnorepinephrine by Dr. Muscholl and Maître and later for dopamine and 5-hydroxydopamine by our group.

Philippu: Dr. Geffen, as I understand, you find a spontaneous release of dopamine-β-hydroxylase which is not accompanied by a release of chromogranin. How do you explain this dissociation of the release of the two proteins?

GEFFEN: I do not believe that our data demonstrate that there is a dissociated release. The antigenicity of dopamine-β-hydroxylase—in our hands at any rate—is much greater than chromogranin A, so we have much more sensitive immunological assay for the dopamine-β-hydroxylase. It may be that some chromogranin A is also present in the perfusate before stimulation but is below the limit of detection. I want to emphasize that in nerves, unlike in the adrenal medulla, one is working very closely to this limit, even with immunological methods which are capable of measuring nanograms of protein.

I did not get an opportunity to reply to Dr. DAHLSTRÖM and I just want to make this point. It is possible that the release of proteins, as she calls it, is an accident of the mechanism but I could also quote that very eminent authority, NEWTON, who said that Nature does nothing in vain. If nature is extremely parsimonious and if the adrenal medulla, as well as nerves, release so much protein then I think that we have first to look for possible functions for chromogranin outside of the neuron.

PHILIPPU: You think, therefore, that the spontaneous release of catecholamines is also due to exocytosis?

GEFFEN: There is spontaneous release in noradrenaline as we know from junction potentials in the absence of nerve impulse. However, the traces of dopamine-β-hydroxylase which we found in the control perfusates may have been released by previous activity in the spleen. I just do not think we can say anything yet. We have not perfused for long enough without stimulation to be certain that it is spontaneously released.

AXELROD: I just want to continue to question the concept of release of false transmitters raised by Dr. THOENEN. Dopamine is considered to be a false transmitter yet it is normally present in sympathetic nerves and it is released together with noradrenaline on nerve stimulation. Not only is noradrenaline and dopamine released but also other normally occurring amines, such as octopamine and perhaps other amines stored in vesicles. Thus, the multiple release of amines from nerves in addition to noradrenaline complicates the picture.

Bayer-Symposium II, 78—88 (1970)

Session 2b

Mechanism of Transmitter Release with Special Reference to the Problem of Exocytosis (II)

Chairman: P. Holtz

Quantal Aspects of the Secretion of Catecholamines and Dopamine-β-hydroxylase from the Adrenal Medulla

N. KIRSHNER and O. H. VIVEROS

With 1 Figure

During the past several years a number of laboratories have provided evidence which has established that secretion of catecholamines from the adrenal medulla occurs by a process in which the soluble contents of the catecholamine storage vesicles are secreted directly to the exterior of the cell leaving the vesicle membranes within. In the first studies, the materials which appeared in perfusates of isolated glands following chemical stimulation were determined and compared to the amounts which were present in the catecholamine storage vesicles. More recently the composition and properties of the storage vesicles which remain in the gland after stimulation have been examined and provide additional information on the mechanism of the secretory process.

DOUGLAS and co-workers [1] measured the appearance of catecholamines (CA) and ATP, or its hydrolytic products, in perfusates of cat adrenal glands. Following stimulation with acetylcholine, nicotine, excess potassium and calcium, they found increased amounts of AMP and CA in the perfusates such that the ratio of CA/AMP in perfusates (6:1) was similar to the ratio of CA/ATP in the gland (4:1). In subsequent work [2], they reported that after neurogenic stimulation the time course of the appearance in perfusates of AMP was the same as that of the CA. ATP perfused through cat adrenal glands was rapidly broken down to AMP, but when ATP hydrolysis was inhibited by perfusing the gland with Ca^{++}-free, Mg^{++}-free Locke's solution containing 1 to 2 mM EDTA, and CA secretion was evoked by Ba^{++}, large amounts of ATP appeared in the perfusates. Under these conditions the CA/ATP ratio in the perfusates was 11:1, but when total adenine nucleotides were considered, the ratio was again close to 4:1 [3]. BANKS [4] reported similar results using perfused bovine adrenal glands. These experiments clearly suggested that secretion occurred by exocytosis.

Banks and Helle [5] found increased amounts of protein, which was not hemoglobin, in perfusates of bovine adrenal glands stimulated with carbamylcholine. Using antisera prepared against the soluble proteins of the storage vesicles, they were able to detect these proteins in perfusates of stimulated glands but could not detect the proteins in perfusates of unstimulated glands. Shortly thereafter, Sage et al. [6] and Kirshner and Sage [7, 8] using a quantitative immunologic method, demonstrated that stimulation of isolated perfused bovine adrenal glands with acetylcholine or nicotine released CA and chromogranin A, the major soluble protein component of the storage vesicle [9 to 17], into the perfusate in the same relative amounts as that found in the isolated intact storage vesicles. They also showed that phenylethanolamine-N-methyl-transferase (PNMT), an anzyme which is present in the soluble fraction of adrenal homogenates, was not present in the perfusates of stimulated glands. Blaschko et al. [18] found that chromogranin A was also released under more physiological conditions. Using a calf anesthetized with pentobarbitone, they were able to measure catecholamines and chromogranin A in adrenal venous plasma following stimulation of the splanchnic nerve, but could not detect chromogranin A in the absence of stimulation. In a subsequent report, Schneider et al. [19], using starch gel electrophoresis and immunological methods, reported the presence of chromogranin A and other soluble proteins of the catecholamine storage vesicles in the perfusates of stimulated glands. In these same experiments, they found that stimulation with carbamylcholine did not increase the amounts of lactate dehydrogenase or lysolecithin in the perfusate and caused only small increases in the amounts of phospholipids, cholesterol and fatty acids.

The appearance of catecholamines, adenine nucleotides or their breakdown products, and intravesicular protein in perfusates of stimulated adrenal glands in the same ratios as that found in storage vesicles provides compelling evidence that the storage vesicles are the direct source of the released products and that these products are released directly to the exterior of the cell. This process can occur by at least two different mechanisms: (a), one in which the entire vesicle is extruded from the cell followed by lysis and liberations of its contents; (b), one in which the storage vesicles interact with the cell membrane in some manner such that the vesicle contents can then be released directly to the exterior. In this latter process, the vesicle membranes may become incorporated into the plasma membrane, or, after release of its contents, may return into the cytoplasm.

Electron microscopic studies of adrenal glands before and after stimulation showed structures which resembled "empty" storage vesicles [20, 21], but the observed changes may have been due to other events which may have occurred during the secretory process. These same studies showed no evidence of structures resembling vesicle membranes in the extracellular space. Although these studies were suggestive that the vesicle membranes were retained, they were not conclusive.

To obtain information on the fate of the storage vesicle membranes following secretion, Trifaro et al. [22] and Schneider et al. [19] examined perfusates of stimulated adrenal glands for the appearance of increased amounts of typical membrane components such as phospholipids, cholesterol and fatty acids during stimulation. Both groups found either no change or only very modest increases of these components, which suggested that the vesicle membranes were not released from the cell during secretion. However, even if the vesicle membranes are released from the cell, they may not readily penetrate the vascular walls or they may not be rapidly broken down into their lipid components. Although electron microscope studies [20, 21] showed no evidence of vesicle membranes in the extracellular space, the adrenal glands in these studies were taken from intact animals where the vesicle membranes, if extruded, may have been cleared very quickly from the gland *in vivo*.

Poisner et al. [23] examined homogenates of perfused adrenal glands following stimulation. Using discontinuous sucrose density gradient centrifugation, they found a decrease in the catecholamine and protein content of the fraction containing the storage vesicles and a slight increase in the lipid content of a membrane fraction which equilibrated at a lower density than the storage vesicles. The increase in the lipid content of the lighter fraction was not quantitatively equal to the decrease in the storage vesicle fraction but was suggestive that vesicle membranes were retained by the gland. Malamed et al. [24] prepared the storage vesicle fractions from stimulated and nonstimulated glands by differential centrifugation and compared their appearance by electron microscopy. The vesicles obtained from stimulated glands appeared more translucent than those from the control glands and again was suggestive that the "empty" vesicles were retained.

Prior to the above reports, we had also initiated a series of studies on the fate of the storage vesicles and their membranes following secretion. Two properties of the vesicles appeared, at that time, to offer means of investigating their fate. One was the ability of the isolated vesicles to incorporate labelled catecholamines into their endogenous stores in the presence of ATP and Mg^{++} [25, 26], and the second was the association of dopamine-β-hydroxylase (DBO) with the particulate fraction of the storage vesicles [27, 28]. Investigations of the incorporation would provide information on the functional integrity of the storage vesicle which remained after secretion while DBO would serve as a marker for the membrane itself.

Preliminary studies of DBO activity in lysed and intact storage vesicles of cat adrenals led to the finding in our laboratory [29], and independently by others [30 to 33], of the presence of potent endogenous inhibitors of the enzyme. The endogenous inhibitors could be inactivated by heavy metals such as Cu^{++}, Hg^{++} and Ag^{+} or by other sulfhydryl reactive reagents such as p-chloromercuribenzoate and N-ethylmaleimide. Inactivation of the endogenous inhibitors revealed the presence of both particulate and soluble DBO in the storage vesicles of the cat, cow [29], rabbit and rat (Table 1). The proportions of soluble and particulate DBO in storage vesicles varies considerably among the species tested but is quite constant within the species.

The presence of soluble DBO within the storage vesicle provided an additional means of testing whether the soluble contents of the storage vesicle were released upon stimulation. Viveros et al. [34] readily confirmed that CA and soluble DBO were released upon stimulation of isolated perfused bovine adrenal glands in the same ratios as found in the soluble fraction of lysed storage vesicles.

The presence of both particulate and soluble DBO in the storage vesicles proved to be a much more valuable tool for studying the secretory process than the presence of either one alone. If secretion occurred by release of the soluble contents but with retention of the vesicle membrane, one would expect to find after secretion a decrease in the CA and soluble DBO content and no decrease in the particulate DBO. Either because of loss of ATP and chromogranins, or because of disruption of the storage vesicles during secretion one would also expect a decrease in the ability of the storage vesicles to take up catecholamines which, on a percent basis, should be comparable to the loss of CA. One should also be able to distinguish neurogenic secretion from pharmacologic depletion induced by agents such as reserpine. After treatment of animals with reserpine, one would expect depletion of the CA, little or no change in DBO and complete, or almost complete, inhibition of the Mg^{++}-ATP stimulated uptake.

Table 1. *Distribution of dopamine-β-hydroxylase in rat and rabbit adrenals*

	DBO % of Total	
	Rabbit	Rat
Homogenate, 0.3 M Sucrose	100 ± 7	100 ± 7
800 g pellet	9 ± 1	—
26,000 g pellet	68 ± 6	80 ± 6
a) pellet particulate	28 ± 3	59 ± 6
b) pellet soluble	39 ± 4	21 ± 1
26,000 g supernatant	30 ± 3	19 ± 1
a) 100,000 g pellet	13 ± 2	6 ± 1
b) 100,000 g supernatant	17 ± 2	13 ± 1

The adrenals were homogenized in 0.3 M sucrose and subjected to differential centrifugation [41]. DBO in each of the fractions was assayed as previously described [41]. DBO activity in rabbit adrenals (n = 7) was 18.5 mμ moles of octopamine formed/hr/gland pair; in rat adrenals (n = 6) DBO activity was 1.2 ± 0.08 mμ-moles of octopamine formed/hr/gland.

In a first series of experiments, Viveros et al. [35, 36] found that neurogenic stimulation by insulin-induced hypoglycemia in rabbits produced a decrease in the CA and DBO of the storage vesicle fraction and a decrease in the ability of the storage vesicles to take up CA. After depletion of the CA with small doses of reserpine (0.25 to 1.0 mg/kg) there was almost complete inhibition of uptake and no change in the DBO content for several days. Five days after the reserpine treatment, there was a significant increase in the DBO content of the glands and at 8 days the levels were approximately twice that of the untreated controls. Large doses of reserpine (5 mg/kg) cause neurogenic stimulation of the adrenal gland and produce effects similar to those observed after insulin treatment [36].

In these same experiments [35, 36] the levels of PNMT and of tyrosine hydroxylase (TH) were also studied in control and insulin-treated animals. Insulin treatment had no effect on PNMT levels, but between 24 and 48 h after insulin treatment, there was an approximate doubling of the TH, and it remained at this level until the end of the experimental period (5 days). de Quattro et al. [37] also reported increases in heart and adrenal TH 24 to 48 h after sino-aortic denervation in dogs. More recently Mueller et al. [38, 39] reported increases in adrenal TH of rats after treatment with 6-hydroxydopamine and Thoenen et al. [40] reported increases in TH in adrenals and cervical sympathetic ganglia after treatment with reserpine.

Viveros et al. [41] confirmed and extended their previous studies and clearly showed that during neurogenic stimulation all of the DBO which was lost came from the soluble DBO of the storage vesicles and that there was no loss of particulate DBO (Fig. 1). The decrease of particulate DBO of the 26,000 g pellet was accounted for by an increase in the 100,000 g sediment. During secretion a portion of the vesicle membrane apparently became fragmented to the extent that it no longer sedimented at 26,000 g. Norman [42] observed fragmentation of storage vesicle membranes after release of neurosecretory granules in the corpus cardiacum of the blow fly, *Calliphora erythrocephala*, and Amsterdam et al. [43] reported similar observations for the zymogen granules of the rat parotid gland. Depletion of the adrenal CA content by small doses of reserpine, or by large doses of reserpine if a long-acting ganglionic blocking agent was administered before the reserpine, resulted in no loss of DBO [41].

These experiments established that neurogenic stimulation results in release of the soluble contents of the storage vesicles and retention of the storage vesicle membranes. However, they do not enable one to determine whether those vesicles which participated in tbe secretory response released their total content or only a portion of their content. By examining the relative amounts of DBO and CA in a population of purified storage vesicles isolated from untreated animals, from animals treated with insulin and from animals treated with reserpine and a ganglionic blocking agent, Viveros et al. [44] have obtained information which indicates that secretion occurs by release of the total contents of the stoarge vesicles.

In rabbit adrenals approximately 50% of the DBO of the storage vesicle is firmly bound to the particulate fraction and the remainder is readily solubilized upon lysis of the vesicles in distilled water [45]. During neurogenic stimulation only the soluble DBO is released in amounts proportionate to that of the CA [41]. Thus if the storage vesicles release only a portion of their contents, one would expect an increase in the DBO/CA ratio of the vesicles. In addition one might reasonably expect that partially depleted vesicles would have a lighter bouyant density than "filled" vesicles and would equilibrate at a different position in an isopycnic density gradient.

On the other hand, if vesicles released their total contents, one would expect the remaining vesicles to have the same DBO/CA ratios and the same bouyant density as vesicles obtained from unstimulated adrenal glands. One would also expect a portion but not all (see discussion of Fig. 1), of the membranes of the empty vesicle to be located in the portion of the gradient isopycnic with membranes from lysed vesicles.

When crude storage vesicle fractions prepared from adrenal glands of control animals, animals treated with insulin, animals treated with chlorisondamine (a long-lasting ganglionic blocking agent) and chlorisondamine plus reserpine were centrifuged through sucrose density gradients and the fractions containing the purified vesicles assayed for CA and DBO the results summarized in Table 2 were obtained [44]. A more detailed analysis of these experiments will be published elsewhere. R_g is a relative measure of the bouyant density and is defined as the distance of the CA and DBO peaks from the top of the gradient divided by the total length of the gradient. The DBO/CA ratios and R_g of the storage vesicles were the same in the control group, the insulin-treated group and the chlorisondamine-treated group, but the DBO/CA ratio of the animals which received chlorisondamine plus reserpine was significantly greater than, and the R_g of the vesicles was significantly less than those of the other three groups. The conclusion

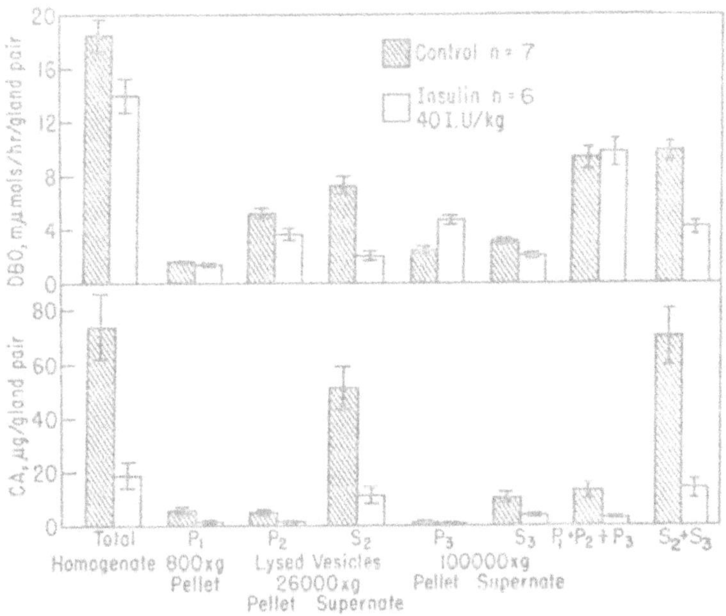

Fig. 1. Effect of insulin treatment on the DBO and CA content and distribution in rabbit adrenals. Glands from control and insulin-treated animals were subjected to differential centrifugation and assayed as described by VIVEROS et al. [41]. Note the decrease of DBO in the P_2 fraction of the insulin-treated animals and the compensatory increase in the P_3 fraction. Insulin-treated animals were killed 4 h after receiving 10 I.U./kg of insulin i.v.

drawn from these experiments is that during neurogenic stimulation, each of the vesicles which participates in the secretory response releases its total contents.

The evidence that the total content of each participating vesicle is released during the secretory response suggests that secretion from the adrenal medulla is quantized, and that the storage vesicle is the quantal unit. This implies that for each stimulus that causes a secretory response, one or more storage vesicles release their total content, and it also implies that the minimal stimulus is sufficient to cause the release of the total content of a storage vesicle. Although our data suggest that the content of each vesicle is released in an "all or none" fashion, they do not rule out the possibility that more than one stimulus is required for the total release from each vesicle. For example, vesicles may become apposed to the plasma membrane at the first stimulus and release only a portion of their contents. They may then remain attached to the membrane and respond to subsequent stimuli until they are completely depleted. If secretion proceeded in this manner, our method would not be sufficiently sensitive to detect 10% of the vesicle population which was partially depleted. Numerous electron microscope studies have shown

Table 2. *Dopamine-β-hydroxylase, catecholamines and R_g of purified storage vesicles after various drug treatments*

Treatment		DBO mμmoles × 100 / 1 hr/gland pair	R_gDBO	CA μg/gland pair	R_gCA	DBO/CA
None	(7)	853 ± 88	0.74 ± 0.01	53 ± 7	0.79 ± 0.01	15 ± 1.6
Insulin, 3 h	(6)	116 ± 16[a]	0.72 ± 0.02	10 ± 1.2[a]	0.77 ± 0.02	12 ± 1.3
Chlorisondamine	(7)	600 ± 16[c]	0.74 ± 0.02	46 ± 5	0.75 ± 0.02	14 ± 1.8
Chlorisondamine + Reserpine	(7)	537 ± 48[b]	0.64 ± 0.01[a]*	17 ± 3.5[a]*	0.68 ± 0.02[a]*	42 ± 9.6[c]*

Animals were treated with the drugs, the crude storage vesicle fraction (26,000 g pellet) prepared, centrifuged through sucrose density gradients and assayed for CA and DBO as previously described [44]. The figures in parentheses are the numbers of animals in each group. Data are means and standard errors of the means: a) $p < .001$; b) $p < .01$; c) $p < .02$ compared to untreated animals. The DBO values of the chlorisondamine-treated group and the chlorisondamine plus reserpine-treated group were not significantly different from each other.

* p values were the same compared to either the untreated animals or the chlorisondamine-treated animals.

Table 3. *Catecholamine secretion from cat adrenal medulla*[1]

1. Release by supramaximal stimulation first splanchnic nerve (50)	1×10^{-8} g/pulse/gland
2. Catecholamine content one adrenal gland (51)	3×10^{-4} g
3. Total adrenal gland weight[1]	1.5×10^{-1} g
4. Mean chromaffin cell diameter (52)	1×10^{-3} cm
5. Mean chromaffin vesicle diameter (52)	2.5×10^{-5} cm
6. Mean catecholamine concentration/vesicle (53)	5×10^{-1} M or 88 g/l
7. Total medulla weight (from 1/3 of 3)	5×10^{-2} g
8. Total chromaffin cell weight (from 2/3 of 7)	3.3×10^{-2} g
9. Chromaffin cell volume (from 4^{-3})	1×10^{-9} cm^3
10. Number of cells in one gland (from 8/9) (Assuming 1 cm^3 = 18)	3.3×10^{7} cells/gland
11. Vesicle volume (from 5 and volume of sphere)	8.2×10^{-18} l
12. Amount of catecholamine/vesicle (from 6 × 11)	7.2×10^{-16} g/vesicle or 2.4×10^{6} molecules/vesicle
13. Total number of vesicles/gland (from 2/12)	4.2×10^{11} vesicles/gland
14. Mean number of vesicles/cell (from 13/10)	1.3×10^{4} vesicles/cell
15. Number of vesicles released/pulse (from 1/12)	1.4×10^{7} vesicle/pulse
16. Number of vesicles released/pulse/cell (from 15/10)	4×10^{-1} vesicles/pulse/cell

[1] This calculation assumes that stimulation of the first splanchnic nerve will depolarize every cell. However, this is not true (50). A conservative estimate would be that 1/2 to 2/3 of the cells are innervated by the first splanchnis nerve.

Viveros, O. H. and Kirshner, N., unpublished (308 ± 12.6 mg/pair of gland, n = 54).

very few if any vesicles closely applied to the plasma membrane in a manner that suggests they were secreting their contents. Quantal release does not require that the entire content of a storage vesicle, whether from the adrenal medulla, from the noradrenergic or cholinergic synapse, or from other secretory cells be released per minimal stimulus—it only requires that, within a statistical range, a fixed number of molecules be released.

The quantum hypothesis for the release of neurotransmitters proposed by DEL CASTILLO and KATZ [46] has been widely accepted and has been confirmed in several different chemical synapses [47]. However, their suggestion that the synaptic vesicles are the quantum subcellular unit still lacks experimental verification. FOLKOW et al. [48] and STJARNE et al. [49] have proposed that sites other than the storage vesicles of the noradrenergic synapse are the immediate source of noradrenaline released by nerve stimulation. They have calculated that only 3 to 10% of the content of a single vesicle would be released from each varicosity on stimulation of the nerves to skeletal blood vessels. However, because of many questionable assumptions, these calculations offer little support for their proposal or against the concept that noradrenergic synaptic vesicles are the subcellular quantal units. Similar calculations (Table 3) for the release of catecholamines from the cat's adrenal medulla for each maximal stimulus applied to the first splanchnic nerve is of the order of one vesicle content per pulse per chromaffin cell.

The release of intravesicular secretory products from different cells and nerve terminals present remarkable similarities [54, 55]. Recent evidence [56 to 58] indicates that chromogranin and DBO may be within the same synaptic storage vesicle as noradrenaline. Studies similar to those reported here may determine whether release from sympathetic varicosities proceeds by exocytosis and whether the synaptic vesicles are the subcellular units of quantal noradrenaline release.

Summary

Several laboratories employing a variety of techniques—electron microscopy, chemical, immunological and enzymatic analysis—have provided evidence that secretion from the adrenal medulla occurs by a process in which the soluble contents of the storage vesicle are secreted directly to the exterior of the cell without first entering the cytosol. This process has been termed exocytosis.

Studies of the distribution of particulate and soluble dopamine-β-hydroxylase in homogenates of adrenal glands obtained from control animals and from insulin-treated animals show that only the soluble dopamine-β-hydroxylase is lost during neurogenic secretion and that all of the particulate enzyme remains within the gland. After reserpine administration, providing that neurogenic input to the adrenal gland is absent or blocked, there is a depletion of the adrenal catecholamines but no loss of either the soluble or particulate dopamine-β-hydroxylase. These studies provide additional evidence for exocytosis as the mode of secretion from the adrenal medulla.

Isopycnic centrifugation through sucrose density gradients of adrenal catecholamine storage vesicles obtained from control animals and from reserpine-treated animals show that vesicles obtained from the reserpine-treated animals have a lower bouyant density and a higher dopamine-β-hydroxylase : catecholamine ratio than do vesicles obtained from the control animals. These results are in concordance with the known direct effect of reserpine on the storage vesicles. Similar studies using storage vesicles obtained from insulin-treated animals show that these vesicles have the same bouyant density and the same dopamine-β-hydroxylase: catecholamine ratio as do vesicles from control animals even though the catecholamine content and the dopamine-β-hydroxylase activity of the glands from the insulin-treated animals were markedly depleted. These observations indicate that during neurogenically stimulated secretion the vesicles release their content in an "all or none" response.

Acknowledgements

The authors wish to acknowledge the following sources of support: United States Public Health Service Grant AM 05427; American Medical Association Education and Research Foundation: NIH Research Career Award K3 GM-15,184 and an NIH International Postdoctoral Fellowship.

References

1. DOUGLAS, W. W., POISNER, A. M., RUBIN, R. P.: Efflux of adenine nucleotides from perfused adrenal glands exposed to nicotine and other chromaffin cell stimulants. J. Physiol. (Lond.) **179**, 130—137 (1965).
2. — — Evidence that the secreting adrenal chromaffin cell releases catecholamines directly from ATP-rich granules, J. Physiol. (Lond.) **183**, 236—248 (1966).
3. — — On the relation between ATP splitting and secretion in the adrenal chromaffin cell: Extrusion of ATP (unhydrolyzed) during release of catecholamines. J. Physiol. (Lond.) **183**, 249—256 (1966).
4. BANKS, P.: The release of adenosine triphosphate catabolites during the secretion of catecholamines by bovine adrenal medulla. Biochem. J. **101**, 536—541 (1966).
5. —, HELLE, K.: The release of protein from the stimulated adrenal medulla. Biochem. J. **97**, 40 c—41 c (1965).
6. SAGE, H. J., SMITH, W. J., KIRSHNER, N.: Mechanism of secretion from the adrenal medulla I. A. microquantitative immunologic assay for bovine adrenal catecholamine storage vesicle proteins and its application to studies of the secretory process. Molec. Pharmacol. **3**, 81—89 (1967).
7. KIRSHNER, N., SAGE, H. J., SMITH, W. G., KIRSHNER, A. G.: Release of catecholamines and specific protein from adrenal glands. Science **154**, 529—531 (1966).
8. — — — Mechanism of secretion from the adrenal medulla II. Release of catecholamines and storage vesicle protein in response to chemical stimulation. Molec. Pharmacol. **3**, 254—265 (1967).
9. HILLARP, N.-A., LAGERSTEDT, S., NILSON, B.: The isolation of a granular fraction from the suprarenal medulla, containing the sympathomimetic catecholamines. Acta physiol. scand. **29**, 251—263 (1953).
10. BLASCHKO, H., HELLE, K. B.: Interaction of soluble protein fractions from bovine adrenal medullary granules with adrenaline and adenosinetriphosphate. J. Physiol. (Lond.) **169**, 120—121 (1963).
11. KIRSHNER, N., HOLLOWAY, C., SMITH, W. J., KIRSHNER, A. G.: Uptake and storage of catecholamines. In: *Mechanisms of release of biogenic amines*, pp. 109—123. Ed. by VON EULER, U. S., ROSELL, S., UVNOS, B. Eds: Oxford: Pergamon Press 1966.
12. BLASCHKO, H., SMITH, A. D., WINKLER, H.: Untersuchungen an Eiweißfraktionen der chromaffinen Granula. Naunyn-Schmiedebergs-Arch. exp. Path. Pharmak. **253**, 23 (1966).
13. SMITH, A. D., WINKLER, H.: Studies of soluble protein from adrenal chromaffin granules. Biochem. J. **95**, 42 P (1965).
14. HELLE, K. B.: Some chemical and physical properties of the soluble protein fraction of bovine adrenal chromaffin granules. Molec. Pharmacol. **2**, 298—310 (1966).
15. SMITH, W. J., KIRSHNER, N.: A specific soluble protein from the catecholamine storage vesicles of bovine adrenal medulla I. Purification and chemical characterization. Molec. Pharmacol. **3**, 52—62 (1967).
16. SMITH, A. D., WINKLER, H.: Purification and properties of an acidic protein from chromaffin granules of bovine adrenal medulla. Biochem. J. **103**, 483—492 (1967).

17. Kirshner, A. G., Kirshner, N.: A specific soluble protein from the catecholamine storage vesicles of bovine adrenal medulla II. Physical characterization. Biochim. biophys. Acta (Amst.) **181**, 219—225 (1969).
18. Blaschko, H., Comline, R. S., Schneider, F. H., Silver, M., Smith, A. D.: Secretion of chromaffin granule protein, Chromogranin, from the adrenal gland after splanchnic stimulation. Nature (Lond.) **215**, 58—59 (1967).
19. Schneider, F. H., Smith, A. D., Winkler, H.: Secretion from the adrenal medulla: Biochemical evidence for exocytosis. Brit. J. Pharmacol. **31**, 94—104 (1967).
20. de Robertis, E. D. P., Vaz Ferreira, A.: Electron microscope study of the excretion of catechol-containing droplets in adrenal medulla. Exp. Cell Res. **12**, 568—574 (1957).
21. Wetzstein, R.: Elektronenmikroskopische Untersuchungen am Nebennierenmark von Maus, Meerschweinchen und Katze. Z. Zellforsch. **46**, 517—576 (1957).
22. Trifaro, J. M., Poisner, A. M., Douglas, W. W.: The fate of the chromaffin granule during catecholamine release from the adrenal medulla I. Unchanged efflux of phospholipid and cholesterol. Biochem. Pharmacol. **16**, 2095—2100 (1967).
23. Poisner, A. M., Trifaro, J. M., Douglas, W. W.: The fate of the chromaffin granule during catecholamine release from the adrenal Medulla II. Loss of protein and retention of lipid in subcellular fractions. Biochem. Pharmacol. **16**, 2101—2108 (1967).
24. Malamed, S., Poisner, A. M., Trifaro, J. M., Douglas, W. W.: The fate of the chromaffin granule during catecholamine release from the adrenal medulla III. Recovery of a purified fraction of electron translucent granules. Biochem. Pharmacol. **17**, 241—246 (1968).
25. Kirshner, N.: Uptake of catecholamines by a particulate fraction of the adrenal medulla. J. biol. Chem. **237**, 2311—2317 (1962).
26. Carlsson, A., Hillarp, N.-A., Waldeck, B.: Analysis of the Mg^{++}-ATP dependent storage mechanism in the amine granules of the adrenal medulla. Acta physiol. scand. **59**, Suppl. **215**, 5—38 (1963).
27. Kirshner, N.: Pathway of noradrenaline formation from Dopa. J. biol. Chem. **226**, 821—825 (1957).
28. Levin, E. Y., Levenberg, B., Kaufman, S.: The enzymatic conversion of 3,4-dihydroxyphenylethylamine to norepinephrine. J. biol. Chem. **235**, 2080—2086 (1960).
29. Duch, D. S., Viveros, O. H., Kirshner, N.: Endogenous inhibitors in adrenal medulla of dopamine-β-hydroxylase. Biochem. Pharmacol. **17**, 255—264 (1968).
30. Creveling, C. R.: Studies on dopamine-β-oxidase. Doctoral Thesis, The George Washington University, Washington, D.C. 1962.
31. Nagatsu, T.: Endogenous inhibitors of dopamine-β-hydroxylase. In: Biological and chemical aspects of oxygenases. Eds.: Bloch, K. E., Hayaishi, O., Tokyo: Maruzen 1966, p. 273.
32. —, Kuzuya, H., Kidaka, H.: Inhibition of dopamine-β-hydroxylase by sulfhydryl compounds and the nature of the natural inhibitors. Biochim. biophys. Acta (Amst.) **139**, 319—327 (1967).
33. Austin, L., Livett, B. G., Chubb, I. W.: Biosynthesis of noradrenaline in sympathetic nervous tissue. Amer. Heart Ass. Monograph **17**, 111—117 (1967).
34. Viveros, O. H., Arqueros, L., Kirshner, N.: Release of catecholamines and dopamine-β-oxidase from the adrenal medulla. Life Sci. **7**, 609—618 (1968).
35. — — — Mechanism of secretion from the adrenal medulla. The fate of the catecholamine storage vesicles. Fed. Proc. **27**, 601 (1968).
36. — —, Connett, R. J., Kirshner, N.: Mechanism of secretion from the adrenal medulla IV. The fate of the storage vesicles following insulin and reserpine administration. Molec. Pharmacol. **5**, 69—82 (1969).
37. de Quattro, V., Maronde, R., Nagatsu, T., Alexander, N.: Altered norepinephrine synthesis and storage in the hypertensive buffer denervated rabbit. Fed. Proc. **27**, 240 (1968).
38. Mueller, R. A., Thoenen, H., Axelrod, J.: Adrenal tyrosine hydroxylase: Compensatory increase in activity after chemical sympathectomy. Science **158**, 468—469 (1969).
39. — — — Increase in tyrosine hydroxylase activity after reserpine administration. J. Pharmacol. exp. Ther. **169**, 74—79 (1969).

40. THOENEN, H., MUELLER, R. A., AXELROD, J.: Increased tyrosine hydroxylase activity after drug-induced alteration of sympathetic transmission. Nature (Lond.) **221**, 1264 (1969).
41. VIVEROS, O. H., ARQUEROS, L., KIRSHNER, N.: Mechanism of secretion from the adrenal medulla V. Retention of storage vesicle membranes following release of adrenaline. Molec. Pharmacol. **5**, 342—349 (1969).
42. NORMANN, T. C.: Experimentally induced exocytosis of neurosecretory granules. Exp. Cell Res. **55**, 285—287 (1969).
43. AMSTERDAM, A., OHAD, I., SCHRAMM, M.: Dynamic changes in the ultrastructure of the acinar cell of the rat parotid gland during the secretory cycle. J. Cell Biol. **41**, 753—773 (1969).
44. VIVEROS, O. H., ARQUEROS, L., KIRSHNER, N.: Quantal secretion from adrenal medulla: All or none release of storage vesicle content. Science **165**, 911—913 (1969).
45. — —, CONNETT, R. J., KIRSHNER, N.: Mechanism of secretion from the adrenal medulla III. Studies of dopamine-β-hydroxylase as a marker for catecholamine storage vesicle membranes in rabbit adrenal glands. Molec. Pharmacol. **5**, 60—68 (1969).
46. DEL CASTILLO, J., KATZ, B.: Quantal components of the end-plate potential. J. Physiol. (Lond.) **124**, 560—573 (1954).
47. MARTIN, A. R.: Quantal nature of synaptic transmission. Physiol. Rev. **46**, 51—66 (1966).
48. FOLKOW, B., HAGGENDAL, J., LISANDER, B.: Extent of release and elimination of nor-adrenaline at peripheral adrenergic nerve terminals. Acta physiol. scand. Suppl. **307** (1967).
49. STJÄRNE, L., HEDGVIST, P., BYGDEMAN, S.: Neurotransmitter quantum released sympathetic nerves in cat's skeletal muscle. Life Sci. 8, 189—196 (1969).
50. MARLEY, E., PROUT, G. I.: Physiology and pharmacology of the splanchnicadrenal medullary junction. J. Physiol. (Lond.) **180**, 483—513 (1965).
51. —, PATON, W. D. M.: The output of sympathetic amines from the cat's adrenal gland in response to splanchnic nerve activity. J. Physiol. (Lond.) **155**, 1—27 (1961).
52. COUPLAND, R. E.: The natural history of the chromaffin cell. London: Longman, Green and Co., Ltd. 1965.
53. HILLARP, N.-A.: Further observations on the state of the catecholamines stored in the adrenal medullary granules. Acta. physiol. scand. **47**, 271—279 (1959).
54. DOUGLAS, W. W.: Stimulus-secretion coupling: The concept and clues from chromaffin and other cells. Brit. J. Pharmacol. **34**, 451—474 (1968).
55. SMITH, A. D.: Biochemistry of adrenal chromaffin granules. In: The interaction of drugs and subcellular components of animal cells, p. 239—292. CAMPBELL, P. W. Ed. London: J. A. Churchill, Ltd. 1968.
56. POTTER, L. T., AXELROD, J.: Properties of norepinephrine storage particles of the rat heart. J. Pharmacol. **142**, 299—305 (1963).
57. LADURON, P., BELPAIRE, F.: Transport of noradrenaline and dopamine-β-hydroxylase in sympathetic nerves. Life Sci. **7**, 1 (1968).
58. LIVETT, B. G., GEFFEN, L. B., RUSH, R. A.: Immunohistochemical evidence for the transport of dopamine-β-hydroxylase and a catecholamine binding protein in sympathetic nerves. Biochem. Pharmacol. 18, 923—924 (1969).

Dr. N. KIRSHNER
Department of Biochemistry and Pathology
Duke University, Medical Center
Durham/North Carolina 27706/USA

Discussion

DE ROBERTIS: It was very interesting to see the difference between the release when one stimulates the adrenal by a neurogenic mechanism or by reserpine. I remember a paper by SJÖSTRAND in which he used reserpine and the catecholamine containing vesicle did not disappear. From our work we know that when the splanchnic nerve was stimulated the total content of the vesicle disappeared. I am happy to see how well your results fit with the views of electron microscopy.

CARLSSON: I must say the data are really first-class and one certainly feels inclined to be absolutely convinced but, of course, it is nice if one could at least try to play the role of the doubting THOMAS and put forward a few questions!

KIRSHNER: But we too have some questions.

CARLSSON: May be they are the same—in one of your first slides where you had the supernatant, the loss of supernatant DBO and loss of supernatant catecholamines following insulin, it seemed to me that the loss of catecholamines was considerably greater than the loss of DBO, whereas later in other experiments the ratios agreed very well. I should appreciate it if you could comment on that. Another thing that came to my mind is that if you stimulate for such a long time, 8 h or so, there is a tremendous synthesis of new catecholamines. Therefore, if the ratio now between the two is constant, as it was in some of the experiments, would you not then have to postulate that DBO is being synthesized at the same rate as the catecholamines, or is this somehow wrong?

KIRSHNER: I do not think the two pieces of data are completely comparable and I still have some questions about the rate of re-synthesis of adrenaline or noradrenaline during stimulation from the adrenal gland. I think if one looks at the data one can find from one laboratory, rates of re-synthesis amounting to from nothing to 40 or 50% of the amount which was initially present and I find this rate of synthesis very hard to believe.

CARLSSON: So you do not think there is any synthesis during the course of the 4 h—any real appreciative synthesis.

KIRSHNER: We have done some studies on isolated perfused bovine adrenal glands and found no evidence of appreciable increases in synthesis, following stimulation with acetylcholine. The amounts of catecholamine recovered in the perfusate was equal to the amount lost from the gland. The insulin data with intact glands indicate, and my own feeling is, that there is no appreciable re-synthesis of noradrenaline during this period of time.

CARLSSON: Would you care to comment upon the apparent discrepancy between the first slide and the second slide; I mean from the point of view of the release of more catecholamines than DBO?

KIRSHNER: Yes, if one calculated the dopamine-β-hydroxylase to catecholamine ratios of a normal gland and what one might expect after secretion, the ratios here were somewhat less than what one would expect, that is, there was not as much release of the soluble DBO as one might have expected from the decrease in the catecholamine content. However, there is a fairly large standard area in these experiments, i.e. there is wide variation from one animal to another that may possibly be an explanation. In other experiments we do find a very good correspondence and very little, if any, significant difference between the DBO to catecholamine ratios of what is present in an intact animal after stimulation and what we would calculate by difference, knowing what is left at the end.

HELLE: Do you have any quantitative data on how much of the granule protein is made up by the enzyme dopamine-β-hydroxylase?

KIRSHNER: We do not have any strict data on the quantitation but Dr. GEFFEN showed a chromatogram this morning—a separation of the protein on DEAE cellulose and we have obtained entirely similar results. It looks like, and it certainly is, less than 10%; probably in the order of only 2 or 3% of the total soluble protein.

Bayer-Symposium II, 91—97 (1970)

Some Aspects of the Quantal Release of the Adrenergic Transmitter

BJÖRN FOLKOW and JAN HÄGGENDAL

In a recent study concerning the noradrenaline (NA) release at the vasoconstrictor nerve endings in the cat calf muscles, the results suggested that the transmitter amount released per stimulus was only about 1/50.000 of the total regional NA content (FOLKOW et al., 1967). Assuming that, first, by far the major part of tissue NA is present in the varicosity granules (vesicles) and, second, the majority of varicosities discharge transmitter when an impulse arrives, it was deduced that some 400 NA molecules would be released per varicosity and impulse. Such an amount, which creates a peak NA concentration of 0.5 to 1 µg/ml if evenly distributed in a junction gap having the size of 1 to 2 μ^2 and a width of 1000 A, would correspond to only some 3% of the NA content of one granule, provided that the varicosity contains about 1000 granules and each granule about 15.000 NA molecules. These figures were based on calculations of the number of varicosities in the rat peripheral adrenergic neuron and the total NA content of the corresponding tissue (DAHLSTRÖM et al., 1966).

It is generally accepted that resting motor nerve fibres release transmitter by means of a random discharge of subthreshold "quantal packets", which become jointly discharged in such large numbers when an impulse arrives that effector excitation ensues (KATZ, 1962; ECCLES, 1964). Experimental evidence suggests that this is the case also for peripheral adrenergic nerves [BURNSTOCK et al., 1962 (1, 2)], where a random quantal release may occur once each 30th to 90th second at the individual junction. The quantal packet is generally assumed to be identical with the transmitter-containing vesicle, though KATZ (1962) has as an alternative proposed that the limitation of the "packet" might be set by the nerve membrane in the form of a restricted number of coupling-discharge sites for the transmitter molecules. In that case the size of the packet may well correspond to only a *fraction* of the vesicle content. Another possibility is that transmitter vesicles in nerve fibre endings are themselves "compartmentalized" so that only a fraction of the total content is available for immediate release, as discussed by e.g. FOLKOW et al. (1967). Thus, assuming that the majority of adrenergic varicosities release transmitter at the arrival of an impulse, the quantal packet of NA may correspond to far less than 400 molecules, i.e. to about one per cent of the granule content at most, in case the release of *several* subthreshold quanta may be necessary to induce an effector cell excitation. This by no means contradicts the wellestablished concept of a quantal transmitter release, but it would no doubt suggest that the quantal packet is far smaller than the content of one granule (vesicle).

The present discussion will be mainly devoted to the problem as to whether there may be alternative ways of explaining the mentioned experimental results, but first some other aspects will be briefly commented upon. Recently Stjärne et al. (1969) performed a similar study but with the modification that they perfused the calf muscles with Krebs Henseleit's solution instead of with blood and used phenoxybenzamine instead of the membrane pump blockers used by us. This was based on the assumption that phenoxybenzamine may more effectively block the membrane reuptake pump and further, that it would facilitate the NA washout by preventing vasoconstriction. With these procedures, the NA release in the early stimulation period corresponded to some 1200 molecules when calculated per varicosity and stimulus as outlined above. Provided that phenoxybenzamine causes a more effective blockade of local routes for NA elimination but does not interfere with its release, such a figure—about three times higher than ours—would be a better reflection of the physiological release of the adrenergic transmitter.

When using phenoxybenzamine in the blood-perfused calf muscles, we have obtained similar figures for NA release as Stjärne et al. (1969), being about 3 times higher than when specific membrane pump blockers were used, even when the samples were taken during the first 2 min of stimulation. The question arises, however, as to whether the *highest* NA figures are a valid reflection of the physiological release. As will be discussed by Dr. Häggendal it is possible that phenoxybenzamine affects also the NA *release* mechanism and in such a direction that abnormally large amounts are released per stimulus. That drugs may interfere with the release process in different ways appears to be only seldom considered in these connections but such interferences are, of course, possible; we observed, for example, that very high concentrations of the membrane pump blockers used reduced the NA release (Folkow et al., 1967).

It is therefore debatable whether the results with phenoxybenzamine more accurately reflect the physiological NA release; the range would be 400 to 1200 NA molecules per stimulus and varicosity where the lower level may after all be the most relevant one. Furthermore it seems justified to make also another correction of quantitative importance for the subsequent considerations. In our study we assumed—based on by then available results—that each varicosity contained 1000 granules as an average. Recent systematic calculations of the number of granules in the varicosities of the rat iris suggest a figure of 300 to 400, though with a considerable scatter (Hökfelt 1968, 1969). If this figure is valid also for the cat the average NA content of one granule would be almost 3 times higher than we predicted, i.e. approaching 40.000 molecules instead of 15.000. Such a figure is not unreasonably high because the approximate percentage weight of NA in one granule, assuming a specific gravity of about 1.3 and a mean diameter of 500 to 550 A, would be only slightly higher than the figure of 6.7% found by Hillarp (1959) in the adrenal medullary granules. The net result of these considerations would be that the average NA release per varicosity and stimulus is still only at most 3% of the NA granule content; it would be closer to 1% if phenoxybenzamine causes an abnormally high NA release. The quantal packet would then make up perhaps only a fraction of 1%, if *several* packets are discharged per impulse and varicosity.

However, the important question arises: May each varicosity, rather than releasing 1 to 3% of the NA content of one granule per impulse, discharge the *entire*

NA content of one or even several granules but correspondingly *less frequently*? One prerequisite for this latter alternative would be that the spontaneous random release of some 40.000 molecules into one junction gap, here creating a peak concentration of as much as 50 to 100 μg/ml if evenly distributed in the gap, would be insufficient to excite the cell; a considerable number of simultaneously discharged packets is no doubt needed [BURNSTOCK et al., 1962 (1, 2)]. This enormous NA concentration should be contrasted to the finding on the rat portal vein that exposure of *all* junction gaps to a sustained NA concentration of only 1 to 3 μg/ml induces 50% of the maximal response (LJUNG, 1969), where the uptake of NA by the "membrane pump" greatly decreases the efficiency of exogenous NA. If this uptake is eliminated, e.g. by nerve degeneration, only about one tenth of this concentration is needed.

Another prerequisite would be that the individual varicosity could discharge NA only about each 30th to 100th impulse, or even less frequently if several "granular" packets are released per varicosity, i.e. at least 97 to 99% of the impulses would leave the varicosity unaffected. Such a type of nerve function would apparently imply a most inefficient "coupling" between the nerve impulse and the transmitter release. As mentioned, a fairly *simultaneous* local depolarization in many junction gaps and in several effector cells seems to be needed to produce excitation and a propagated response in smooth muscle [BURNSTOCK et al., 1962 (1)]. A system where enormous concentrations of transmitter are created, but only in a minor fraction of available junctions, would appear as an inefficient and uneconomical mode of transmitter turnover. For such reasons this view may seem unlikely though of course not theoretically impossible.

Let us therefore consider these difficulties in relation to the organization of the adrenergic neuro-effectors to see whether they may be only apparent, or whether they make the hypothesis that the NA quantal packet is identical with one granule unlikely. In the densely innervated "multiunit" type of smooth muscle, such as the vas deferens or the iris, "unit bundles" are formed, about 1 mm in length and some 100 μ in diameter (MERRILLEES, 1967; see also BENNET et al., 1968). Apparently cell-to-cell propagation occurs *within* the individual bundle, but *not* between adjacent bundles, so that each bundle may function as a syncytium that is isolated from the surrounding ones. A joint activation of such bundles therefore calls for a nervous integration, in analogy to the situation when different motor units of a skeletal muscle are activated simultaneously.

Assuming an interstitial space of up to 40%, a cell length of about 150 μ and an average cell diameter of some 3 μ, such a "unit bundle" of multi-unit smooth muscle would contain approximately 3000 to 4000 cells. The number of varicosities in the rat iris and vas deferens (DAHLSTRÖM et al., 1966) may be used for estimating the number of varicosities each bundle would contain, giving a figure of some 8000; i.e. 2 to 3 varicosities per smooth muscle cell. If these 3000 to 4000 muscle cells within the unit bundle are electrically closely coupled to allow a complete bundle activation if only a limited number of them are simultaneously excited (BENNETT et al., 1968), there would be available 8000 varicosities for such an excitation. At a fibre discharge of one impulse/sec, corresponding to the tonic sympathetic activity during "rest" (FOLKOW, 1955), transmitter release would as a mean occur within the bundle from only 80 to 300 of its 8000 varicosities per impulse (1 to 3%

per imp.), even if the contents of only *one* granule were discharged from each of these excited varicosities.

It is known from recordings of the resting discharge of 1 to 2 imp/sec in bundles of sympathetic nerve axons that these bursts of impulses are not strictly synchronized in time but scattered over a period of perhaps 0.3 sec (e.g. Bronk et al., 1940). This scatter in time of the impulses may be expected to be still larger once the effector cells are reached, which means that these are probably never affected quite simultaneously by the transmitter release during physiological circumstances. The miniature potentials in adrenergic effectors seem, however, to be fairly prolonged (about 150 msec) and so are the junction potentials produced by direct nerve stimulations [Burnstock et al., 1962 (1)]; further, there is a considerable spread of electrotonus within smooth muscle bundles (Holman, 1968). These observations therefore suggest that even a fairly marked scatter in time and space of neurogenically induced subthreshold potentials (where each is tentatively supposed to be caused by some 40.000 NA molecules) may nevertheless fuse and be able to cause bundle excitation. Consequently, no theoretically insurmountable difficulties are encountered so far in reasoning along these lines. However, the concentrations of NA produced in a very limited number of junctions would then be enormous compared with what is really needed if instead a substantial number of junctions are exposed to NA release. It would almost be like shooting at a flock of birds with occasional artillery grenades instead of using the no doubt less expensive and probably more efficient way of firing shots from a shotgun.

Further, these figures for assumed granule release during a fibre discharge of one impulse per second, which is known to cause almost 50% of the maximal response in such effectors as the nictitating membrane, the systemic veins or the spleen, must be related to those produced by the spontaneous random release during fibre rest, where the NA concentrations are clearly subthreshold. Analyses of intracellular recordings of spontaneously occurring miniature potentials in the vas deferens [Burnstock et al., 1962 (1, 2)] have led to the proposal that a quantal packet of NA might be discharged into the individual junction gap as seldom as once each 30th to 90th second. If this proposal is even remotely correct it follows, however, that granules would be discharged spontaneously *each second* from some 100 to 300 of the 8000 varicosities present within the unit bundle. The situation would in this respect probably not be very different in blood vessels, though fewer varicosities seem to be present per unit number of muscle cells because the varicosities reach only the outermost muscle layer of the vessels. However, this outer layer seems to be as richly innervated as the cells of the "unit bundle" in multiunit types of smooth muscle and the inner muscle layers of vascular walls appear to be effectively excited by means of cell-to-cell propagation because of their "syncytial" type of arrangement (Johansson et al., 1968).

Thus, the hypothesis that one NA granule should be identical with the quantal packet appears to end in the paradox that the random subthreshold discharge of NA from resting nerve fibres would be *largely the same* per second as that produced when the fibres discharge once per second, in which case almost 50% of the maximal response is reached in many adrenergic neuro-effector units. It should further be stressed that a great many studies—including our own results—indicate that the NA release from resting nerve fibres is only a minor fraction of that released during, say, one impulse/sec. The hypothesis would further have the consequence that an adrenergic fibre varicosity, when reached by the nerve impulse, would be capable of discharging transmitter only once each 30th to 100th impulse, at the most. Finally, during fibre rest smooth muscle cells would not be activated if the enormous NA concentrations of 50 to 100 μg/ml were created at random and almost simultaneously in quite a number of junction gaps per unit bundle, while only a minor fraction of this concentration is known to cause 50% of maximal response in e.g. the portal vein if *all* junction gaps are reached.

On the basis of such considerations concerning the amounts of released transmitter during rest and activity, and concerning the relationship between fibre varicosities and subordinated smooth muscle effectors, it appears difficult to maintain the view that the NA quantal packet should be identical with one granule. However, all such difficulties are easily solved, without any principal disagreement with the concept of quantal release of transmitter, if the assumption is accepted that only a *small fraction* of the granular content is immediately available for release. This would mean that the transmitter granules (vesicles) of nerve fibres may be "compartmentalized". This would be in contrast to the situation in the adrenal medulla where some evidence indicates that the entire granule content can be released. Such a proposal for differentiation may not be unrealistic; there is, after all, a marked difference between neurons and adrenal gland cells. Thus, in a neuron the transport routes for granules from the source of production to the site of release may be about a 100.000-fold longer. In return, the neuron has the ability to deliver the active substance in immediate contact with the effector cell which greatly decreases the transmitter amounts needed to reach an effective local concentration. By a compartmentalization of the neuronal transmitter granules, combined with the ability for NA reuptake into the granules and synthesis of new transmitter, the serious transport problems and unnecessary excesses of release would be effectively met with. In fact, a great number of physiological and pharmacological studies suggest the presence of a small pool of readily available adrenergic transmitter (cf. CARLSSON, 1965; IVERSEN, 1967). Further, the life span of the adrenergic granules appears to be several weeks (DAHLSTRÖM and HÄGGENDAL, 1966).

Such aspects should also be considered in the light of the fact that "tonic" motoneurons, which can command up to 200 muscle fibres, may for long periods excite all the muscle fibres at rates of upto 20/sec with the release of perhaps hundreds of quanta per endplate and impulse. An apparently impossible axonal transport problem (or local resynthesis problem) would here arise if entire vesicles (or their total contents) constitute the quantal packet. Their turnover could then be towards a million per second and each minute the cell soma of such a neuron would have to produce 10% of its own weight in transmitter vesicles, if these are considered to be discharged and therefore lost at the nerve endings; not to mention the enormous flow of vesicles along the axons that would have to occur if it is not assumed that vesicles are formed in the nerve terminals. In fact, concerning the cholinergic transmission, from which the concept of quantal release emerges by the brilliant studies of KATZ and coworkers, a number of experimental findings seems to be best compatible with the idea that the quantal packet corresponds to only a *fraction* of the vesicle content of transmitter (cf. FOLKOW et al.,

1967). For example, the figures for ACh quantal release that can be deduced from the results of Emmelin et al. (1956) seem to be most satisfactorily explained by the hypothesis that the transmitter vesicles are compartmentalized and release only a few per cent of their transmitter content, if it is assumed that the ACh vesicle contains roughly the same number of transmitter molecules as the NA granule. Such an assumption is not unreasonable as the two types of vesicles (granules) and transmitter molecules are of about equal size; further, the principles of specific binding within the vesicles (granules) may be expected to be about the same.

Summary

On the basis of measurements of the NA release from vasoconstrictor nerve endings, quantitative histochemical data, and parallels to the ACh release from motoneurons, it is discussed whether the quantal packet of the adrenergic transmitter is constituted by the entire content of one varicosity granule or by only a small fraction of this content.

It seems difficult to harmonize available experimental data with the hypothesis that one granule should constitute the quantal packet as it appears that only a few per cent of the NA content of one transmitter granule corresponds to the minimal quantum released. In comparison to the situation in adrenal medullary cells, the transport routes for the neuronal transmitter granules from their site of production are enormous, but in return the site of release is very close to the effector cell, allowing for a marked reduction in the amount of transmitter release needed per impulse.

It is proposed that this situation has led to the development of a *compartmentalization of the neuronal granules (vesicles)* so that they contain different pools of transmitter. A small pool, perhaps corresponding to only a few percent of the content, would contain readily releasable transmitter that constitutes the true quantal packet while the greater part of the NA granule content would be held in store. From this large store the readily releasable transmitter pool would be maintained, to which the processes of transmitter synthesis and reuptake would contribute.

Acknowledgements

The results discussed in the present paper have been supported by the Air Force Office of Scientific Research through the European Office of Aerospace Research OAR United States Air Force under Contract F 61052-68-C-0044, by a grant from US Public Health Service (HE-05675-08) and from the Swedish Medical Research Council by grants B 70-14-X-16-06 B and B 70-14 X-166-06 B.

References

Bennet, M. R., Burnstock, G.: Electrophysiology of the innervation of intestinal smooth muscle. Handbook of physiology, section 6, vol. IV, 1709—1732 (1968).

Bronk, D. W., Pitts, R. F., Larrabee, M. G.: Role of hypothalamus in cardiovascular regulation. Ass. Res. nerv. Dis. Proc. **20**, 323—341 (1940).

Burnstock, G., Holman, M. E.: (1) Spontaneous potentials at sympathetic nerve endings in smooth muscle. J. Physiol. (Lond.) **160**, 446—460 (1962).

— — (2) Effect of denervation and of reserpine treatment on transmission at sympathetic nerve endings. J. Physiol. (Lond.) **160**, 461—469 (1962).

Carlsson, A.: Drugs which block the storage of 5-hydroxytryptamine and related amines. Handbook of experimental pharmacology **19**, 529—592 (1965).

DAHLSTRÖM, A., HÄGGENDAL, J.: Some quantitative studies on the noradrenaline content in the cell bodies and terminals of a sympathetic adrenergic neuron system. Acta physiol. scand. 67, 271—277 (1966).
— —, HÖKFELT, T.: The noradrenaline content of the nerve terminal varicosities of sympathetic adrenergic neurons in the rat. Acta physiol. scand. 67, 289—294 (1966).
ECCLES, J. C.: The physiology of synapses. Berlin-Göttingen-Heidelberg-New York: Springer 1964.
FOLKOW, B.: Nervous control of blood vessels. Physiol. Rev. 35, 629—663 (1955).
—, HÄGGENDAL, J., LISANDER, B.: Extent of release and elimination of noradrenaline at peripheral adrenergic nerve terminals. Acta physiol. scand. Suppl. 307 (1967).
HILLARP, N.-Å.: Further observations on the state of the catechol amines stored in the adrenal medullary granules. Acta physiol. scand. 47, 271—279 (1959).
HOLMAN, M. E.: Introduction to electrophysiology of visceral smooth muscle. Handbook of physiology, section 6, vol. IV, 1665—1708 (1968).
HÖKFELT, T.: Electron microscopic studies on peripheral and central monoamine neurons. From Dept. of Histology, Karolinska Institutet, Stockholm. Stockholm 1968.
— Distribution of noradrenaline storing particles in peripheral adrenergic neurons as revealed by electron microscopy. Acta physiol. scand. 76, 427—440 (1969).
IVERSEN, L. L.: The uptake and storage of noradrenaline in sympathetic nerves. Cambridge: Univ. Press 1967.
JOHANSSON, B., LJUNG, B.: Role of myogenic propagation in vascular smooth muscle response to vasomotor nerve stimulation. Acta physiol. scand. 73, 501—510 (1968).
KATZ, B.: The transmission of impulses from nerve to muscle and the subcellular unit of synaptic action. Proc. roy. Soc. B. 155, 455—479 (1962).
LJUNG, B.: Local transmitter concentrations in vascular smooth muscle during vasoconstrictor nerve activity. Acta physiol. scand. 77, 212—222 (1969).
MERRILLEES, N. C. R.: The nervous environment of individual smooth muscle cells reconstructed from serial sampling with the electron microscope. J. Cell Biol. 37, 794—817 (1968).
STJÄRNE, L., HEDQVIST, P., BYGDEMAN, S.: Neurotransmitter quantum released from sympathetic nerves in cat's skeletal muscle. Life Sci. 8, 189—196 (1969).

Prof. Dr. B. FOLKOW
Department of Physiology
University of Göteborg
Göteborg/Sweden

Discussion

KIRSHNER: I would like to ask Dr. FOLKOW what limits of reliability he places on the estimates of the release of the transmitter for stimulus on the estimates of the number of vesicles per varicosity and on the content of each vesicle?

FOLKOW: We have the following general background for this estimation: We use the figures deduced by HÄGGENDAL, DAHLSTRÖM and HÖCKFELT concerning the number of varicosities as related to number of granules per varicosity and total NA content of the preparation studied. We assume, on the basis of the mentioned findings on the iris and the vas deferens, which mutually agree very well, that by far the major part of the catecholamines are *inside* the varicosities and here in the granules (or vesicles, if one prefers that name). One then ends up with a figure for an *average* NA content per varicosity and per granule. These data we relate to the NA content of the preparation we have studied and to the fraction of NA released per impulse from this tissue. Now, of course, one begins to come into troubles because errors tend to get bigger and bigger with each step in such deductions but, on the whole, we have tested our data from different angles and nevertheless we end up with about the same figures concerning the NA content per granule. Of course, some of these granules are bigger and will then probably contain proportionally more of NA and some are smaller, but I think as an *average* our figures are not too far from the truth. I freely admit that such calculations may include several errors but I think they are, nevertheless, most useful because one may then begin to consider what are reasonable theories for the extent and mode of transmitter release. These attempts to a quantitative estimation make it difficult for us to accept the view that one granule, with its large amount of NA content, should be identical with the quantal packet, as this would mean, among other things, that this large NA amount (30,000 to 40,000 molecules) should be subthreshold to the effector cells. Calculations of this type thus lead to an appraisal, whether current theories of transmitter release may be reasonable or not from the quantitative point of view. For any theory concerning transmitter release debit and credit have to match.

Geffen: Professor Folkow has freely admitted the errors involved both in the calculations made in the iris and the vas deferens which the original authors agreed were only approximations.

I am not particularly happy about the calculations because of the compounding of errors as one proceeds. The most important point for me is that the calculations of the number of varicosities and vesicles are not made in the same organs as one measures the release of transmitter from. In fact, in the vas deferens and iris it is extremely difficult to collect any released transmitter at all. One can also critisize the calculation of the concentrations in the synaptic gap. We have no idea of what the volume of the synaptic gap is and secondly we have no idea of whether every single smooth muscle cell must be in contact with an appropriate concentration of amine. From the work of Holman, Burnstock and Bennet it seems that there are certain key cells which are exposed to the high concentrations of amine and that they are electrically coupled to the remaining cells in that unit. I personally do not think we are anywhere near in a position to make any calculations about the effective concentration of amines in the synaptic gap required to fire either a key cell or every single cell. The last point I want to make is that from electrophysiological recordings of smooth muscle cells, one cannot yet say anything about quantal release. One can talk of spontaneous release of transmitter but the size of the spontaneous junction potentials is extremely variable. This may depend upon the position of the electrode with respect to the site of release and on the geometry of the tissue. I do not think that Dr. Holman is yet prepared to talk of quantal release in the sense of a fixed size packet from electrophysiological data.

Folkow: I agree with you on many of these points and am grateful for the points you take up for consideration. However, I do *not* think it matters very much for this type of calculation how many varicosities there are per unit number of smooth muscle cells; the important thing is that the adrenergic nerve *varicosities* appear to be fairly equal in size and content of granules, whether they are found in the vascular walls or in the vas deferens. It is further likely that the size and composition of the granule (vesicle) is about the same in such tissues.

As I said, there *are* considerable differences between these two effector organs in many other respects but you must remember that we, of course, take our figures for *noradrenaline content* from the same tissue in which we have measured the release. If we now assume that the varicosities are constructed in the same way in all tissues and each varicosity contains largely the same amounts of granules and NA, which seems reasonable, then the figures will still be valid for that very tissue; in this particular case the calf muscles of the cat.

I fully agree with you that the junction gaps vary quite a lot. It seems to be so, especially in vascular smooth muscle, where the gaps are in some places up to 2000 Å, but in other, well-innervated vessels of the order of 500 to 1000 Å; we have used the latter figure as an average and we have used the commonly accepted size for a varicosity to get some reasonable figure for the volume of the junction gap at the adrenergic nerve contact with the blood vessels which we have studied. However, as far as I understand shrinkage of tissues studied in the electron microscope may rather lead to an *over-estimation* of the size of the junction gaps. If so, the NA concentrations we deduced may in reality be still higher.

You drew up another question about the "miniature" subthreshold potentials —whatever they are, they are certainly, as far as I can see, caused by a "spontaneous" noradrenaline release. I think that Dr. HOLMAN, even if she is very careful in her statements, would agree to this so far. Then comes the question: What is the *amount* of NA released in connection with each such potential? Well, I would think this might vary somewhat from junction to junction and perhaps from moment to moment. However, what I object against—at least at the present moment though I will readily give in if someone comes with a very good argument—is that this spontaneous subthreshold release should always be *identical with the entire NA content of one granule*. It is *this* very question which is the key point in our deductions and arguments. As far as we can see our deductions, even if they were wrong by factor of ten, would still seriously question the assumption that the entire NA content of one granule should be identical with the quantal packet.

We simply try to use what quantitative figures that are available to see whether this or that theory for transmitter release appears the most probably one. As we have discussed in our original paper even some of the studies of cholinergic transmission appear to be best compatible with the view that the quantal packet corresponds to only *part* of the vesicle (granule) content of ACh. To settle these, in some respects the more important questions, one must use quantitative evaluations and somewhere one has to start and I think it is worth the effort.

I also agree to the fact that there are so-called "key cells" and in blood vessels only few of the muscle cells are directly innervated, usually the outer layer, the others being excited by cell to cell propagation; as judged from studies on the portal vein by Dr. JOHANSSON and Dr. LJUNG in our department. However, if so this does not basically change the possible validity of our deductions. Really, if one considers these problems from *all* points of view one appears to run into several serious problems concerning the concept that the minimal amount of transmitter released is identical with the *total* content of one vesicle (granule). After all, this concept is no dogma; it can be questioned and should, of course, be challenged if available information is better compatible with another, modified view. I am simply trying to provide some evidence that may justify a debate for an alternative view.

Bayer-Symposium II, 100—109 (1970)

Some Further Aspects on the Release of the Adrenergic Transmitter

JAN HÄGGENDAL

With 5 Figures

It often seems assumed that under normal conditions the amount of transmitter, released at neuro-effector junctions per impulse, is largely constant at a given physiological rate of discharge, being composed of a certain number of transmitter "quanta" (cf. ECCLES, 1964). Under experimental conditions, however, the amount of acetylcholine in a quantum can be modified (ECCLES, 1964). Furthermore, a deficency of acetylcholine in the presynaptically formed quanta is discussed at myastenia gravis (ELMQVIST, 1965). However, variations in the transmitter content in tissues (after blockade of synthesis) or in the overflow of the transmitter into a tissue perfusate during e.g. sympathetic nerve activity, are mostly ascribed to either a changed rate of impulse flow, or to variations in the inactivation of the amount of noradrenaline (NA) released per nerve impulse.

Released NA appears to be inactivated in principally three different ways at the nerve terminal-receptor level: (1) by means of enzymatic destruction, (2) by active reuptake by the so-called membrane pump, and (3) by diffusion into the blood stream (for ref. see e.g. CARLSSON, 1965; FOLKOW et al., 1967; ANDÉN et al., 1969).

It is known from many studies that administration of different α-adrenergic blocking agents, e.g. phenoxybenzamine (PBZ), dibenamine, azapetine and Hydergine, considerably increases the output of NA upon sympathetic stimulation of. e.g. the spleen, the intestines and the skeletal muscles (e.g. BROWN, 1960; THOENEN et al., 1964; BOULLIN et al., 1967; GEFFEN, 1965; SALZMANN et al., 1968; HEDQVIST et al., 1969). This increase has generally been ascribed to blockade of the membrane pump, and/or of local enzymatic destruction, and also to the maintained high blood flow during the sympathetic stimulation due to the receptor blockade. Thus it has been suggested that after these α-blocking drugs almost the entire amount of transmitter, released to the synaptic gap, has escaped into the blood stream.

ad 1: However, quantitatively the local enzymatic destruction appears to be of only small importance (ROSELL et al., 1963; see also FOLKOW et al., 1967, IVERSEN, 1967). An inhibition of this metabolic pathway by α-receptor blockers may thus be of limited importance for the net overflow of transmitter. In agreement with this view, it was found in recent experiments on field stimulation of rat's portal vein, preincubated with ^{3}H-NA, that the total outflow of radioactivity (^{3}H-NA and metabolites taken together) was increased after PBZ (HÄGGENDAL, JOHANSSON, JONASON and LJUNG, to be published).

ad 2: Reuptake by the membrane pump may under normal physiological conditions be an important inactivating mechanism (for ref. and results see e.g. Folkow et al., 1967). However, when the adrenergic vasoconstrictor fibres to the muscles of the cat's hind limb were stimulated (Folkow and Häggendal, to be published; for methods see Folkow et al., 1967), after administration of membrane pump blockers [desipramine (DMI) or 1-phenyl-1-(3-methyl-aminopropyl)3,3-dimethyl-phthalane (Lu 3-010)] in appearantly optimal doses, the amount of NA released per nerve impulse was considerably increased when PBZ was added. Larger doses of membrane pump blockers than those used do not appear to increase the NA output further. The increase of the NA output observed after PBZ is thus unlikely

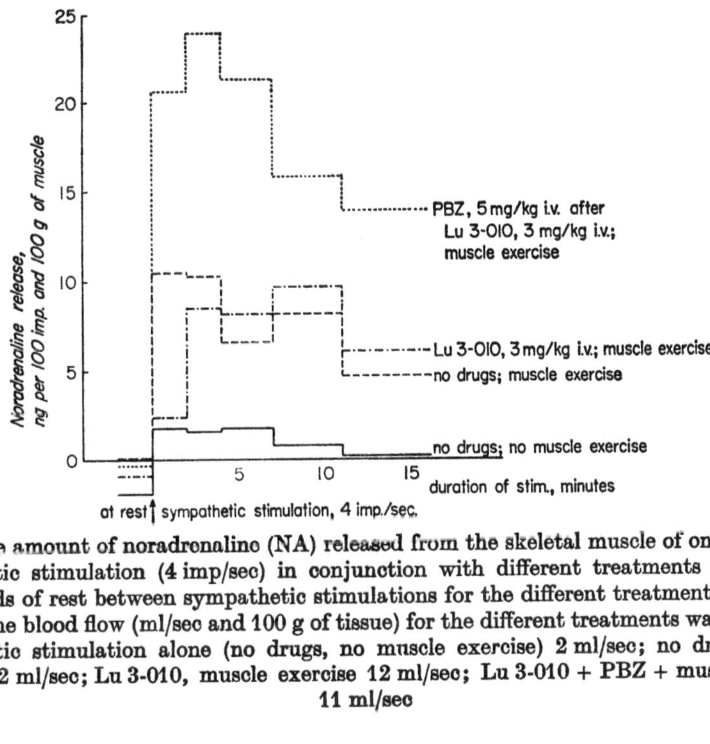

Fig. 1. The amount of noradrenaline (NA) released from the skeletal muscle of one cat during sympathetic stimulation (4 imp/sec) in conjunction with different treatments (see below). The periods of rest between sympathetic stimulations for the different treatments were 12 to 15 min. The blood flow (ml/sec and 100 g of tissue) for the different treatments was as follows: sympathetic stimulation alone (no drugs, no muscle exercise) 2 ml/sec; no drugs, muscle exercise 12 ml/sec; Lu 3-010, muscle exercise 12 ml/sec; Lu 3-010 + PBZ + muscle exercise 11 ml/sec

to be due to a further increased blockade of the membrane pump caused by PBZ. An example is shown in Fig. 1, where Lu 3-010 was used. Muscle exercise was performed in this experiment and the blood flow was about the same when the drugs were given. Similar results are also found during stimulation of the spleen. This is in agreement with the results reported by Geffen (1965), who found that DMI did not significantly increase the NA overflow, but the subsequent addition of PBZ produced a 10-fold rise in the NA output.

Whether PBZ is a "true" membrane pump blocker has in fact been questioned (Malmfors, 1965). In any case, the possible membrane pump blocking effect of PBZ appears to be of short duration, while the receptor blocking effect and the capacity to increase NA output is long-lasting (for discussion see Boullin et al., 1967). Thus it seems unlikely that the main explanation for the increased NA output after PBZ is that the drug blocks the membrane pump.

ad 3: The blood flow has been shown to be of importance for the NA overflow. During markedly reduced blood flow the NA output is less than at increased blood flow (Rosell et al., 1963; Carlsson et al., 1964). Since α-receptor blockers are able to prevent the vasoconstriction during adrenergic stimulation the maintained blood flow after these drugs may favour the NA output.

However, in experiments where the blood flow was maintained at a rather high level (by muscle exercise), both before and after administration of PBZ there was

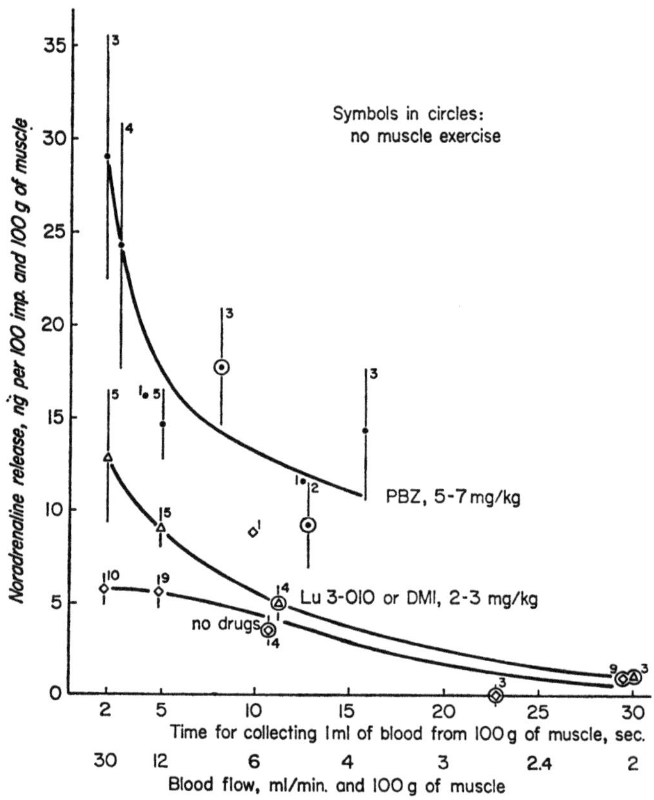

Fig. 2. The amount of noradrenaline (NA) released from cat's skeletal muscle during sympathetic stimulation (4 to 6 imp/sec) correlated to the time for collecting 1 ml of blood (sec) or blood flow (ml/min). Observations from 20 cats. The values are omitted, where the time for stimulation was less than 60 sec or more than 900 sec when half the sample was collected. In the phenoxybenzamine (PBZ) group some values are included where the animals had received both PBZ and desipramine (DMI) or Lu 3-010. These values did not appear to differ from those obtained after PBZ alone. The mean NA values ± s.e.m. (vertical bars) and the number of observations (small figures) are indicated. Symbols in circles: no muscle exercise

still a clearcut increase of the NA output after the α-receptor blocker as illustrated in Fig. 1. In Fig. 2 results from several observations (Folkow and Häggendal, to be published) are taken together. It can be seen that at the same blood flow, the highest NA output is found after PBZ or PBZ + membrane pump blockers, while membrane pump blockers alone give lower values and no drug treatment gives the lowest NA output. (No clearcut difference in the NA output was observed when PBZ was given alone or together with membrane pump blockers.)

Thus, it seems rather unlikely that the increased NA output after PBZ (or other α-receptor blockers) is mainly due to the effects of the blockers on the discussed inactivating mechanisms; see also BOULLIN et al. (1967).

It is possible that the α-receptor blockers may have effects on other, still mainly unknown mechanisms for inactivation of the released NA, such as binding to the tissues. However, if this is the case, it seems unlikely that a blockade of the inactivation of these mechanisms would be of great quantitative importance in explaining the significant increase of NA output as found after e.g. PBZ.

All the discussed explanations of the increased NA output after α-receptor blockers assume that the amount of NA released into the synaptic gap per nerve impulse is about constant; an assumption which may be incorrect since difficulties arise with all the proposed explanations of the increased NA output. Hence it would be of interest to discuss the possibility that the amount of transmitter, *released to the synaptic gap per nerve impulse*, may vary. An increased release of transmitter per impulse may be due to either A) direct effect of PBZ on the nerve membrane or B) some feed-back mechanism connected to the effector cell response.

A. The increased NA output, after PBZ and other α-receptor blockers, may be due to direct effects of the drugs on the nerve terminals.

Some observations do indicate that e.g. PBZ may affect the nerve terminals. Thus, it has been observed *in vitro* that the uptake-concentration mechanism of amine granules is inhibited by PBZ (CARLSSON et al., 1963; EULER et al., 1964). If the increased NA output per nerve impulse after PBZ is due to a direct effect on the nerve terminals, this would imply that an unphysiologically high NA output is obtained per nerve impulse after administration of the drug. On the other hand the nerve terminals appear not to be markedly affected by PBZ (in the doses used), since the transmitter leaking out spontaneously at impulse rest (the outflow of NA) appears to be about the same as normally (FOLKOW and HÄGGENDAL, to be published).

It cannot be excluded that at least to some degree the increased NA output may be due to a direct effect on the nerve terminals of one or several of the α-receptor blockers. However, it is noticeable that all these chemically different drugs which increase the NA output have one feature in common—they affect the α-receptors of the effector cell.

B. It will therefore be discussed if the effector cell may influence the transmitter release via some feed-back mechanism (briefly outlined at the XIII Scandinavian Congress of Physiology, Aug. 1969, HÄGGENDAL, 1969; also discussion in DAHLSTRÖM, 1969). If this is the case, it seems reasonable to suggest that more transmitter will be released when the effector cell is difficult to excite, e.g. after α-receptor blockers, than when the effector cell is supra-excitable.

The state of activity of the effector cells (the smooth muscle cells in the peripheral vascular bed) during sympathetic stimulation may be considered to be reflected by the peripheral resistance

$$\text{PRU} = \frac{\text{blood pressure (in mmHg)}}{\text{blood flow (from 100 g of tissue in ml/min)}} \cdot$$

The NA output per stimulus from cat's skeletal muscle has thus been correlated to PRU (Fig. 3). No drugs were given and the blood flow was relatively high due to a moderate degree of muscular exercise (2 imp/sec). In the group of 5 observations showing a high NA output the *blood flow* was as a mean the same as in the group with the lower values for NA output, since the blood pressure was on the average lower in the former group having lower PRU values. It can be seen that the group with the highest PRU values showed the lowest NA output. Thus, the difference in NA output at different peripheral resistances is apparently not mainly due to changes in the blood flow.

In Fig. 4 the NA output per stimulus is correlated to PRU after administration of membrane pump blockers and after PBZ. In the cases with increased PRU the NA output tends to be reduced. However, the blood flow is also generally reduced

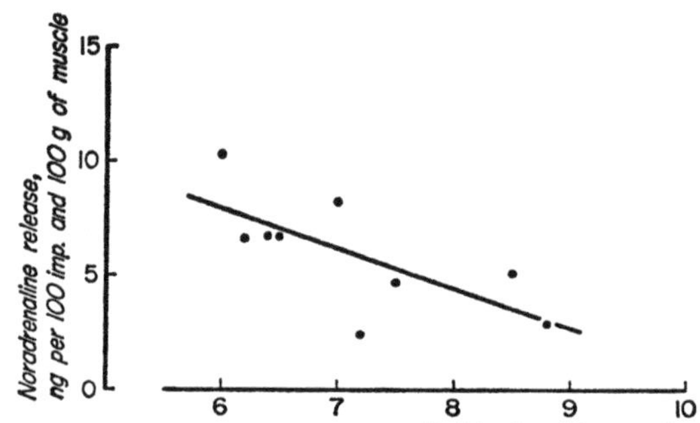

Fig. 3. The noradrenaline (NA) release from cat's skeletal muscle during sympathetic stimulation, and simultaneous muscular exercise related to peripheral resistance units

$$\left(\text{PRU} = \frac{\text{Blood pressure in mm Hg}}{\text{Blood flow from 100 g of tissue in ml/min}}\right).$$

No drugs were given. Individual values: Correlation coefficient = 0.70; $p < 0.05$. The mean values for the blood flow time (sec/ml) were for the 5 highest NA values 4.7 ± 0.34, and for the 4 lowest NA values 4.9 ± 0.66; the corresponding blood pressures were 82 ± 4.6 and 103 ± 14.8 mmHg, respectively

at this situation, which may offer better possibilities for the released NA to be inactivated locally. Observe, however, that with the dose of membrane pump blockers used, the reuptake by the membrane pump is likely to be blocked rather efficiently (cf. FOLKOW et al., 1967). Consequently, it is unlikely that the reduced NA output is due to an increased local inactivation, at least with respect to inactivation by the membrane pump. After PBZ the results are even more significant. If for a moment the view is accepted, that the increased NA output after PBZ is due to withdrawal of quantitatively important local NA inactivation, it then becomes difficult to explain why the NA output in the presence of PBZ should drop when the blood flow is reduced.

After ergotamine the α-receptors are blocked but the effector cells are at the same time stimulated, showing increased tonus. In recent experiments (FOLKOW and HÄGGENDAL, to be published) the output of NA was studied in cats after administration of ergotamine (500 μg/kg body weight i.v. or 300 μg/kg body weight i.a., after Lu 3-010, 3 mg/kg i.v.). The PRU was found to be increased at rest—due to direct stimulation of the effector cells (see e.g. SALZMANN et al., 1968). The effect of adrenergic stimulation was almost completely, but not totally blocked in most of the cases. The degree of α-receptor blockade was similar to that obtained after PBZ in a dose of 5 mg/kg body weight i.v. The output of NA per stimulus was about the same or even somewhat reduced after ergotamine as compared to when membrane pump blockers were given. The NA output after ergotamine [being 6.7 $\pm$ 2.8 ng per 100 imp. and 100 g of muscles at PRU of about 5.5 (12 observations in 4 cats)], thus showed a clearcut decrease as compared to the output after PBZ (see Fig. 4 for PBZ values). It is likely that the difference in the NA

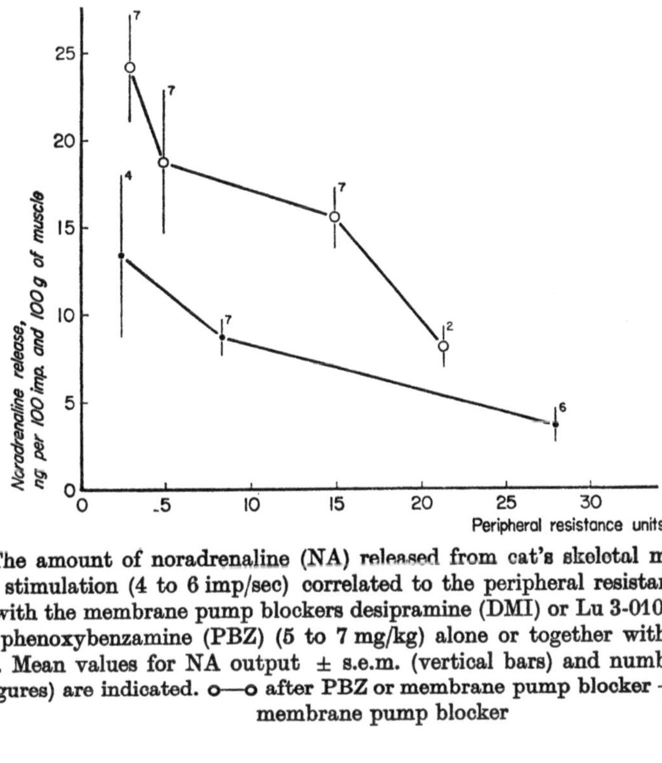

Fig. 4. The amount of noradrenaline (NA) released from cat's skeletal muscle during sympathetic stimulation (4 to 6 imp/sec) correlated to the peripheral resistance. The cats were treated with the membrane pump blockers desipramine (DMI) or Lu 3-010 (2 to 3 mg/kg i.v.) or with phenoxybenzamine (PBZ) (5 to 7 mg/kg) alone or together with membrane pump blockers. Mean values for NA output $\pm$ s.e.m. (vertical bars) and number of observations (small figures) are indicated. o—o after PBZ or membrane pump blocker + PBZ. ●—● after membrane pump blocker

output is related to the stimulating effect of ergotamine on the effector cell, because after ergotamine the effector cell is more active than after PBZ, the degree of α-receptor blockade being about the same.

The results discussed appear to indicate that the amount of transmitter released per nerve impulse may depend on the reactive state of the effector cell, since more transmitter than normally is found released when the effector cell is "depressed" by e.g. α-receptor blockers. Also the release of transmitters appears to be lower when the effector cell is excited.

It then seems reasonable to suggest that some sort of feed-back machanism may operate via the synaptic gap, between the effector cell and the nerve terminal.

The nature of such a *hypothetical* local feed-back mechanism is certainly obscure. However, *tentatively* its role in a schematic working model for transmitter release may be the following (Fig. 5):

1. At nerve impulse rest the membrane pump (mp) is pumping back most of the small amounts of NA leaking out at random (perhaps in the form of very small quanta). NA is available to be released from pools in the granules and perhaps also from "release points" on the varicosity membrane.

2. Following a nerve impulse, depolarization of the varicosity membrane occurs. During release of NA into the synaptic gap (from the release points on the membrane and/or from the granules directly) the membrane pump is switched off. For the mechanisms of transmitter release the ionic shift occurring at the depolarization may be of importance. A particularly important role may be played by the Ca^{++}-ions (for ref. see Iversen, 1967).

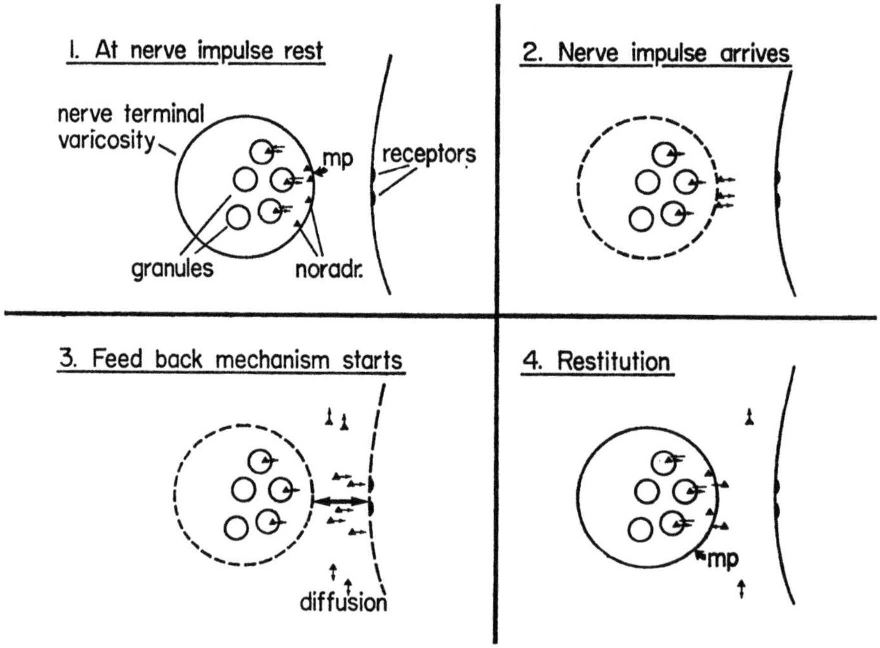

Fig. 5. Hypothetical model of mechanisms of transmitter release at the nerve terminal effector cell level. For further explanation see text

3. The transmitter will continously be released into the synaptic gap to reach the receptors of the effector cell until a sufficient number of receptors are activated to evoke the effector cell response—or until the immediately available pools for transmitter release are exhausted. At response of the effector cell the hypothetical feed-back mechanism may operate by some tentative mechanisms, e.g.:

The ionic-shift at the depolarization may via the synaptic gap affect the depolarized nerve terminal membrane and stop the NA release.

Also some particular ions, for instance Ca^{++}-ions, may play a key-function. Thus, after the depolarization of the nerve terminal, Ca^{++} may enter the varicosity and take part in the NA release mechanisms. If Ca^{++}-ions also enter the effector cell at depolarization and only small amounts of Ca^{++}-ions are available in the synaptic gap, there may be a competition for the Ca^{++}-ions between the effector cell and the nerve terminals and the flux of Ca^{++}-ions into the nerve terminals may be shifted. Ca^{++}-ions will then be withdrawn from the release mechanisms and the transmitter release switched off.

Some recent data indicate that a certain compound may be released from the effector cell at nerve activity and influence the amount of NA released per impulse. Thus prostaglandin E_2 has been discussed as being of importance for a negative feed-back mechanism at transmitter release [HEDQVIST, 1969 (1, 2)]. Prostaglandin has been shown to be released from stimulated spleen. When infused into the spleen it decreases the release of NA including the increased NA output found at PBZ administration.

Whatever the precise nature of the hypothetical feed-back mechanism may be, the presence of such a mechanism connected to the response of the effector cell would imply that more NA is released when the effector cell is depressed than when the effector cell is in a normal or easily excitable state.

4. When the feed-back mechanism has influenced the nerve terminal the latter is repolarized and the membrane pump starts to recapture transmitter that has not been utilized or otherwise inactivated by e.g. disappearing into the circulation. When large amounts of transmitter are released into the synaptic gap a blockade of the membrane pump would be of larger quantitative importance than when only small amounts are released (compare the results in Fig. 2, without and after membrane pump blockers). The suggested release points on the nerve terminal membrane will be reloaded and the varicosity will be fit for NA release at the next nerve impulse.

If the working model for transmitter release can be shown to be true with respect to the presence of a local feed-back mechanism, this would mean that the figure of 400 molecules of NA released per nerve impulse and varicosity (FOLKOW et al., 1967, 1970) will be too high for many physiological conditions. Consequently, it seems even more unreasonable that one transmitter quantum should correspond to the total NA content of one amine storage granule.

It may be pointed out that, irrespective of the mechanisms involved in the output of NA per nerve impulse after some drugs (e.g. after PBZ), this output may vary. Consequently it may be difficult to assume that the observed variations, both in the output of NA and NA stores in innervated tissue, necessarily involve a change in the rate of nerve impulse flow.

Summary

At stimulation of adrenergic sympathetic fibres many authors have shown an increased outflow of NA to the blood after different α-receptor blockers. Mainly based on recent results on the NA output per nerve impulse, when stimulating the vascoconstrictor fibres of the cat's skeletal muscle, different mechanisms for the increased outflow of NA after PBZ is discussed in the present paper. Neither the effects of the drug on the main inactivating mechanisms for the released transmitter at the nerve terminal receptor level nor the magnitude of the blood flow appears to account for the observed increase. Instead the possibility is discussed that the amount of transmitter, released per nerve impulse to the synaptic gap, may vary. A direct effect of the drug on the nerve terminals may give an increased release. Of more importance is that the released amount per nerve impulse appears to be correlated to the effector cell response. Tentatively a local feed-back mechanism operating via the synaptic gap is therefore suggested. Different possibilities for such a mechanism are briefly outlined, e.g. as due to either ionic changes in the synaptic gap, occurring at effector cell response, and/or to the production of some key compound, maybe prostaglandin E_2.

Acknowledgements

For stimulating discussions I am most indebted to Professor BJÖRN FOLKOW. The results discussed in the present paper have been supported by the Air Force Office of Scientific Research through the European Office of Aerospace Research OAR United States Air Force under Contract F 61052-68-C-0044, by a grant from US Publich Health Service (HE-05675-08) and from the Swedish Medical Research Council by grants B 70-14 X-16-06 B (B. FOLKOW) and B 70-14 X-166-06 B and by "Wilhelm och Martina Lundgrens Vetenskapsfond" (J. HÄGGENDAL).

References

ANDÉN, N.-E., CARLSSON, A., HÄGGENDAL, J.: Adrenergic mechanisms. Ann. Rev. Pharmacol. **9**, 119—134 (1969).

BOULLIN, J., COSTA, E., BRODIE, B. B.: Evidence that blockade of adrenergic receptors causes overflow of noradrenaline in cat's colon after nerve stimulation. J. Pharmacol. exp. Ther. **157**, 125—134 (1967).

BROWN, G. L.: Release of sympathetic transmitter by nerve stimulation. Adrenergic mechanisms, p. 116—124. VANE, WOLSTENHOLME, O'CONNOR, Eds. London: J. & A. Churchill Ltd. 1960.

CARLSSON, A.: Drugs which block the storage of 5-hydroxytryptamine and related amines. Handbook of experimental pharmacology, XIX, 529—592. Berlin-Heidelberg-New York: Springer 1965.

— FOLKOW, B., HÄGGENDAL, J.: Some factors influencing the release of noradrenaline into the blood following sympathetic stimulation. Life Sci. **3**, 1335—1341 (1964).

— HILLARP, N.-Å., WALDECK, B.: Analysis of the Mg^{++}-ATP dependent storage mechanism in the amine granules of the adrenal medulla. Acta physiol. scand. **59**, Suppl. 215 (1963).

DAHLSTRÖM, A.: Adrenergic neurons. Vth Internat. Symp. on Neurosecr. in Kiel. Aug. 1969. Z. Zellforsch. (in press).

ECCLES, J. C.: The physiology of synapses. Berlin-Göttingen-Heidelberg-New York: Springer 1964.

ELMQVIST, D.: Neuromuscular transmission with special references to myasthenia gravis. Acta physiol. scand. **64**, Suppl. 249 (1965).

VON EULER, U. S., STJÄRNE, L., LISHAJKO, F.: Effects of reserpine, segontin and phenoxybenzamine on the catecholamines and ATP of isolated nerve and adreno medullary storage granules. Life Sci. **3**, 35—40 (1964).

FOLKOW, B., HÄGGENDAL, J.: Some aspects on the quantal release of the adrenergic transmitter—this volume—1970.

— — LISANDER, B.: Extent of release and elimination of noradrenaline at peripheral adrenergic nerve terminals. Acta physiol. cand. Suppl. 307 (1967).

GEFFEN, L. B.: The effect of desmethylimipramine upon the overflow of sympathetic transmitter from the cat's spleen. J. Physiol. (Lond.) 181, 69—70 P (1965).

HÄGGENDAL, J.: On release of transmitter from adrenergic nerve terminals at nerve activity. Acta physiol. scand. Suppl. **330**, 29 (1969).

HEDQVIST, P.: (1) Modulating effect of prostaglandin E_2 on noradrenaline release from the isolated cat spleen. Acta physiol. scand. **75**, 511—512 (1969).

— (2) Antagonism between prostaglandin E_2 and phenoxybenzamine on noradrenaline release from the cat spleen. Acta physiol. scand. **76**, 383—384 (1969).

IVERSEN, L. L.: The uptake and storage of noradrenaline in sympathetic nerves. Cambridge: Univ. Press 1967.

MALMFORS, T.: Studies on adrenergic nerves. Acta physiol. scand. **64**, Suppl. 248 (1965).

ROSELL, S., KOPIN, I. J., AXELROD, J.: Fate of ^{3}H-norepinephrine in skeletal muscle before and following sympathetic stimulation. Amer. J. Physiol. 205, 317—321 (1963).
SALZMANN, R., PACHA, W., TAESCHLER, M., WEIDMANN, H.: The effect of ergotamine on humoral and neuronal action in the nictitating membrane and the spleen of the cat. Naunyn-Schmiedebergs Arch. Pharmak. exp. Path. 261, 360—378 (1968).
STJÄRNE, L., HEDQVIST, P., BYGDEMAN, S.: Neurotransmitter quantum released from sympathetic nerves in cat's skeletal muscle. Life Sci. 8, 189—196 (1969).
THOENEN, H., HÜRLIMANN, A., HAEFELY, W.: Wirkung von Phenoxybenzamin, Phentolamin und Azapetin auf adrenergische Synapsen der Katzenmilz. Helv. physiol. pharmacol. Acta 22, 148—161 (1964).

Dr. J. HÄGGENDAL
Department of Pharmacology
University of Göteborg
Göteborg/Sweden

Discussion

THOENEN: Dr. HÄGGENDAL, you assume that phenoxybenzamine has virtually no effect on the uptake of norepinephrine into the adrenergic nerve endings. I should like to ask you how you explain the following findings: If you add norepinephrine to the arterial inflow of the isolated perfused spleen—perfused with Krebs-Henseleit or blood—you find only 30 to 40% of the infused norepinephrine in the venous outflow. By addition of phenoxybenzamine the recovered amount of norepinephrine is increased to about 80%. This arterio-venous difference of norepinephrine cannot be further decreased by the addition of cocaine, a potent and generally accepted inhibitor of the norepinephrine uptake into the adrenergic nerve endings. I certainly would not exclude other effects of phenoxybenzamine on the adrenergic nerve endings, such as an interference with the storage mechanism or an increased amount of norepinephrine liberated per single nerve impulse, as is the case in the presence of tetraethyl-ammonium. However, I doubt very much that phenoxybenzamine is devoid of any effect on the uptake of norepinephrine into the adrenergic nerve endings.

HÄGGENDAL: There has indeed been much discussion as to whether phenoxybenzamine has a direct effect on the membrane pump or not. But for the present discussion on the released amounts of transmitter, a membrane pump blocking effect of phenoxybenzamine seems not to be a very likely explanation of the increased transmitter release after the drug, since, if the membrane pump blocker is first given followed by phenoxybenzamine, a clearcut increase in the amount of released transmitter is observed. Some interesting observations indicate that phenoxybenzamine may have an effect on the membrane pump. This effect, however, disappears rather fast, apparently within about 18 h after drug administration. But still there is an α-receptor blocking effect and also an increased amount of transmitter released per nerve impulse.

THOENEN: I completely agree that there is a dissociation between the α-adrenergic blocking action and the effect on the norepinephrine uptake. This dissociation is not only detectable several hours after the administration of phenoxybenzamine but also in the very first stage of the drug action observed *in vitro*. In the isolated perfused spleen of the cat the addition of phenoxybenzamine to the perfusion fluid produces a virtually complete blockade of the effect of sympathetic nerve stimulation at a stage when the norepinephrine output is not yet increased by inhibition of reuptake.

I have a further question: Is your preparation decentralized or is the reflex arch still intact? In the latter case one could assume that after administration of a mere blocker of the membrane pump—devoid of any α-blocking effects—the additional increase of the norepinephrine output, evoked by phenoxybenzamine, results from a reflex increase in sympathetic nerve activity, due to the blockade of the α-receptors.

Dahlström: I just want to make a comment on the possible effect of phenoxybenzamine on the membrane pump by referring to experiments by Dr. Hamberger. He used irides from reserpine-pretreated rats for *in vitro* studies on the uptake-mechanism of the nerve membrane. During incubation of the irides in a medium containing noradrenaline or α-methylnoradrenaline, the nerve terminals in the irides take up amine. If DMI, protriptyline or some other "membrane pump-blocker" was added to the medium before the amine, no amine was taken up into the nerve terminals. The tissue was studied in the fluorescence microscope. The addition of phenoxybenzamine to the medium instead of a "membrane pump-blocker" had no effect on the uptake of amines as could be observed in the microscope. Dr. Malmfors has performed the same type of experiment *in vivo* and obtained similar results. The results of these experiments thus indicate that phenoxybenzamine has little, if any, blocking effect on the amine uptake in the nerve membrane.

Glowinski: I would like to know if phenoxybenzamine is affecting norepinephrine synthesis in this situation.

Häggendal: An increased synthesis of noradrenaline has been shown after phenoxybenzamine. It has also been shown that if the adrenergic fibres are stimulated by an increased impulse flow, an increased synthesis of noradrenaline may occur, probably due to the increased loss of transmitter. I think this is the explanation also in the present situation. After phenoxybenzamine there is an increased release of transmitter and following this the synthesis will increase.

Muscholl: Dr. Häggendal, I wonder whether you have tried amphetamine in your experiments. Amphetamine is another drug which causes an increase in output much higher than what we expected from its membrane blocking property.

Häggendal: I have had no experience of this in the present system, but I hope it will be followed up.

von Euler: It might be taken into account, too, that phenoxybenzamine, like other α-blockers, inhibits re-uptake in the storage granules so if synthesis is taking place in the usual way then some of it will not be taken up by the granules. This will, of course, tend to result in a greater output. I would also like to put a question to Dr. Häggendal about the shutting off of the re-uptake system: I was a little shocked that it could be shut off so easily because I have been told that re-uptake is the most important way of inhibiting or ending the action and also important for transmitter economy. One then might ask at which stimulation frequency is this re-uptake shut off?

Häggendal: I think the membrane pump may be shut off during the period of depolarisation of the nerve varicosity, when the transmitter is released.

When the varicosity membrane is polarised again after the transmitter has been released, the membrane pump is likely to pump back available transmitter from the synaptic gap. This transmitter may thus be reutilized. Some of the released transmitter must, however, have disappeared into the circulation.

von Euler: But then, is it not strange when you give an uptake inhibitor like DMI in the exercise experiments that at a flow of, I think it was something between 6 and 2 ml/min, there was practically no difference in output between the normal and inhibited preparation. This would tend to show that the re-uptake mechanism was effectively shut off during this time and I am sure that in the experiments carried out together with Dr. Folkow, the stimulation frequency was not in any way unappropriately high, so there should have been time for re-uptake and then one should have expected a greater outflow in the presence of DMI—do you not think so?

Häggendal: In the experiments carried out together with Dr. Folkow the impulse frequency was 4 to 6 impulses/sec. After a membrane pump blocker, I think there is a clearcut tendency towards increased amounts of released transmitter when the blood flow is high, but much less marked when the flow is low, as compared to the released amounts when no drugs are given. A tentative explanation may be based on the discussed hypothetical mechanism for transmitter release. The blood flow may thus be low because the capacity of the effector cells in the blood vessels to respond is high. In such a situation only a little transmitter will be needed because the effector cells will respond promptly to the released transmitter. Thus there will only be small amounts of transmitter which can be pumped back again by the membrane pump or which can escape into the circulation if the membrane pump is blocked.

The situation is different when the blood flow is high in spite of the sympathetic stimulation. The effector cells do not respond in the same way as in the earlier situation. More transmitter will be needed to be released into the synaptic gap. When the membrane pump is blocked in this situation, rather more released transmitter, which would have been taken up by the membrane pump, will now escape into the blood stream.

Bayer-Symposium II, 112—127 (1970)

Quantal or Graded Secretion of Adrenal Medullary Hormone and Sympathetic Neurotransmitter

L. Stjärne

With 7 Figures

Introduction: The Concept of "Quantal" Neurotransmitter Secretion

One and a half decades have gone by since the concept of quantal secretion of neurotransmitter was originally introduced (for refs. see Katz, 1969). The initial work was done on the skeletal neuromuscular endplate, but it appears that the postjunctional membrane potential (MP) in essentially all neuro-effector junctions so far investigated shows extensive similarities to that of the endplate. Thus in the absence of nerve impulses there are spontaneous subthreshold spike-like fluctuations in the MP, similar to the MEPP (miniature endplate potentials) of the neuromuscular endplate, and on nerve stimulation the MP does not rise continuously but in steps, to a level above the resting potential in some instances apparently corresponding to an integral number of the amplitude of a single MEPP. Since focal application of exogenous acetylcholine to the neuromuscular endplate does not produce a stepwise, but a smoothly graded rise in MP, the important conclusion has been drawn that the chemoreceptors of the postjunctional membrane are not enough sensitive to detect, or to signal the detection of, single transmitter molecules. It has thus been postulated that the spontaneous miniature potentials are not due to diffuse leakage of transmitter, but to the impact on the postjunctional membrane of focally secreted multimolecular packets of transmitter, of preset size. The electronmicroscopical findings of vesicular structures accumulating in nerve terminals, and the subsequent demonstration by biochemical techniques that these "synaptic vesicles" are loaded with neurotransmitter, led to the assumption that the transmitter quanta are stored in and secreted from the vesicles, or "transmitter granules", and that the size of the quantum is possibly determined by their total content of transmitter.

These and other observations have led to the theory that miniature potentials are due to random collision of critical structures on the transmitter vesicles with specific reactive sites in the axonal membrane, each resulting in focal secretion of a single transmitter quantum. The effect of arrival of propagated nerve impulses is postulated to consist merely in quantitative acceleration of this process, involving a sudden enormous increase in probability for critical collisions, by a factor of a few hundred thousand for a very short time, thus causing the synchronous secretion of hundreds of discrete quanta, which produce the stepwise rise in MP triggering the motor response of the postjunctional cells.

The experiments behind this concept of neurotransmitter secretion have been carried out at such a high level of technical sophistication, and the results appear so conclusive that there can be no doubt about the "quantal" behaviour of the MP of the innervated postjunctional structures. However, the evidence concerning the prejunctional events, the "quantal" nature of transmitter secretion, is still incomplete. Firstly it must be kept in mind that the conclusions concerning the secretory events in the prejunctional nerve endings are based on indirect evidence, i.e. postjunctional recording of changes in MP, and not on actual measurement of the amount of transmitter secreted. Moreover, the crucial question remains whether iontophoretic application of exogenous acetylcholine to the junction can mimick the local secretion of endogenous acetylcholine so accurately that the observed smooth gradation of the changes in postjunctional MP induced by this technique finally rules out the possibility that the spontaneous or nerve stimulation induced "quantal" jumps in MP could be at least to some extent due to properties of the postjunctional membrane itself, and not exclusively to "quantal" jumps in the amount of transmitter secreted. Furthermore, it seems to be difficult to explain, in terms of this hypothesis, the fact that the spontaneously occurring miniature potentials are not dependent on the presence of Ca^{++}, which is an obligatory requirement for the further increase in postjunctional MP which triggers the motor response in the postjunctional cells. And finally it appears for economical reasons highly unlikely that the postulated transmitter quantum equals the total transmitter content of individual transmitter vesicles, simply because there are too few such vesicles present in the secretory units, at least of noradrenergic nerve terminals (STJÄRNE, 1966; FOLKOW et al., 1968; HÖKFELT, 1969).

Thus in conclusion, the circumstantial evidence in favour of the general concept of quantal secretion of neurotransmitter is very strong, but still incomplete. What is then the present evidence concerning the applicability of the quantal concept to catecholamine (CA) secretion from adrenal medulla and sympathetic nerves?

Is the Adrenal Medulla a Good Model System for Study of Sympathetic Neurotransmitter Secretion?

While our knowledge of several other aspects of the physiology of the adrenal medulla and sympathetic nerves, such as mechanisms of synthesis, uptake and storage of CA, has rapidly improved over the last two decades, the mechanisms of secretion of CA have proved to be more difficult to elucidate. Only during the last few years has real progress been made in this respect, particularly in the study of mechanisms of secretion from the adrenal medulla (cf. DOUGLAS, 1968). For a number of reasons this tissue is easier to study than the sympathetic nerves. In view of the fact that the two systems are homologous and make the same or closely related secretory material, which they store in somewhat similar ways, there has been a widespread tendency to use the adrenal medulla as a model organ for the study of sympathetic nerves, and to discuss the physiology of the nerves on the basis of extrapolation from the experimental results obtained with adrenal medulla (cf. STJÄRNE, 1964). This has stimulated and directed research concerning the mechanisms of secretion of the sympathetic neurotransmitter, and has recently led to some very intriguing observations of similarities between the two tissues in events accompanying CA secretion (cf. below). However, it must be kept in mind that we are dealing with two different systems, which although they have much in common, including ontogenetic origin, also differ in several respects. Thus their closely related precursor cells have become differentiated for quite different physiological purposes. The adrenal medullary cell has become specialized for secretion into the general circulation, to affect distant receptors, while the sympathetic nerves have been specialized for focal secretion, to affect a strictly limited number of nearby receptors. The anatomical differentiation for the two functions involves enormous differences logistically, in terms of length of the supply lines of the secretory units. Thus although one has reason to expect certain basic qualitative similarities in secretory mechanisms, there is also every reason to expect marked differences in the details of these mechanisms in adrenal medullary cells and in sympathetic neurons.

In view of this it is interesting that the electrophysiological characteristics of the cell membranes of the two secretory systems appear to be widely different. There does not seem to be any good reason to assume that the MP of sympathetic nerve terminals should differ from that of other neurons, in terms of resting amplitude or reversal and overshoot on adequate stimulation. The potential difference across the membrane of the adrenal medullary cell, on the other hand, exhibits the low value characteristic of glandular cells (POULSEN and PETERSEN, 1966; DOUGLAS, KANNO and SAMPSON, 1967), and stimulation does not lead to reversal and overshoot and thus to a propagated action potential, but only to minimal local depolarization (KANNO and DOUGLAS, 1967). However, these marked electrophysiological differences between the two tissues need not necessarily imply basic differences in their secretory mechanisms, since there seems to be good evidence, in sympathetic nerves and adrenal medulla as well as in muscle, that the phasic shifts in MP, which essentially reflect alteration in sodium and/or potassium conductance, are not obligatory requirements for the secretory (DOUGLAS and RUBIN, 1963) or mechanical (SCHILD, 1960) motor responses triggered by the same stimuli.

While these differences between adrenal medulla and sympathetic nerves with respect to the electrical properties of the cell membranes need thus not imply differences in secretory mechanisms, it may be more significant that the two tissues are similar in that their secretion is strongly dependent on the presence of Ca^{++} (DOUGLAS and RUBIN, 1961; HUKOVIC and MUSCHOLL, 1962; KIRPEKAR and MISU, 1967). An additional similarity is the common secretory stimulus, acetylcholine, secreted from the splanchnic and preganglionic sympathetic nerves, respectively. Moreover, according to recent experimental evidence acetylcholine may in addition, by a direct effect on the nerve terminals, affect the process of secretion of the sympathetic transmitter (LINDMAR et al., 1968; LÖFFELHOLZ and MUSCHOLL, 1969). The possible physiological relevance of these pharmacological findings (cf. BURN and RAND, 1965) is supported by the histochemical demonstration of cholinesterase activity in the axonal membrane of noradrenergic nerves (BURNSTOCK and ROBINSON, 1967), as well as by the evidence that noradrenergic sympathetic and cholinergic parasympathetic nerve terminals run together in the autonomic ground plexus of certain tissues such as the iris, enclosed in the same Schwann cell sheath (EHINGER et al., 1969).

In this connection it may be pertinent to point out that the adrenal medulla and sympathetic nerves differ *in vivo* in CA turnover rate as well as in their sensitivity to drugs (cf. STJÄRNE, 1964). Moreover, while the two tissues show many similarities in their intracellular CA storage mechanisms, there are also striking differences in the properties of the isolated amine storage vesicles in a number of respects, such as in size, in resistance to physical changes in the environment, in response to various drugs and in rate of spontaneous amine release and uptake on incubation *in vitro* (cf. STJÄRNE, 1964).

Thus there are extensive similarities as well as marked differences between adrenal medulla and sympathetic nerves. The use of one as a model for the other may be completely justified, but only as long as the limitations of the analogy are clearly realized.

Quantal or Graded Secretion from Adrenal Medulla?

Is there then any experimental evidence indicating whether secretion from the adrenal medulla is quantal or graded in nature? This is a problem, which until quite recently appeared difficult to study experimentally. However, the dramatic progress during the last few years in our understanding of the cellular mechanisms of CA secretion from the adrenal medulla (cf. DOUGLAS, 1968) has made it technically feasible to study this issue. On the basis of recent experiments involving determination of the CA and soluble and insoluble dopamine-β-oxidase (DBO) remaining in the CA storage vesicles of rabbit adrenal medulla after insulin treatment (VIVEROS et al., 1969), the conclusion was drawn that adrenal medullary secretion is indeed quantal in nature, and that the size of the secretory quantum is determined by the entire CA content of individual vesicles. Furthermore, the estimation was made that the amount of CA secreted per splanchnic nerve impulse corresponds to about 3×10^{-5} of the total organ content (KIRSHNER and VIVEROS, 1970). This is remarkably close to the recent estimate made at a different laboratory using different techniques, that in the cat adrenal the CA secretion per planchnic nerve impulse amounts to about 3.5×10^{-5} of the organ content (FOLKOW et al., 1968). Since the cat adrenal medullary cell has been reported to contain an average of 13,000 CA vesicles (KIRSHNER and VIVEROS, 1970), the implication is that if all adrenal medullary cells respond to each splanchnic nerve stimulus, the size of the postulated secretory quantum could not be larger than one third of the CA content of a single vesicle. Alternatively, if the above mentioned concept that the amine vesicles secrete all-or-none is correct, these considerations imply that only one out of three adrenal medullary cells responds to each splanchnic nerve impulse.

The above-mentioned results appear to imply a striking constancy in the CA secretory response of the gland to splanchnic nerve stimulation. However, one might equally well emphasize the opposite aspect, namely the equally striking variability in the response to splanchnic nerve stimulation, under other experimental conditions. The CA output from the cat adrenal gland per splanchnic nerve impulse is thus strongly dependent both on nerve stimulation frequency and on the total number of pulses, i.e. duration of stimulation (MARLEY and PATON, 1961; KIRPEKAR and CERVONI, 1963; MARLEY and PROUT, 1965). On stimulation of short duration the CA output per splanchnic nerve impulse shows progressive, and apparently linear, increase with frequency of stimulation throughout the physiological range (Fig. 1, STJÄRNE and HEDQVIST, to be published). This observation indicates that processes of facilitation and/or summation are involved in determining the secretory response of the adrenal medulla to nerve stimulation. On prolonged stimulation, the secretory response also exhibits distinct fading, which may reduce the CA output per splanchnic nerve stimulus quite considerably (MARLEY and PATON, 1961). The early onset of fading, as well as the observation that fading occurs in the presence of apparently unchanged sensitivity of the adrenal medullary cells to direct stimulation by nicotinic agents implies that it is not primarily a post-junctional phenomenon, due to e.g. CA depletion. Since the splanchnic nerve-adrenal medullary preparation consists of two series-coupled secretory systems, the cholinergic nerve terminals and the adrenal medullary cells,

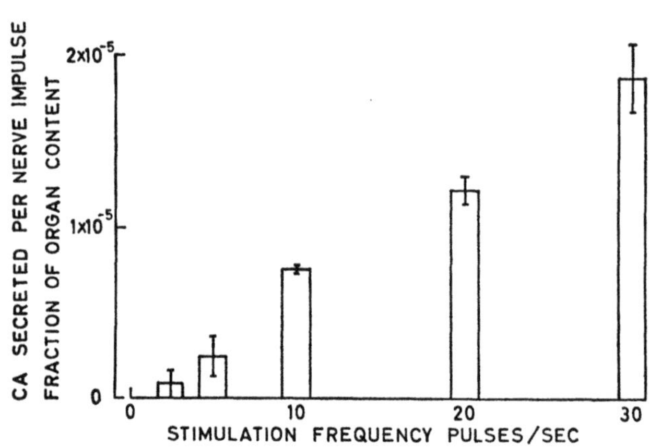

Fig. 1. Typical experiment showing CA secretion per supramaximal splanchnic nerve stimulus. Cat adrenal medulla. Ordinate: CA secretion expressed as fraction of total organ content. Mean and range of two observations in same experiment. From: STJÄRNE and HEDQVIST, to be published

it seems quite possible that both facilitation/summation and fading may be partly or exclusively dependent on prejunctional events. In that case it appears that these phenomena cannot be used as strong arguments either for or against the concept of quantal secretion of CA from the adrenal medulla.

Quantal or Graded Secretion of Sympathetic Neurotransmitter?

What then is the evidence for quantal or for graded secretion of the sympathetic neurotransmitter? The sympathetic neuro-effector junction has the advantage over the adrenal medulla that it allows the application of essentially the same electrophysiological techniques as those utilized in the study of the neuromuscular endplate. The results obtained are also largely similar (cf. BURNSTOCK and HOLMAN, 1966). Thus in the absence of nerve stimulation the MP of sympathetically innervated smooth muscle exhibits spontaneous subthreshold spike-like fluctuations ("spontaneous excitatory junction potentials", SEJP), and on nerve stimulation the further rise in MP is discontinouus ("excitatory junction potentials", EJP). The amplitude of the spontaneous miniature potentials, SEJP, is much larger than that of the MEPP in skeletal muscle, but evidently within the predicted range in view of the differences in caliber of the cells. There is also a much greater variability in size of the individual SEJP than of MEPP, but this difference is apparently predictable in view of the much wider range in size of the junctional gap, and thus of diffusional distances, in the sympathetic neuro-effector junctions. Thus one has to conclude that if the electrophysiological data from the skeletal neuro-muscular endplate is accepted as satisfactory evidence that transmitter secretion in that system is quantal in nature, then the same conclusion might apply to the secretion of the sympathetic neurotransmitter as well.

The direct determination of sympathetic neurotransmitter secretion, resting and in response to nerve stimulation, is much more difficult than that of adrenal medullary secretion, for two reasons. Firstly, the amounts secreted are low, close to the lower limit of sensitivity of the available methods for NA assay. Secondly, and even more importantly, there is strong evidence that a large proportion of the neurotransmitter secreted is recaptured into the nerve terminals and reused (cf. CARLSSON, 1965). Several tricks have been used to overcome this difficulty, such as the use of supraphysiological stimulation frequencies (HÄGGENDAL and MALMFORS, 1969), which apparently minimize recapture, or pharmacological blockers of either the reuptake process itself (cocaine-like drugs), or of reuptake plus the mechanical effect caused by the transmitter secreted (mainly phenoxybenzamine and Hydergin).

In studies concerning this issue, which have in some cases been explicitly aimed at estimation of the size of the postulated secretory quantum in relation to the total transmitter content of the whole organ, and thus by calculation, to that of individual transmitter granules, several research groups have emphasized the relative constancy in NA secretion per nerve stimulus. This is true as long as the experimental conditions are maintained constant, although there are wide variations in this respect between different experimental series, in which different tissues and techniques have been utilized. Thus the relative NA secretion per nerve stimulus from the sympathetic nerves of cat skeletal muscle was estimated to from 2×10^{-5} (perfusion with blood, variable flow, cocaine-like pharmacological blockers of NA reuptake, intermittent collection starting 6 min after beginning of supramaximal nerve stimulation at 6 per second, FOLKOW et al., 1968) to 6×10^{-5} (peak value on perfusion with Krebs-Henseleit medium, constant flow, phenoxybenzamine, continuous collection, same type of nerve stimulation, STJÄRNE et al., 1969) of the total organ content of NA. Estimation of the NA secretion per supramaximal nerve stimulus in the cat spleen in the presence of phenoxybenzamine or cocaine plus Hydergin has given higher values, ranging from about 8×10^{-5} to 3×10^{-4} of the total organ content of NA (STJÄRNE and HEDQVIST, to be published). The output per stimulus under these conditions appears to be to some extent frequency dependent, although the results reported from different laboratories are variable in this respect. Thus the output per stimulus was found to reach a maximum already at a frequency of 2 per second and to maintain this level up to 8 per second (HAEFELY et al., 1965), or to show a maximum at 10 per second with progressive decrease both at lower and higher frequencies (DAVIES and WITHRINGTON, 1968), or to have its maximum at 1 per second with progressive decline up to 10 per second (Fig. 2, STJÄRNE and HEDQVIST, to be published). The differencies in results are probably due to variations in the techniques used.

Thus in general, depending on where one prefers to put the emphasis, the secretion of NA per stimulus from sympathetic nerves may be regarded as relatively constant, or as somewhat variable, possibly involving e.g. tissue differences. However, the results agree on one point: Since the probable secretory unit of the sympathetic nerve terminals, the varicosity, contains less than 1000 transmitter vesicles (Hökfelt, 1969), it follows that if each nerve impulse invades all the nerve terminals, and if this leads to NA secretion from each varicosity, the size of the postulated secretory quantum is at least less than the total NA content of a single vesicle, ranging from 2 to 30% of that amount. On the other hand, if it is assumed that the nerve vesicles secrete all-or-none, as suggested for the adrenal medullary vesicles (cf. above), it follows that not more than one out of from

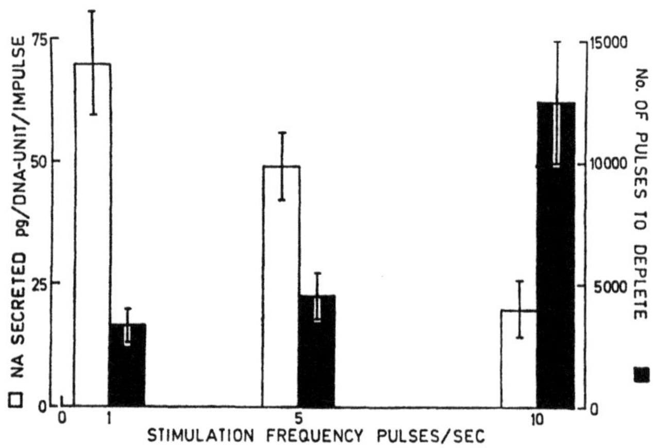

Fig. 2. Peak NA secretion in response to supramaximal splenic nerve stimuli. Cat spleen perfused with Krebs-Henseleit medium containing phenoxybenzamine 10 μg/ml, or cocaine 8 μg/ml + Hydergin 0.3 μg/ml. Left ordinate: NA secretion per nerve stimulus relative to DNA content of spleen. Right ordinate: Number of pulses required to deplete the organ in case the peak secretion could be maintained. Means ± SEM, n = 4. From: Stjärne and Hedqvist, to be published

3 to 30 varicosities secretes in response to each propagated nerve impulse (cf. Folkow et al., 1968).

Returning to the electrophysiological evidence, the successive increase in EJP with each nerve impulse indicates that the transmitter secretion in response to nerve stimuli is not constant at all, but rather subject to facilitation and/or summation processes (Burnstock and Holman, 1961). This has recently been confirmed in studies of the effect of short trains of nerve stimuli on the transmitter secretion and on the resulting vasoconstrictor response in the isolated perfused cat spleen (Stjärne and Hedqvist, to be published). It turned out that the mechanical response to 5 to 100 supramaximal nerve shocks delivered at frequencies ranging from 2 to 10 per second was not constant but showed a progressive increase with the stimulation frequency. This was true both in the presence and in the absence of phenoxybenzamine (Fig. 3). The variation in mechanical effect with stimulation frequency might obviously represent summation of the contractile response due to suitable shortening of the intervals between secretion of possibly constant amounts of neurotransmitter. However, direct assay indicated that the amount of NA secreted per nerve impulse showed the same frequency dependence. Thus transmitter secretion in response to propagated nerve impulses does not appear to be constant, but rather subject to modification by facilitation and/or summation, largely in the same manner as described above for the splanchnic nerve-adrenal medullary system.

It is well known that prolonged nerve stimulation at different frequencies is accompanied by fading of the mechanical response. The onset and time course of fading has been found to be frequency dependent. Thus at 1 per second the vasoconstrictor response reached its maximum after a considerable latency and

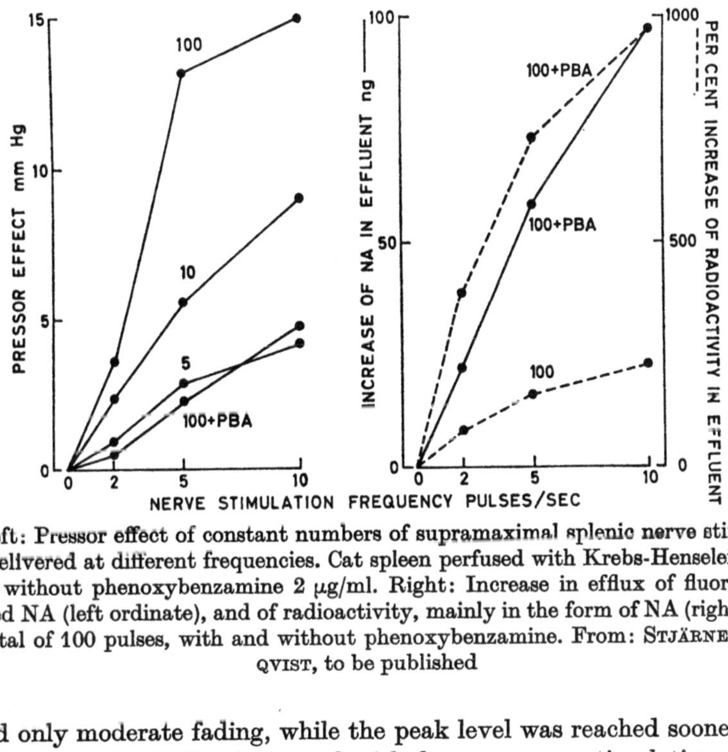

Fig. 3. Left: Pressor effect of constant numbers of supramaximal splenic nerve stimuli (5, 10 or 100), delivered at different frequencies. Cat spleen perfused with Krebs-Henseleit medium, with and without phenoxybenzamine 2 µg/ml. Right: Increase in efflux of fluorimetrically determined NA (left ordinate), and of radioactivity, mainly in the form of NA (right ordinate) after a total of 100 pulses, with and without phenoxybenzamine. From: Stjärne and Hedqvist, to be published

exhibited only moderate fading, while the peak level was reached sooner and the degree of progressive fading increased with frequency, on stimulation at 10 and 30 per second (Fig. 4, Stjärne and Hedqvist, to be published). The early onset of fading, as well as direct determination of the NA remaining in the spleen indicate that fading occurs in spite of an essentially unchanged NA content. It is still conceivable that it could be due to depletion of a small separate "available transmitter pool". However, recent studies of the effect of prostaglandins on the sympathetic neuro-effector junction indicate that the fading process may represent active regulation of transmitter secretion by a negative feed-back mechanism, triggered by the mechanical response to nerve stimulation. This concept is based on the observation that low concentrations of exogenous prostaglandin E_2, a substance which is known to occur in the spleen and to be secreted in large amounts in response to splenic nerve stimulation, markedly inhibit the sympathetic neuro-effector junction both pre-junctionally, by reducing the amount of transmitter secreted per nerve impulse (Fig. 5), and postjunctionally, by antagonizing the effector response to transmitter secreted (Hedqvist, 1969, and personal communication).

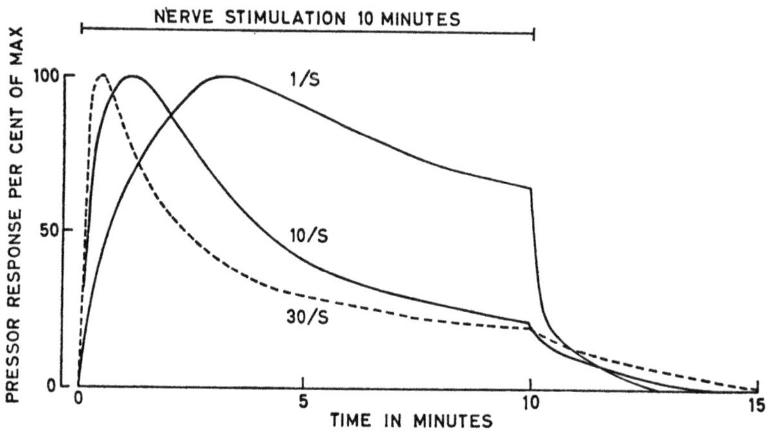

Fig. 4. Pressor effect of prolonged supramaximal splenic nerve stimulation at different frequencies. Cat spleen. Ordinate: Pressor effect expressed as per cent of maximum response at each frequency. From: Stjärne and Hedqvist, to be published

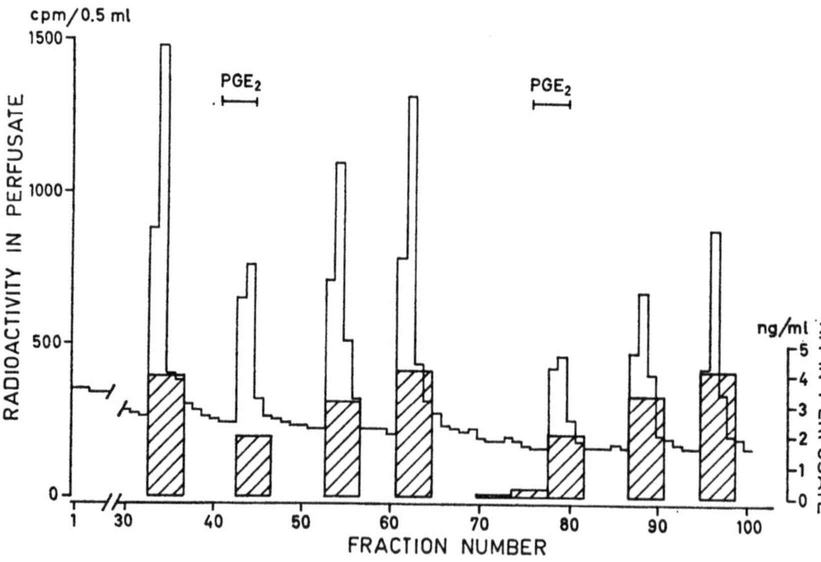

Fig. 5. NA in effluent from perfused cat spleen, resting and in response to trains of 200 supramaximal splenic nerve stimuli, delivered at 10/sec. Effect of infusion of prostaglandin E_2 (PGE_2). Final concentration 1.6×10^{-6} M. Left ordinate: Radioactivity. Increase over baseline mainly NA. Right ordinate: NA fluorimetrically determined. From Hedqvist: Acta physiol. scand. 75, 511—512 (1969)

The frequency dependence of the facilitation/summation phenomena in transmitter secretion might possibly be related to variation in the degree of hyperpolarization of the sympathetic C-fibers (cf. Burnstock et al., 1964), or could possibly be secondary to a progressive build-up of available intraaxonal Ca^{++} (KATZ and MILEDI, 1967). In the latter case it is tempting to speculate on the possibility that the inhibitory effect of prostaglandins on transmitter secretion might be related to interference with Ca^{++} availability (for ref. see HORTON, 1969).

The observations of facilitation, summation and fading processes in secretion of NA from sympathetic nerves need of course not be incompatible with the concept of quantal secretion of neurotransmitter, since they can be to a certain extent explained in terms of variations in the degree of "recruitment" of nerve terminals invaded by each nerve impulse (KRNJEVIC and MILEDI, 1958), or "recruitment" of secreting varicosities, or finally "recruitment" of possibly quantally secreting units in each varicosity. However, the results appear to be equally compatible with semi-quantal or graded secretion of neurotransmitter.

Cellular Mechanisms of CA Secretion from Adrenal Medulla and Sympathetic Nerves

Dramatic progress has been made during the last few years in the study of the cellular mechanisms involved in stimulus-secretion coupling, particularly in the adrenal medulla (cf. DOUGLAS, 1968). At present all the evidence seems to indicate that secretion from the medulla may occur by exocytosis. The experimental evidence for this concept is the following (Fig. 6): The CA secreted in response to splanchnic nerve stimulation, or to acetylcholine, are not accompanied by any cytoplasmic material, nor by insoluble vesicular components, but by all the soluble compounds of the vesicles, which appear together with the CA in the effluent from the stimulated gland, in the same proportions in which they occur in the vesicles (BANKS and HELLE, 1965; DOUGLAS et al., 1965; DOUGLAS, 1966; BLASCHKO et al., 1967; KIRSHNER et al., 1967; VIVEROS et al., 1968).

To what extent does this model apply to secretion of sympathetic neurotransmitter? This question has been difficult to answer for several reasons, including the impurity of the nerve vesicle preparations available. However, even if the chemical composition of these vesicles is far less well known than that of the adrenal medullary vesicles, it is by now well established that they in addition to NA contain ATP (SCHÜMANN, 1958; STJÄRNE, 1964), the enzyme DBO (STJÄRNE, 1966; POTTER, 1966; STJÄRNE and LISHAJKO, 1968) and also protein immunologically indistinguishable from the chromogranin A of adrenal medullary vesicles (BANKS et al., 1968). As in the adrenal medulla Ca^{++} is required for the secretion induced by nerve stimulation (cf. refs. above). This does not lead to increased efflux of axoplasmic compounds, such as of NA-metabolites (ROSELL et al., 1963) or of the enzyme dopa decarboxylase. On the other hand, according to recent evidence, soluble DBO as well as chromogranin accompany the NA appearing in the effluent (DE POTTER et al., 1969). The close similarity to the events accompanying secretion from adrenal medulla makes these observations look like convincing arguments in favour of transmitter secretion by the same mechanism: exocytosis. However, although at this early stage the information is too incomplete to allow final conclusions, it appears that there are striking quantitative differences between the secretory events in adrenal medulla and sympathetic nerves. Thus while in the secretion from adrenal medulla the relative proportions of CA to soluble vesicular compounds are maintained in the effluent, there appears to be a large excess of NA in relation to protein in the effluent from stimulated sympathetically innervated tissue (de Potter et al., 1969). Similarly, while stimulation by acetylcholine of CA secretion from the adrenal medulla has been found to cause parallel efflux of labelled adenine nucleotide material, there was no detectable efflux of nucleotides accompanying the NA secreted from stimulated nerve tissue (Stjärne et al., 1970).

As previously mentioned the basis for the postulate that adrenal medullary secretion is quantal in nature is the evidence interpreted as indicating that the exocytosis process is an all-or-none phenomenon (Viveros et al., 1969), each time leading to the expulsion of all soluble compounds in individual vesicles. The

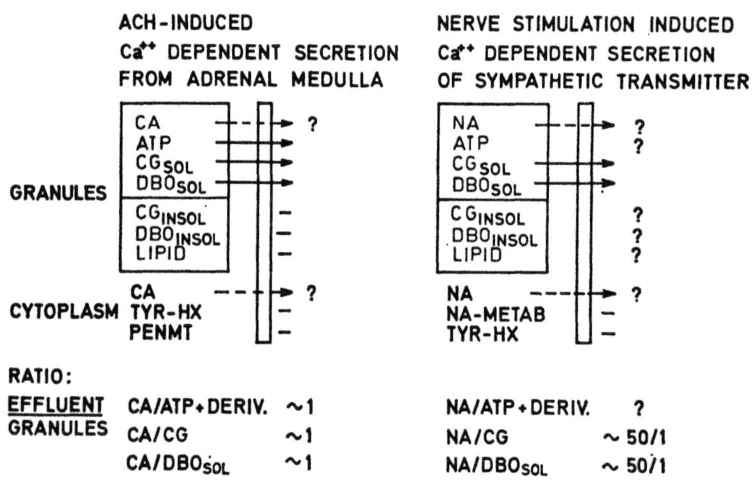

Fig. 6. Diagram to show present evidence concerning secretion by exocytosis in adrenal medulla and sympathetic nerves (see text). Abbreviations not explained in text: CG_{SOL}, CG_{INSOL} = Soluble and insoluble chromogranin; TYR H-X = Tyrosine hydroxylase; PENMT = Phenylethanolamin N-methyltransferase

insoluble material, lipids and insoluble chromogranin and DBO, apparently remain in the cell (Poisner et al., 1967; Malamed et al., 1968; Serck-Hansen, 1969; Viveros et al., 1969). The implication appears to be that the adrenal medullary vesicles are completely disposable structures, used for secretion only once and then probably discarded. If this is correct the "reserve capacity" of the adrenal medulla would correspond to about 13,000 secretory quanta per cell (cf. Kirshner and Viveros, 1970). In view of the facilitation/summation and fading phenomena discussed above this figure might not be unreasonable in relation to the available figures for CA turnover in this tissue under conditions such as hypoglycemic stress: less than 24 h (Bygdeman et al., 1960).

Concerning the mechanisms of secretion of sympathetic neurotransmitter, there does at present not seem to be any good way of deciding whether the small amount of specific protein appearing together with NA in the effluent from stimulated nerve tissue is to be regarded as unspecific leakage of waste products from deteriorating vesicles, or as evidence that exocytosis may be an obligatory requirement at least for the initiation of transmitter secretion. It is not even possible to know with certainty if the transmitter is secreted directly from the vesicles, or possibly from some other source, which depends on the vesicles for refilling with transmitter. According to the quantum theory secretion is focal and not diffuse. This seems to be generally accepted, and at present most workers seem not to consider secretion in terms of acceleration of "leakage" of "free" transmitter, but rather tend to look for structural candidates, possibly related to the axonal membrane, for specific "transmitter secretion compartments" (cf. von Euler, 1970, this Symposium).

Summing up: The present evidence is compatible with the concept of quantal secretion by exocytosis of adrenal medullary hormone. The size of the quantum

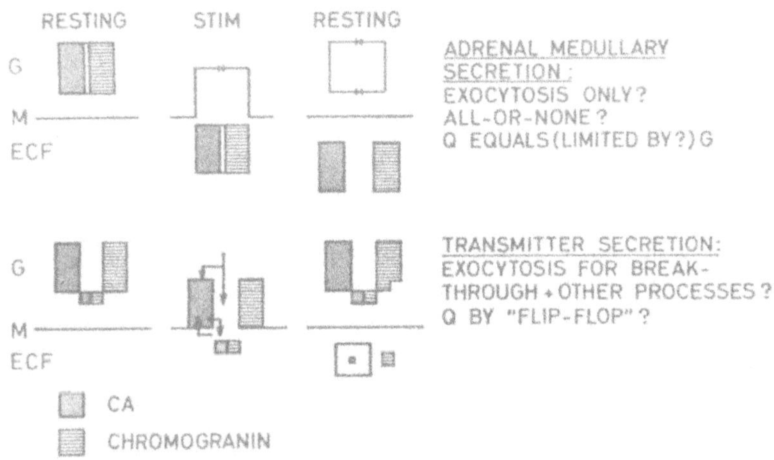

Fig. 7. Diagram to show working hypothesis for cellular mechanisms of CA secretion in adrenal medulla and sympathetic nerves (see text). G = granules. M = cell membrane. ECF = Extra cellular fluid

may be determined by the total CA content of single vesicles, since exocytosis in this tissue may be all-or-none in character (Fig. 7).

Sympathetic neurotransmitter secretion appears to be more complex. The present evidence seems to be best explained by the following working hypothesis: The nerve vesicle obtains its full supply of protein, including chromogranin, in the cell body. The chromogranin in the vesicles does not turn over, and peripheral synthesis is insignificant. The chromogranin exists in the vesicle in two "states" or "compartments", one of very small and one of large capacity. The vesicle may similarly have two NA compartments, and also has the enzymatic equipment for NA synthesis (DBO) as well as the capacity to take up and concentrate preformed NA from the surrounding medium, provided that ATP and Mg^{++} are available (Euler and Lishajko, 1963). In contrast to chromogranin the NA is constantly turning over and most of the NA synthesis occurs peripherally. A small proportion of the NA is bound in relation to chromogranin in the small "compartment". Only NA thus bound in relation to specific protein is "available" for secretion by exocytosis, and this may be an absolute requirement for the initial breakthrough of the axonal membrane. Once this is done, secretion of NA unrelated to chromogranin is maintained by other processes, involving NA in the big "compartment", newly formed NA (Kopin et al., 1968), possibly extravesicular NA and NA related to nearby vesicles, and finally NA secreted, recaptured and ready for reuse. NA secretion may be quantal or "semi-quantal", the relative constancy of the amount of transmitter secreted not necessarily depending on packets of preset size, but possibly on a "flip-flop"-mechanism (Katz and Miledi, 1965): an "opening" of standard size in the axonal membrane, exposed to a steady "transmitter diffusion pressure". On return to resting, the vesicle becomes detached from the axonal membrane and both NA compartments become refilled, by recapture and local synthesis, but a small proportion of the chromogranin is permanently lost. The reloaded vesicle is ready for reuse, and the cycle is repeated until the supply of chromogranin is used up. The relative size of the fraction of the total chromogranin lost in each secretory cycle determines the life span of the nerve vesicles.

Summary

Results obtained in studies of postjunctional potentials in various tissues indicate that neurotransmitter secretion does not occur by acceleration of diffuse leakage, but by focal secretion of packets of transmitter, of preset size. The paper deals with the applicability of the theory of quantal secretion to adrenal medulla and sympathetic nerves. The technical advantages and disadvantages in the study of the two tissues are discussed and it is emphasized that the use of one as a model organ for the other may be justified only as long as one clearly realizes the important differences between them. Recent evidence which may indicate that secretion from adrenal medulla and sympathetic nerves may be quantal in nature is presented. Facilitation/summation and fading in CA secretion in the two tissues are discussed and the possibility of negative feed-back regulation of the secretion of NA from sympathetic nerves is briefly mentioned. The cellular mechanisms of CA secretion from the two tissues are discussed in terms of the exocytosis theory, and it is emphasized that although exocytosis may be involved in both, there are large quantitative differences between CA secretory events in nerve and medulla. Finally a working hypothesis for hormone and neurotransmitter secretion is presented.

Acknowledgements

This work was supported in part by a grant from the Swedish Medical Research Council, Project No. B70-14X-2705-02, which is hereby gratefully acknowledged.

References

Banks, P., Helle, K.: The release of protein from the stimulated adrenal medulla. Biochem. J. **97**, 40c (1965).

— — Mayor, D.: Evidence for the presence of a chromogranin-like protein in bovine splenic nerve granules. Molec. Pharmacol. **5**, 210—212 (1969).

Blaschko, H., Comline, R. S., Schneider, F. H., Silver, M., Smith, A. D.: Secretion of chromaffine granule protein, chromogranin, from the adrenal gland after splanchnic stimulation. Nature (Lond.) **215**, 58—59 (1967)

Burn, J. H., Rand, M. J.: New interpretation of the adrenergic nerve fiber. Advanc. Pharmacol. **7**, 1 (1965).

BURNSTOCK, G., HOLMAN, M. E.: The transmission of excitation from autonomic nerve to smooth muscle. J. Physiol. (Lond.) **155**, 115—133 (1961).

— — Junction potentials at adrenergic synapses. Pharmacol. Rev. **18**, 481—493 (1966).

— — KURIYAMA, H.: Facilitation of transmission from autonomic nerve to smooth muscle of guinea-pig vas deferens. J. Physiol. (Lond.) **172**, 31—49 (1964).

— ROBINSON, P. M.: Localization of catecholamine and acetylcholin-esterase in autonomic nerves. Circulat. Res. **XX—XXI**, 43—55 (1967).

BYGDEMAN, S., v. EULER, U. S., HÖKFELT, B.: Resynthesis of adrenaline in the rabbit's adrenal medulla during insulin-induced hypoglycemia. Acta physiol. scand. **49**, 21—28 (1960).

CARLSSON, A.: Drugs which block the storage of 5-hydroxytryptamine and related amines. Handbook of experimental pharmacology, XIX, 529—592. Berlin-Heidelberg-New York: Springer 1965.

DAVIES, B. N., WITHRINGTON, P. G.: The release of noradrenaline by the sympathetic post-ganglionic nerves to the spleen of the cat in response to low frequency stimulation. Arch. int. Pharmacodyn. **171**, 185—196 (1968).

DE POTTER, W. P., DE SCHAEPDRYVER, A. F., MOERMAN, E. J., SMITH, A. D.: Evidence for the release of vesicle-proteins together with noradrenaline upon stimulation of the splenic nerve. J. Physiol. (Lond.) **1969**, 52 p.

DOUGLAS, W. W.: The mechanism of release of catecholamines from the adrenal medulla. 2nd Catecholamine Meeting, Milano, July. Pharmacol. Rev. **18**, 471—480 (1965).

— Stimulus-secretion coupling: the concept and clues from chromaffin and other cells. The first gaddum memorial lecture. Brit. J. Pharmacol. **34**, 451—474 (1968).

— KANNO, T., SAMPSON, S. R.: Effects of acetylcholine and other medullary secretagogues and antagonists on the membrane potential of adrenalin chromaffin cells: an analysis employing techniques of tissue culture. J. Physiol. (Lond.) 188, 107—120 (1967).

— POISNER, A. M., RUBIN, R. P.: Efflux of adenine nucleotides from perfused adrenal glands exposed to nicotine and other chromaffin cell stimulants. J. Physiol. (Lond.) **179**, 130—137 (1965).

— RUBIN, R. P.: The role of calcium in the secretory response of the adrenal medulla to acetylcholine. J. Physiol. (Lond.) **159**, 40—57 (1961).

— — The mechanism of catecholamine release from the adrenal medulla and the role of calcium in stimulus-secretion coupling. J. Physiol. (Lond.) **167**, 288—310 (1963).

EHINGER, B., FALCK, B., PERSSON, H., ROSENGREN, E., SPORRONG, B.: Localization of acetylcholine in the feline iris. Acta physiol. scand. **58** P, Suppl. 330 (1969).

v. EULER, U. S., LISHAJKO, F.: Effect of adenine nucleotides on catecholamine release and uptake in isolated adrenergic nerve granules. Acta physiol. scand. **59**, 454—461 (1963).

FOLKOW, B., HÄGGENDAL, J., LISANDER, B.: Extent of release and elimination of noradrenaline at peripheral adrenergic terminals. Acta physiol. scand. **72**, Suppl. 307 (1968).

HAEFELY, W., HÜRLIMAN, A., THOENEN, H.: Relation between the rate of stimulation and the quantity of noradrenaline liberated from sympathetic nerve endings in the isolated perfused spleen of the cat. J. Physiol. (Lond.) **181**, 48—58 (1965).

HÄGGENDAL, J., MALMFORS, T.: The effect of nerve activity on the uptake of noradrenaline into the adrenergic nerve terminals. Acta physiol. scand. **75**, 33—38 (1969).

HEDQVIST, P.: Modulating effect of prostaglandin E_2 on noradrenaline release from the isolated cat spleen. Acta physiol. scand. **75**, 511—512 (1969).

HÖKFELT, T.: Distribution of noradrenaline storing particles in peripheral adrenergic neurons as revealed by electron microscopy. Acta physiol. scand. **76**, 427—440 (1969).

HORTON, E. W.: Hypotheses on physiological roles of prostaglandins. Physiol. Rev. **49**, 122—161 (1969).

HUKOVIC, S., MUSCHOLL, E.: Die Noradrenalin-Abgabe aus dem isolierten Kaninchenherzen bei sympathischer Nervenreizung und ihre pharmakologische Beeinflussung. Naunyn-Schmiedebergs Arch. exp. Path. Pharmak. **244**, 81—96 (1962).

KANNO, T., DOUGLAS, W. W.: Effect of rapid application of acetylcholine or depolarising current on transmembrane potentials of adrenal chromaffin cells. Proc. Canad. Fed. biol. Soc. **10**, 39 (1967).

Katz, B.: The release of neutral transmitter substances. The Sherrington Lectures X. Liverpool: University Press 1969.
— Miledi, R.: The quantal release of transmitter substances. Studies in Physiology, pp. 118 to 125. Berlin-Heidelberg-New York: Springer 1965.
— — The role of calcium in neuromuscular facilitation. J. Physiol. (Lond.) **195**, 481—492 (1968).
Kirpekar, S. M., Cervoni, P.: Effect of cocaine, phenoxybenzamine and phentolamine on the catecholamine output from spleen and adrenal medulla. J. Pharmacol. exp. Ther. **142**, 59—70 (1963).
— Misu, Y.: Release of noradrenaline by splenic nerve stimulation and its dependence on calcium. J. Physiol. (Lond.) 188, 219—234 (1967).
Kirshner, N., Sage, H. J., Smith, W. J.: Mechanism of secretion from the adrenal medulla. II. Release of catecholamines and storage vesicle protein in response to chemical stimulation. Molec. Pharmacol. **3**, 254—265 (1967)
— Viveros, O. H.: Quantal Aspects of the secretion of catecholamines and dopamine-β-hydroxylase from the adrenal medulla. This volume, p. 78.
Kopin, I. J., Breese, G. R., Krauss, K. R., Weise, V. K.: Selective release of newly synthesized norepinephrine from the cat spleen during sympathetic nerve stimulation. J. Pharmacol. exp. Ther. **161**, 271—278 (1968).
Krnjevic, K., Miledi, R.: Failure of neuromuscular propagation in rats. J. Physiol. (Lond.) **140**, 440—461 (1958).
Lindmar, P., Löffelholz, K., Muscholl, E.: A muscarinic mechanism inhibiting the release of noradrenaline from peripheral adrenergic fibres by nicotinic agents. Brit. J. Pharmacol. **32**, 280—294 (1968).
Löffelholz, K., Muscholl, E.: Die Hemmung der Noradrenalin-Abgabe durch Acetylcholin am sympatisch gereizten, isolierten Kaninchenherzen. Arch. int. Path. Pharmak. **263**, 236 (1969).
Malamed, S., Poisner, A. M., Trifaro, J. M., Douglas, W. W.: The fate of the chromaffin granule during catecholamine release from the adrenal medulla. III. Recovery of a purified fraction of electrontranslucent structures. Biochem. Pharmacol. **17**, 241—246 (1968).
Marley, E., Paton, W. D. M.: The output of sympathetic amines from the cat's adrenal gland in response to splanchnic nerve activity. J. Physiol. (Lond.) **155**, 1—27 (1961).
— Prout, G. I.: Physiology and pharmacology of the splanchnic adrenal medullary junction. J. Physiol. (Lond.) **180**, 483—513 (1965).
Poisner, A. M., Trifaro, J. M., Douglas, W. W.: The fate of the chromaffin granule during catecholamine release from the adrenal medulla. II. Loss of protein and retention of lipid in subcellular fractions. Biochem. Pharmacol. **16**, 2101—2108 (1967).
Potter, L. T.: Storage particles in noradrenergic tissues. Pharmacol. Rev. 18, 425—432 (1966).
Poulsen, J. H., Petersen, O. H.: Resting and secretory transmembrane potentials in the submandibular gland of the cat. Acta physiol. scand. **68**. Suppl. **277**, 166 (1966).
Rosell, S., Kopin, I. J., Axelrod, J.: Fate of H^3-noradrenaline in skeletal muscle before and following sympathetic stimulation. Amer. J. Physiol. **205**, 317—321 (1963).
Schild, H. O.: Effect of adrenaline on depolarized smooth nuscle. Ciba Foundation Symposium on Adrenergic Mechanism, pp. 288—292. (Vane, J. R., Wolstenholme, G. E. W., O'Connor, M., Eds.) London: Churchill Ltd. 1960.
Schümann, H. J.: Über den Noradrenalin- und ATP-Gehalt sympatischer Nerven. Naunyn-Schmiedebergs Arch. exp. Path. Pharmak. **233**, 296—300 (1958).
Serck-Hanssen, G.: Chromogranin, a constituent of the chromaffin granule membrane. Acta physiol. scand. **63** P, Suppl. 330 (1969).
Stjärne, L.: Studies of catecholamine uptake, storage and release mechanism. Acta physiol. scand. **62**, Suppl. 228 (1964).
— Storage particles in noradrenergic tissues. Pharmacol. Rev. 18, 432 (1966).
— Hedqvist, P.: Facilitation/summation and fading in catecholamine secretion from adrenal medulla and sympathetic nerves. To be published.
— — Bygdeman, S.: Neurotransmitter quantum released from sympathetic nerves in cat's skeletal muscle. Life Sci. 8, 189—196 (1969).

Stjärne, L., Hedqvist, P., Lagercrantz, H.: Catecholamines and adenine nucleotide material in effluent from stimulated adrenal medulla: a study of the exocytosis hypothesis for hormone secretion and neurotransmitter release. Biochem. Pharmacol. (In press.)

— Lishajko, F.: Localization of different steps in noradrenaline synthesis to different fractions of a bovine splenic nerve homogenate. Biochem. Pharmacol. **16**, 1719—1728 (1967).

Viveros, O. H., Arqueros, L., Kirshner, N.: Release of catecholamines and dopamine-oxidase from the adrenal medulla. Life Sci. **7**, 609—618 (1968).

— — — Quantal secretion from adrenal medulla: All-or-none release of storage vesicle content. Science **165**, 911—913 (1969).

Dr. L. Stjärne
Department of Physiology
Faculty of Medicine
Karolinska Institutet
Stockholm 60/Sweden

Discussion

Geffen: I think I have said enough about possible mechanisms of secretion this morning.

I should like to know what predictive value your model has since this is the usefulness of models—could you perhaps give an idea of what experiments it suggests.

Stjärne: If we are going to interpret the role of the granules for neurotransmitter secretion, we have to have an idea of their life span. And if you look at the estimations of life span available in the literature, you will find that there is one very crucial assumption in those calculations, which may or may not be correct, namely, that the noradrenaline content of the nerve granules in the nerve trunk is identical to that of the nerve granules from the terminals. There is not really any conclusive experimental evidence in favour of this. We have some rather disturbing evidence which might point to the contrary and which might indicate that, as I wanted to show by this model originally, perhaps the noradrenaline you have in the granules in the nerve trunk represents only a small pool, or compartment, in the granules. This noradrenaline compartment shows very distinct features, like having a "ceiling", i.e. a maximum capacity, ATP-Mg^{++} facilitation of amine uptake and inhibitory effects of reserpine on uptake, as well as on release. So there could be reason to assume that things might happen to the nerve granules after their arrival at the terminals. After today's discussion I realize that this is all right—things might happen, but it seems very difficult to know precisely what. But any model serving to propose mechanism for transmitter secretion would have to be constructed in a way which would allow the understanding of the re-cycling, i.e. rapid turnover, of noradrenaline in the presence of essentially no turnover of the chromogranin.

As far as I can understand, fractional secretion of chromogranin could be best explained in terms of different pools of chromogranin in the granules. I think I would ask Dr. Helle to make her point on this in this connection.

Geffen: I agree this is an important thing to determine as a result of the model.

HELLE: In 1964 in your dissertation you postulated a model for the noradrenaline-storing granule of the sympathetic nerve as one with a storage complex of very low solubility. Certainly this postulation agrees very well with that I have found for the noradrenaline-rich granule fraction containing chromogranin. I can only detect chromogranin immunologically when the granule sediment has been solubilized in detergent.

This observation will fit in with your model of a rather insoluble complex of lipoprotein nature in the nerve granule compared with a highly water-soluble complex of chromogranin in the chromaffin granule of the adrenal medulla. Also what I think favours your model is the observation by Professor VON EULER in 1946 where he described an affinity for noradrenaline of the lipids. I favour the idea of a fraction of noradrenaline bound to a chromogranin-lipid complex in the nerve of very low solubility in water. There may, perhaps, be room for some cations like calcium in your model which might display the noradrenaline from the rather insoluble complex of chromogranin in the nerve granules. This would fit your model. I do not know whether this idea is readily discardable by other workers in this field, as far as I can see, this idea would agree with my preliminary results.

DAHLSTRÖM: I presume that the calculations of the life-span of granules you refer to are the ones that Dr. HÄGGENDAL and I did in 1966. It is quite true that the presumptions for these calculations we did were that the granules in the nerve terminals contained about the same amount of noradrenaline as those in the axons. During the last months we have re-investigated this problem, especially concerning the amounts of noradrenaline which can be stored in the granules, in the nerve terminals and in those in the axons. Using the results from this study we have re-calculated the life-span of the storage granules of rat and arrived at a figure of about 3 weeks instead of the 4 to 5 weeks previously calculated in 1966. However, with the present data and information we cannot obtain a figure below these 3 weeks. I wonder—you mentioned some experiments, the results of which were rather discouraging in this field—whether you have time to say something about these experiments.

STJÄRNE: Well, perhaps I could just very briefly state that we have been looking at the noradrenaline content of isolated bovine splenic nerve trunk granule pellets and we have been disturbed in finding that their amine content is very low. As Dr. FOLKOW pointed out, it could very well be that terminal nerve granules have an amine content, which is essentially the same as that of chromaffin granules —nearly 100 mg/g because the average amine concentration of the varicosity is of the order of at least 2 to 3 mg/g. But the amine content of our splenic nerve trunk granule pellets could not possible be that high. Which ever calculation we use we cannot get more than about 50 μg/g. This could, of course, reflect all sorts of things, including contamination—and we are aware of the fact that there is contamination. However, since the splenic nerve trunk consists almost exclusively of sympathetic fibres, one would hesitate to think of contamination of this preparation of the order of 1000-fold. So perhaps the trunk granules become loaded with amine on arrival at the terminals. What has happened more recently is that figures have become available concerning the noradrenaline content of the nerve trunk granules in relation to chromogranin. These are just utterly confusing at present since there is such great disagreement between the results of those who have determined this. I cannot see that it is possible to interpret them as yet.

von Euler: Dr. Helle very graciously mentioned—I had almost forgotten—some of the observations we made quite some time ago. I may add that this phospholipid complex, whatever it is, is very readily split up by, for instance, polyvalent anions like phosphate. Even 2 to 3 millimolar phosphate solution will immediately break it up, while in water it does not seem to break it up at all.

Dahlström: I just want to ask if you have any data of your own on the amine content of a pellet of granules derived from the nerve terminals in the spleen. You have it from the nerve trunk, but do you have it from the nerve terminals too—treated in the same way?

Stjärne: No, we do not, and I am rather pessimistic about the possibilities, at least within the near future, of getting any. The reason is that the density of sympathetic nerves in the innervated end organ, such as the heart or the spleen, is so low—of the order of fractions of a fraction of a per cent. This means that when you homogenize these tissues and try to prepare your amine granules you will have to purify tremendously in order to get rid of contaminating particles. I think that with the techniques presently available, this would be difficult. Of course, we know that Dr. de Potter succeeded in getting what was quoted as a 1000-fold purification, but I do not think that he gave the amine content of those granules. I think that many of us felt very pessimistic when Dr. Michaelson and co-workers made the ultimate experiment in this field. They tried to purify the nerve granules from the heart and compared the chronically denervated, and thus nor-adrenaline free, heart to the normal, innervated heart. After gradient centrifugation of material from both types of heart, they spun down the pellets and looked upon them in the electron microscope. Nobody could tell which pellet was which. So I think that satisfactory purification of terminal granules is not as yet technically feasible.

Bayer-Symposium II, 130—142 (1970)

Session 3

Chemical Sympathectomy

Chairman: P. Holtz

Chemical Sympathectomy with 6-Hydroxydopamine

H. Thoenen, J. P. Tranzer, and G. Häusler

With 10 Figures

The removal or destruction of an organ to elucidate its physiological role is one of the oldest and most commonly used methods in biological research. This principle has found extensive application in the investigation of the physiology and pharmacology of the sympathetic nervous system [Cannon and Rosenblueth, 1949; Trendelenburg, 1963 (1, 2)]. The surgical denervation is still the method of choice for the denervation of organs such as the nictitating membrane, iris or salivary gland which are innervated from single, easily accessible autonomic ganglia. However, for general sympathectomy or for the denervation of an organ with a more complex or less easily accessible innervation the surgical procedure is extremely cumbrous, time consuming and in small animals virtually not practicable. This explains the great interest in methods which provide the possibility to eliminate the sympathetic nervous system by other means. Immunosympathectomy was the first such procedure which became available for destroying or better preventing the development and differentiation of a great part of the pre- and paravertebral sympathetic ganglia. It consists in the administration to newborn animals of an antibody against a protein essential for the development of sympathetic and sensory ganglia of various species (Cohen, 1960; Levi-Montalcini and Booker, 1960; Levi-Montalcini and Angeletti, 1966). More recently, an even simpler procedure became available which consists in the selective destruction of the adrenergic nerve endings by 6-hydroxydopamine [6-HO-DA] [Tranzer and Thoenen, 1967 (1), 1968]. The peculiar properties of 6-HO-DA were detected in the course of studies designed to localize ultramorphologically trihydroxyphenylethylamines which act as false sympathetic transmitters. In previous studies it had been shown that 5-hydroxydopamine [5-HO-DA] and its β-hydroxylated and/or O-methylated metabolites are stored in sympathetic nerve endings and liberated as false transmitters by electrical nerve stimulation (Thoenen et al., 1967). After administration of 5-HO-DA or its metabolic precursor 5-hydroxydopa in doses which reduced the norepinephrine content of peripheral sympathetically innervated organs of cats and rats to less than 10% of control values all the vesicles of the adrenergic nerve endings were completely filled with dense osmiophilic material, whereas the vesicles of the cholinergic nerve endings remained empty (Fig. 1). The dense osmiophilic material represents the ultramorphologic

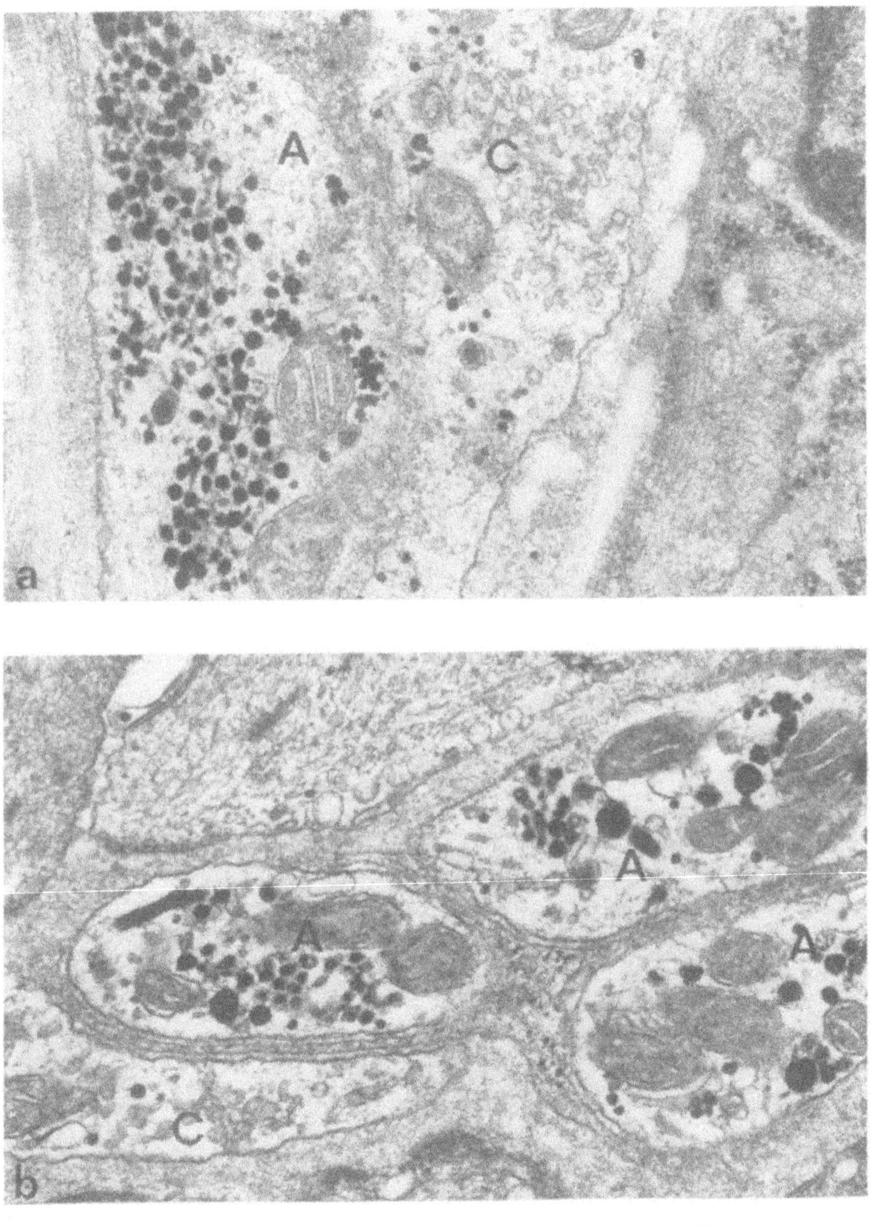

Fig. 1 a and b. Iris of a cat after treatment with 5-HO-DA. All the vesicles of the adrenergic nerve endings (A) contain strongly osmiophilic material which represents the stored 5-HO-DA, whereas the vesicles of the cholinergic nerve endings (C) appear empty. × 50000

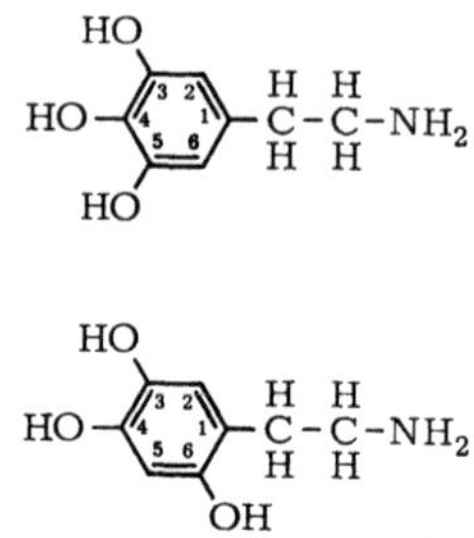

5-Hydroxydopamine

6-Hydroxydopamine

Fig. 2

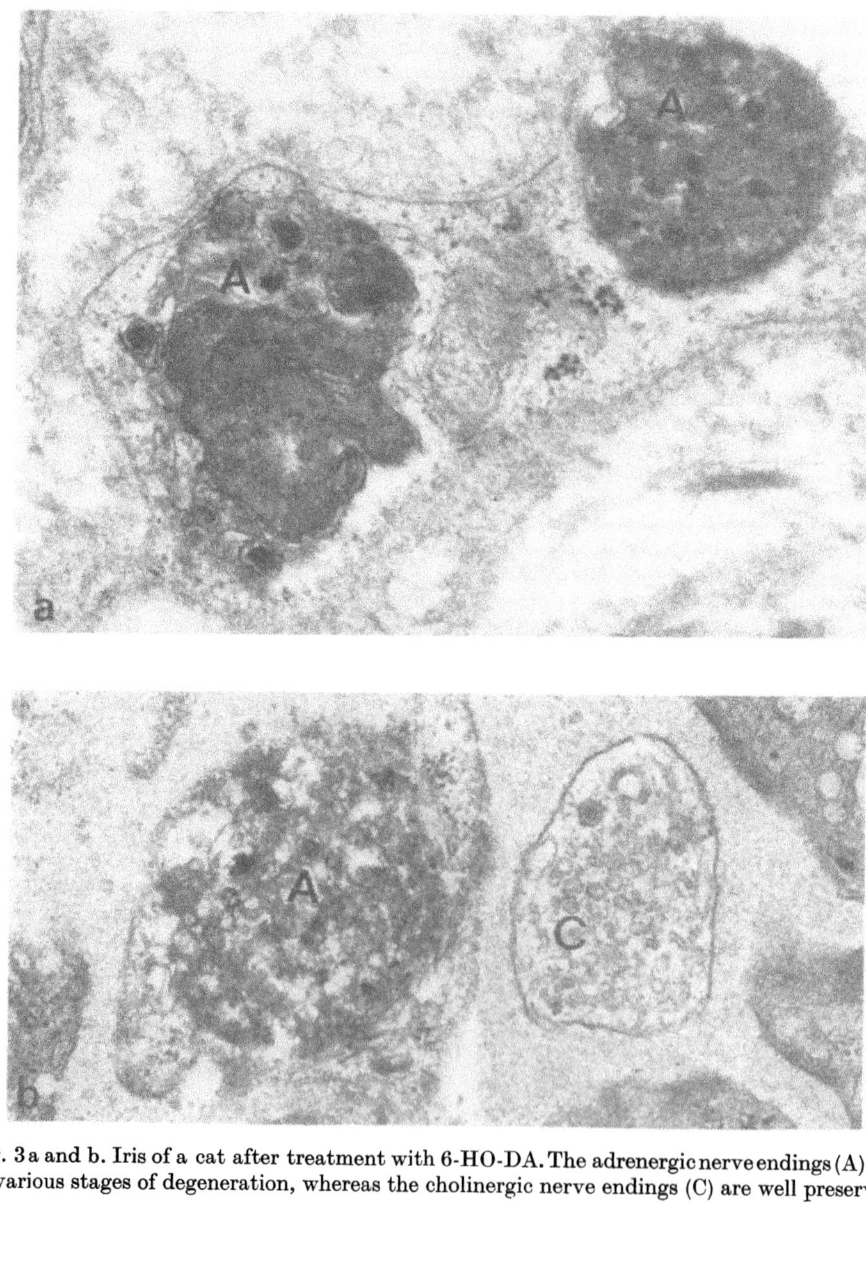

Fig. 3a and b. Iris of a cat after treatment with 6-HO-DA. The adrenergic nerve endings (A) are in various stages of degeneration, whereas the cholinergic nerve endings (C) are well preserved visualization of a false transmitter [Tranzer and Thoenen, 1967 (2)]. In addition, the administration of 5-HO-DA proved also to be a valuable tool for the identification of adrenergic nerve endings in regions such as the central nervous system where the preservation of the electron dense material representing the physiological transmitter norepinephrine has been without success so far (Tranzer et al., 1969) (Fig. 2).

The ultramorphological changes produced by the chemical isomer 6-HO-DA were completely different. Two to three days after the administration of this amine at a dosage which reduced the norepinephrine content to about the same extent as 5-HO-DA, the adrenergic nerve endings were in various stages of degeneration (Fig. 3). The surrounding Schwann cells, the smooth muscle cells and especially the cholinergic nerve endings revealed no ultramorphological alterations. In con-

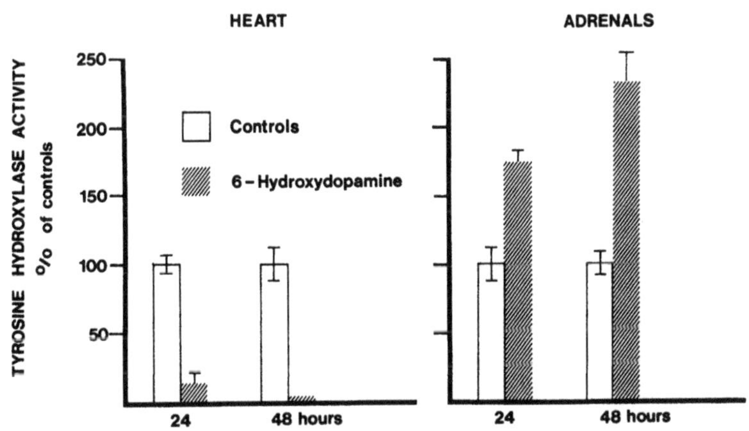

Fig. 4. The effect of 6-HO-DA on tyrosine hydroxylase activity in the rat heart and adrenals. The animals were given two doses of 100 mg/kg of 6-HO-DA i.v. at 8-hour intervals and killed 24 and 48 h after the first dose. Tyrosine hydroxylase activity in heart and adrenals was assayed according to Mueller et al. (1969)

trast to the nerve terminals the cell bodies of sympathetic ganglia were devoid of ultramorphological changes (Tranzer and Thoenen, 1968). Ten days after the administration of 6-HO-DA virtually all the adrenergic nerve terminals in peripheral sympathetically innervated organs had disappeared. The biochemical correlates to this ultramorphological change are a long lasting norepinephrine depletion (Thoenen and Tranzer, 1968) and a marked reduction of the tyrosine hydroxylase activity (Mueller et al., 1969) in sympathetically innervated organs as shown in Fig. 4. The latter finding is consistent with the view that this enzyme is selectively located within the adrenergic neurons (Sedvall and Kopin, 1967).

The time course of norepinephrine depletion in the heart and spleen of the rat after administration of single doses of 6-HO-DA showed that a critical dose of 6-HO-DA is necessary to produce the characteristic long lasting norepinephrine depletion and that these doses vary from organ to organ (Thoenen and Tranzer, 1968). The administration of 1 mg/kg 6-HO-DA produced only a transient reduction of the norepinephrine content both of heart and spleen of the rat and control levels were reached again after 24 h (Fig. 5). In contrast, already 3 mg/kg caused a reduction of the norepinephrine content in the heart which showed no tendency to recover for up to one week. In the spleen, however, this effect was evident only after 30 mg/kg. That the long lasting depletion of the norepinephrine content represents a reliable measure for the destruction of the sympathetic nerve terminals is supported by the observation that 3 mg/kg 6-HO-DA reduced the tyrosine hydroxylase in the rat heart by 35%, whereas 1 mg/kg was without effect. The assumption that the destruction of the adrenergic nerve terminals depends on the

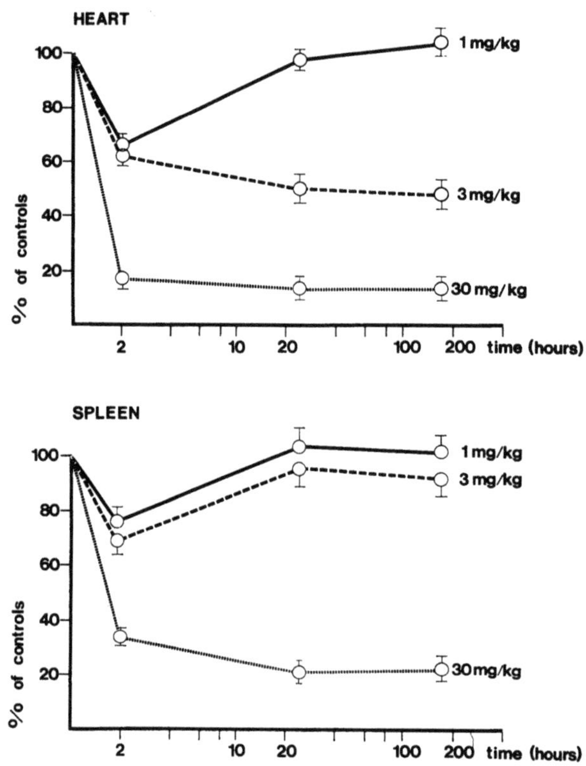

Fig. 5. The effect of single doses of 6-HO-DA on the norepinephrine content of heart and spleen of the rat. Doses of 1, 3 or 30 mg/kg of 6-HO-DA were dissolved in 0.001 N HCl and injected into the tail vein. The norepinephrine content is expressed in % of controls which were injected with 0.001 N HCl only

uptake of a critical amount of 6-HO-DA is further supported by the observation that 30 min after intravenous injection of ^{3}H-6-HO-DA the amounts of ^{3}H-amines retained in the rat heart correspond to the height of the dose administered (Fig. 6). However, after 2 h and even more impressively after 24 h the ^{3}H-amines retained were inversely related to the dose initially given. In other words, in low doses 6-HO-DA is taken up and stored in sympathetic nerve endings and can be liberated as a false adrenergic transmitter as shown in the isolated perfused spleen of the cat (THOENEN and TRANZER, 1968). Higher doses of 6-HO-DA, however, lead to a destruction of the adrenergic nerve terminals and thereby of their own storage sites. The uptake of 6-HO-DA into the storage vesicles of the adrenergic nerve terminals is not a prerequisite for its destroying effect, since pretreatment with reserpine did neither prevent the ultramorphological changes in adrenergic nerve terminals of the cats iris and spleen nor the reduction of tyrosine hydroxylase activity in the rat heart.

After this clear ultramorphological and biochemical evidence for a selective destruction of adrenergic nerve terminals by 6-HO-DA was established we studied

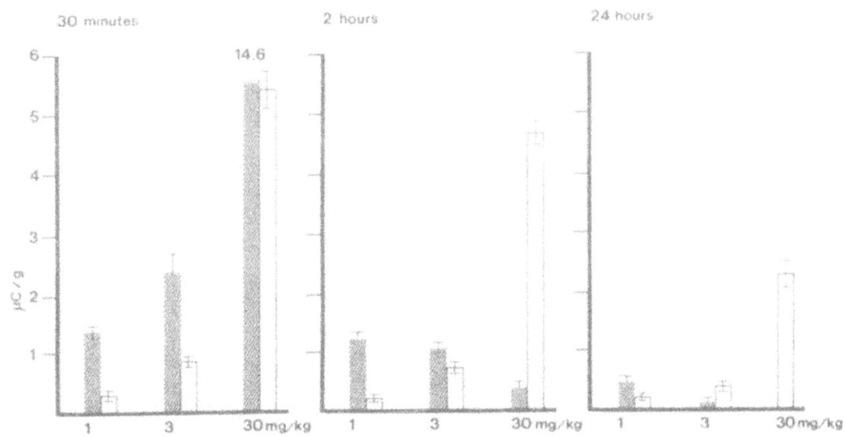

Fig. 6. Uptake and retention of H^3-6-HO-DA in the rat heart. The animals were injected with 1, 3 or 30 mg/kg of H^3-6-HO-DA i.v. and sacrificed 30 min, 2 h or 24 h later. The radioactivity present in the supernatant of hearts homogenized in 0.4 N perchloric acid was separated into acidic and basic (amines) fractions on Dowex 50 WX 4 columns. Open columns = acidic and neutral fraction; shaded columns = amine fraction

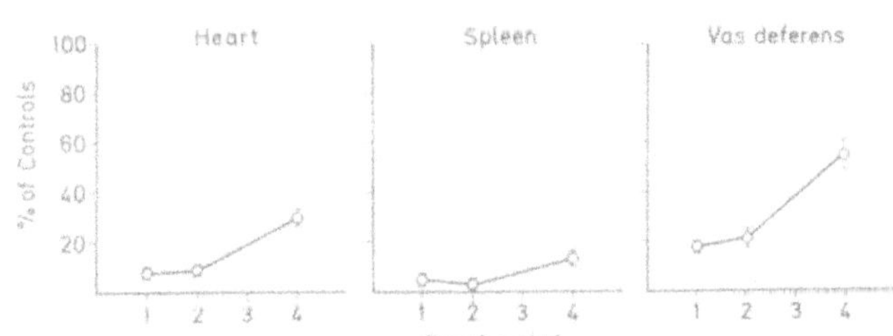

Fig. 7. Time course of the recovery of the norepinephrine content in heart, spleen and vas deferens of rats pretreated with 2 × 50 and 2 × 100 mg/kg 6-HO-DA i.v. (for details see THOENEN and TRANZER, 1968). The norepinephrine content was assayed 1, 2 and 4 weeks after the last dose of 6-HO-DA and compared with that of corresponding controls

to what extent these particular properties of this amine could be used for general chemical sympathectomy. Among various treatment schedules designed to achieve the most complete destruction of the peripheral adrenergic nerve terminals, the following procedure proved to be optimal both in rats and cats: Two doses of 6-HO-DA were given i.v. within 24 h and were followed by two further doses one week later (for details see THOENEN and TRANZER, 1968). One week after the last dose of 6-HO-DA the norepinephrine content of the rat heart was reduced to 7.6% of that of controls, in the spleen to 5.1% and in the vas deferens to 18.8%.

The catecholamine content of the adrenals was not significantly reduced [$P > 0.05$] (Table 1). Two weeks after the last injection of 6-HO-DA there was at most an insignificant trend for recovery of the norepinephrine content (Fig. 7). However, after 4 weeks there was a consistent rise in all organs studied. The rise in the

Table 1. *Chemical sympathectomy of rats with 6-HO-DA*

Organ	Norepinephrine content		
	Controls	6-HO-DA[a]	% of controls
Heart	1.02 ± 0.07 μg/g	0.076 ± 0.008 μg/g	7.6%
Spleen	0.86 ± 0.03 μg/g	0.044 ± 0.008 μg/g	5.1%
Vas deferens	8.80 ± 0.50 μg/g	1.66 ± 0.16 μg/g	18.8%
Adrenal gland	7.31 ± 0.53 μg/organ[b]	6.99 ± 0.51 μg/organ	95.7%

[a] The animals were pretreated with 2 × 50 and 2 × 100 mg/kg of 6-HO-DA. They were sacrificed 1 week after the last dose.

[b] Sum of norepinephrine and epinephrine.

Table 2. *Chemical sympathectomy of cats with 6-HO-DA*

Organ	Norepinephrine content		
	Controls	6-HO-DA[a]	% of controls
Heart	2.04 ± 0.17 μg/g	0.058 ± 0.01 μg/g	2.8%
Spleen	4.28 ± 0.55 μg/g	0.135 ± 0.04 μg/g	3.1%
Nictitating membrane	9.31 ± 0.77 μg/g	0.788 ± 0.14 μg/g	8.3%
Iris	0.52 ± 0.08 μg/organ	0.015 ± 0.004 μg/organ	2.8%

[a] The animals were pretreated with 2 × 20 and 2 × 50 mg/kg of 6-HO-DA. They were sacrificed 1 week after the last dose.

norepinephrine content was accompanied by a corresponding reappearance of the adrenergic nerve endings in electronmicroscopic pictures (Tranzer et al., 1969). Similar results were obtained after analogous treatment with 6-HO-DA in cats (Table 2).

In general the chemical sympathectomy with 6-HO-DA compares favorably with immunosympathectomy or surgical denervation (Cooper, 1966; Hertting and SCHIEFTHALER, 1964; IVERSEN et al., 1966; JELLINEK et al., 1964; LEVI-MONTALCINI and ANGELETTI, 1966). The main advantage of this procedure is its great simplicity and its general effect. However, not only immunosympathectomy but also chemical sympathectomy reveals considerable organ differences (Table 1 and 2). These result most probably from differences in blood supply both with respect to organ mass and especially to their norepinephrine content which reflects the density of the adrenergic innervation. The denser the sympathetic innervation of an organ the smaller is the share of 6-HO-DA per single nerve ending delivered

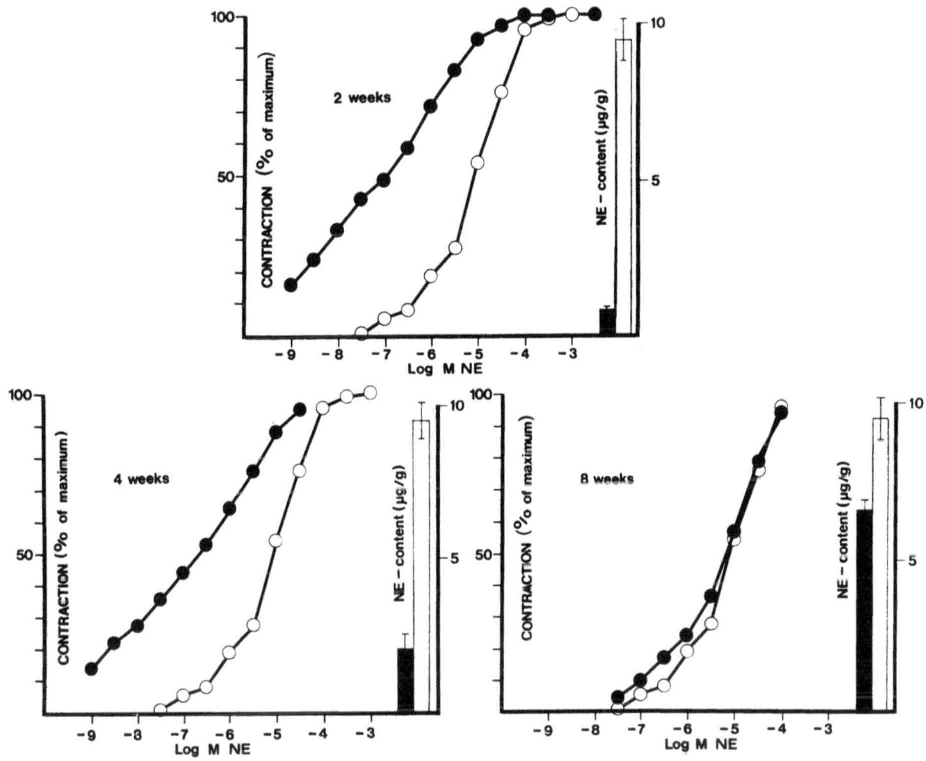

Fig. 8. Norepinephrine content and sensitivity to norepinephrine of the isolated medial smooth muscle of the cat nictitating membrane at various intervals after injection of 6-HO-DA. Open symbols = controls; black symbols = pretreatment with 6-HO-DA

by the blood stream. In this context the recent observation of HÄUSLER et al. (1970) is remarkable that the adrenergic nerves of the rat mesenteric vessels are poorly affected by 6-HO-DA. Their norepinephrine content is only reduced by approximately 50% after the administration of doses of 6-HO-DA which achieve a reduction to less than 10% in heart and spleen. This poor effect on the adrenergic nerves of blood vessels may be due to their particular localization, i.e. they are mainly located in the adventitio-medial junction of the vessels. Intima and media represent a remarkable diffusion barrier between the vascular lumen and the nerve terminals in the adventitia and the blood supply to the adventitia by vasa vasorum is negligible in rats (WOLINSKY and GLAGOV, 1967). This poor effect on the vascular sympathetic nerves could also offer a reasonable explanation for the failure of 6-HO-DA to influence materially the development of the salt-DOCA-hypertension in uninephrectomized rats (MUELLER and THOENEN, unpublished observation) a form of experimental hypertension which results from an increased activity of the sympathetic nervous system (DE CHAMPLAIN et al., 1969).

After chemical sympathectomy of cats the response of the isolated perfused heart to sympathetic nerve stimulation was completely abolished and that of the

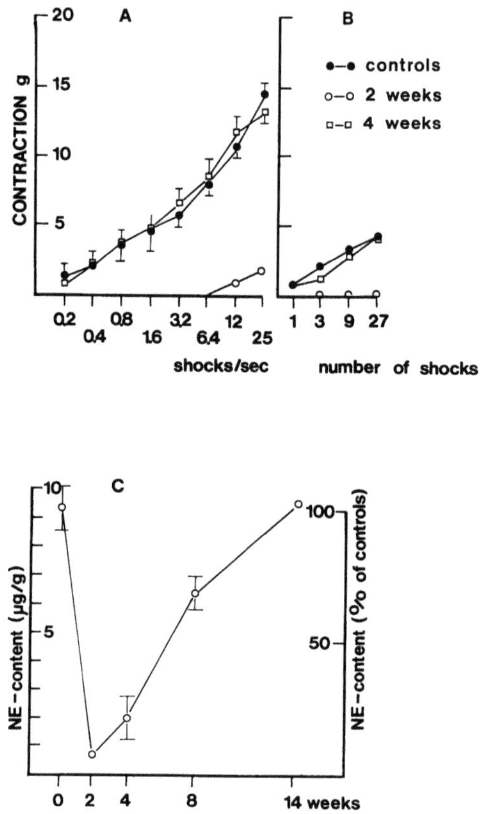

Fig. 9. Contractile response of the isolated nictitating membrane to sympathetic nerve stimulation in comparison to its norepinephrine content at various times after treatment with 6-HO-DA. A = frequency response curve; B = stimulus-number response curve; C = time course of norepinephrine content

nictitating membrane greatly reduced (HÄUSLER et al., 1968, 1969). As does surgical denervation pretreatment with 6-HO-DA provoked a supersensitivity of the nictitating membrane to exogenous norepinephrine (Fig. 8) which partially results from the lack of inactivation of norepinephrine by sympathetic nerve endings and partially from an unspecific postjunctional supersensitivity [TRENDELENBURG, 1963 (1)]. In the heart, this latter type of supersensitivity seems to be absent after chemical sympathectomy (HÄUSLER et al., 1968, 1969).

From a teleological point of view the supersensitivity to the physiological transmitter can be considered as a compensatory mechanism which helps to maintain the homeostasis of autonomic functions after destruction of a greater part of the sympathetic nervous system. The efficiency of this compensation is evident from the fact that the response of the cat nictitating membrane to sympathetic nerve stimulation had returned to control values when the norepinephrine

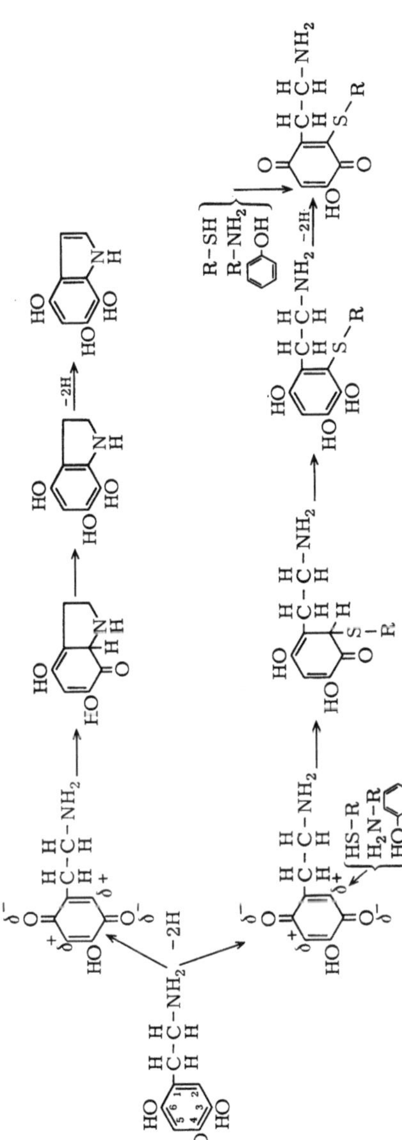

Fig. 10. Oxidation of 6-HO-DA to a para-quinone derivative. Possible further transformation into a trihydroxyindole derivative or interaction with nucleophilic groups of biological macromolecules

content amounted only to 20% of controls (Fig. 9). In addition to these compensatory mechanisms located within each single organ the adrenal medulla which is not directly affected by 6-HO-DA seems to take over—at least partially—the function of the destroyed sympathetic nerves by an augmented delivery of catecholamines to the circulation promoted by an increased reflex activity of the splanchnic nerves. This enhanced splanchnic activity increases the synthesis of adrenal catecholamines and is accompanied by an induction of tyrosine hydroxylase (MUELLER et al., 1969; THOENEN et al., 1969), the enzyme responsible for the rate limiting step in catecholamine synthesis (LEVITT et al., 1965).

As to the mechanism of the destruction of the adrenergic nerve terminals, two properties of 6-HO-DA seem to be of particular importance: a) the efficient accumulation of 6-HO-DA in adrenergic nerve endings (MALMFORS and SACHS, 1968; STONE et al., 1964), b) the extreme susceptibility of 6-HO-DA to non-enzymatic oxidation. Since the formation of H_2O_2 from 6-HO-DA *in vitro* has been shown (BIGLER, unpublished observation), it could be assumed that tissue damage would most readily occur at the sites of the highest concentration, i.e. in the adrenergic nerve endings. However, an alternative explanation seems to be more probable which implies that an oxidation product of 6-HO-DA undergoes covalent binding with biological structures.

Trihydroxyphenols with the structural features of 6-HO-DA are easily oxidized to quinones which beside the formation of indoles can undergo covalent binding with a wide variety of nucleophilic groups, such as SH, NH_2 and phenolic OH (Fig. 10). These covalent bindings might well be responsible for irreversible alteration of biological structures. This assumption is supported by the observation that after administration of ^{3}H-6-HO-DA 30 to 35% of the total radioactivity present in sympathetically innervated tissues cannot be extracted with perchloric acid using a procedure which otherwise leads to the extraction of 99% of norepinephrine administered in an identical way (THOENEN and TRANZER, 1968). That both extractable and non-extractable radioactivity mainly originate from sympathetic nerve endings can be deduced from the fact that in rats previously sympathectomized with non-tritiated 6-HO-DA the total radioactivity retained in the tissues was greatly reduced, but the ratio between extractable and non-extractable radioactivity remained the same.

It seems that the electrophilic attack of the oxidation products of 6-HO-DA on biological macromolecules is rather unspecific and that the high selectivity of the site of destruction is only a consequence of the efficient accumulation of 6-HO-DA in adrenergic nerve endings.

Summary

Electronmicroscopic studies revealed that intravenous administration of high doses of 6-hydroxydopamine produces a selective destruction of peripheral adrenergic nerve terminals. The biochemical correlates to these ultramorphological changes are an efficient and extremely long lasting norepinephrine depletion in peripheral sympathetically innervated organs, a marked decrease in their ability to take up ^{3}H-norepinephrine and a drastic reduction of their tyrosine hydroxylase, an enzyme exclusively located within adrenergic neurons.

By use of adequate dosage schedules 6-hydroxydopamine provides a valuable tool for general peripheral sympathectomy. Applications and limitations of this new method of chemical sympathectomy are discussed, together with a hypothetical mechanism of action.

References

CANNON, W. B., ROSENBLUETH, A.: The supersensitivity of denervated structures. New York: Macmillan Co. 1949.

COHEN, S.: Purification of a nerve growth promoting protein from the mouse salivary gland and its neuro-cytotoxic antiserum. Proc. nat. Acad. Sci. (Wash.) **46**, 302—311 (1960).

COOPER, T.: Surgical sympathectomy and adrenergic function. Pharmacol. Rev. **18**, 611—618 (1966).

DE CHAMPLAIN, J., KRAKOFF, L., AXELROD, J.: Interrelationship of sodium intake, hypertension and norepinephrine storage in the rat. Circulat. Res. **24**, 75—92 (1969).

HÄUSLER, G., HAEFELY, W., HÜRLIMANN, A.: Effect of surgical and chemical adrenergic denervation on vascular responses. Proceedings of the Symposium on the Physiology and Pharmacology of Vascular Neuroeffector Systems, Interlaken 1969. S. Karger, Basle (1970) (in press).

— — THOENEN, H.: Chemical sympathectomy of the cat with 6-hydroxydopamine. J. Pharmacol. exp. Ther. **170**, 50—61 (1969).

— THOENEN, H., HAEFELY, W.: Chemische Sympathektomie der Katze mit 6-Hydroxydopamin: Veränderungen von Sympathicusreizeffekten und Noradrenalinempfindlichkeit. Helv. physiol. pharmacol. Acta **26**, CR 223—225 (1968).

HERTTING, G., SCHIEFTHALER, T.: The effect of stellate ganglion excision on the catecholamine content and the uptake of H^3-norepinephrine in the heart of the cat. Int. J. Neuropharmacol. **3**, 65—69 (1964).

IVERSEN, L. L., GLOWINSKI, J., AXELROD, J.: The physiologic disposition and metabolism of norepinephrine in immunosympathectomized animals. J. Pharmacol. exp. Ther. **151**, 273—284 (1966).

JELLINEK, M. M. P., KAYE, M. P., NIGH, C. A., COOPER, T.: Alterations in chemical composition of canine heart after sympathetic denervation. Amer. J. Physiol. **206**, 971—974 (1964).

LEVI-MONTALCINI, R., ANGELETTI, P. U.: Immunosympathectomy. Pharmacol. Rev. **18**, 619—628 (1966).

— BOOKER, B.: Destruction of the sympathetic ganglia in mammals by an antiserum to a nerve-growth protein. Proc. nat. Acad. Sci. (Wash.) **46**, 384—391 (1960).

LEVITT, M., SPECTOR, S., SJOERDSMA, A., UDENFRIEND, S.: Elucidation of the rate-limiting step in norepinephrine biosynthesis in the perfused guinea-pig heart. J. Pharmacol. exp. Ther. **148**, 1—8 (1965).

MALMFORS, T., SACHS, C.: Degeneration of adrenergic nerves produced by 6-hydroxydopamine. Europ. J. Pharmacol. **3**, 89—92 (1968).

MUELLER, R. A., THOENEN, H., AXELROD, J.: Adrenal tyrosine hydroxylase: compensatory increase in activity after chemical sympathectomy. Science **158**, 468—469 (1969).

SEDVALL, G. C., KOPIN, I. J.: Influence of sympathetic denervation and nerve impulse activity on tyrosine hydroxylase in the rat submaxillary gland. Biochem. Pharmacol. **16**, 39—46 (1967).

STONE, C. A., PORTER, C. C., STAVORSKI, J. M., LUDDEN, C. T., TOTARO, J. A.: Antagonism of certain effects of catecholamine-depleting agents by antidepressant and related drugs. J. Pharmacol. exp. Ther. **144**, 196—204 (1964).

THOENEN, H., HÄFELY, W., GEY, K. F., HÜRLIMANN, A.: Diminished effecr of sympathetic nerve stimulation in cats pretreated with 5-hydroxydopa; formation and liberation of false adrenergic transmitters. Naunyn-Schmiedebergs Arch. Pharmak. exp. Path. **259**, 17—33 (1967).

— MUELLER, R. A., AXELROD, J.: Transsynaptic induction of adrenal tyrosine hydroxylase. J. Pharmacol. exp. Ther. **169**, 249—254 (1969).

— TRANZER, J. P.: Chemical sympathectomy by selective destruction of adrenergic nerve endings with 6-hydroxydopamine. Naunyn-Schmiedebergs Arch. Pharmak. exp. Path. **261**, 271—288 (1968).

TRANZER, J. P., THOENEN, H.: (1) Ultramorphologische Veränderungen der sympathischen Nervenendigungen der Katze nach Vorbehandlung mit 5- und 6-Hydroxy-Dopamin. Naunyn-Schmiedebergs Arch. Pharmak. exp. Path. **257**, 343—344 (1967).

— — (2) Electronmicroscopic localization of 5-hydroxydopamine (3,4,5-trihydroxyphenylethylamine), a new "false" sympathetic transmitter. Experientia (Basel) **23**, 743—745 (1967).

Tranzer, J. P., Thoenen, H.: An electronmicroscopic study of selective, acute degeneration of sympathetic nerve terminals after administration of 6-hydroxydopamine. Experientia (Basel) **24**, 155—156 (1968).

— — Snipes, R. L., Richards, J. G.: Recent developments on the ultrastructural aspect of adrenergic nerve endings in various experimental conditions. Progr. in Brain Res. **31**, 33—46 (1969).

Trendelenburg, U.: (1) Time course of changes in sensitivity after denervation of the nictitating membrane of the spinal cat. J. Pharmacol. exp. Ther. **142**, 335—342 (1963).

— (2) Supersensitivity and subsensitivity to sympathomimetic amines. Pharmacol. Rev. **15**, 255—276 (1963).

Wolinsky, H., Glagov, S.: Nature of species differences in the medial distribution of aortic vasa vasorum in mammals. Circulat. Res. **20**, 409—421 (1967).

Dr. H. Thoenen
Department of Experimental Medicine
F. Hoffmann-La Roche & Co. Ltd.
CH-4002 Basle, Switzerland

Discussion

Holtz: Did I understand you correctly that, in artificially renal hypotensive rats, 6-hydroxydopamine lowers the high blood pressure?

Thoenen: In animals with established hypertension, 6-hydroxydopamine lowers the blood pressure only temporarily and also treatment during the development of hypertension is not very effective. This is most probably due to the poor effect of 6-hydroxydopamine on the adrenergic nerve endings supplying the blood vessels.

Holtz: It would have a parallel to the immunosympathectomy in which you can produce renal hypertension.

Thoenen: I have to mention that the kind of experimental hypertension studied was not a mere renal hypertension. It was a DOCA-salt hypertension in uninephrectomized rats, which is known to be accompanied by increased activity in the peripheral sympathetic nervous system.

Blaschko: I have a brief question in relation to the lacking effect of 6-hydroxydopamine on the adrenal medulla. Is it that the chromaffin cells of the adrenal medulla are less susceptible to the destructive effect of 6-hydroxydopamine than the adrenergic neurons or is it due to a less efficient uptake of this amine into the adrenal medulla?

Thoenen: First of all, the blood supply to the adrenal medulla seems to be relatively small, if considered in relation to the catecholamine content. The amount of 6-hydroxydopamine delivered to the adrenal medulla has to be shared by a large number of chromaffin cells. Furthermore, it has not yet been established—as far as I know—whether the transport of catecholamines through the cell membrane is as efficient in the adrenal medulla as it is in the adrenergic nerve terminals.

de Robertis: I suppose that the 6-hydroxydopamine does not penetrate the blood brain barrier but I wonder if, by injecting it intercerebrally, the effects on the non-adrenergic or dopaminergic neurons may be studied. Have you any evidence of this? I was very interested to hear of this effect and would also like to know if in the pineal gland, where the sympathetic nerve contains 5-HT in addition to norepinephrine, the destructive effect is the same.

Thoenen: As to the pineal gland I must admit that I did not study the effect of 6-hydroxydopamine on this organ. Anyhow I should prefer the surgical procedure for the denervation of the pineal gland since you can easily accomplish a complete denervation by removal of both superior cervical ganglia. As to the effect of 6-hydroxydopamine on the central nervous system, this amine does not cross the blood brain barrier and by intravenous injection does not affect the central noradrenergic or dopaminergic neurons. However, after repeated intraventricular administration the norepinephrine content is reduced to about 10%, whereas the serotonin content is virtually not affected, as shown by Uretsky and Iversen. As in the periphery the reduction of the norepinephrine content is accompanied by a corresponding reduction of the tyrosine hydroxylase activity.

Corrodi: What happens to the cell bodies?

Thoenen: As far as we have studied the effect of 6-hydroxydopamine in the cat, there are no ultramorphological changes in the superior cervical ganglion. However, as in the adrenal medulla we do not know whether the absence of this effect is due to a relatively small supply of 6-hydroxydopamine to the ganglionic cells or whether the amine-uptake mechanism is less efficient in the adrenergic cell body than in the nerve terminals. However, we do have some biochemical evidence that after extremely high doses of 6-hydroxydopamine there is also damage to the cell bodies in the rat superior cervical and stellate ganglion, since the tyrosine-hydroxylase activity of these ganglia is somewhat reduced and the norepinephrine content of the heart and the salivary gland does not return to control levels even after several months. After intracerebral and intraventricular administration of 6-hydroxydopamine it appears that not only a destruction of the adrenergic nerve endings occurs but also a destruction of the cell bodies, as shown in very recent ultramorphological studies by Richards.

Glowinski: Recently, Jouvet in Lyon locally injected 6-hydroxydopamine in the locus coeruleus of the cat in order to study the effect of such lesion on sleep mechanisms. It appears that this injection in the cell bodies of noradrenergic neurons was affecting the content of norepinephrine in the nerve endings. Diminished concentrations of norepinephrine were found in the cortex of these animals, as well as changes in the characteristics of paradoxical sleep.

Dahlström: I wonder if it is impossible that the 6-hydroxydopamine could enter the granules and be stored in the granules by some reserpine-resistant mechanism?

Thoenen: I cannot answer this question definitely. We have direct ultramorphological evidence that 6-hydroxydopamine is taken up into the storage vesicles of the adrenergic nerve terminals. After depletion of norepinephrine by metaraminol, the administration of 6-hydroxydopamine restitutes the electron-dense material in the storage vesicles. So far we have not studied the possibility whether this refilling of the vesicles could be prevented by pretreatment with reserpine. However, reserpine seems to interfere with the storage mechanism of all phenethylamines accumulated in the vesicles.

Bayer-Symposium II, 144—158 (1970)

Session 4a

Effects of Drugs on Uptake and Release of Catecholamines (I)

Chairman: H. Blaschko

Effect of Some Metabolic Factors and Drugs on Uptake and Release of Catecholamines in vitro and in vivo

U. S. von Euler

With 10 Figures

1. Action of Metabolic Factors and Inhibitors on Transmitter Uptake and Release

Since the demonstration of ATP in relatively large amounts in adrenal medullary granules (Hillarp, Högberg and Nilson, 1955) and in nerve granules (Schümann, 1958) this nucleotide has been implicated in the amine storage function of these organelles.

The facilitation by ATP of catecholamine uptake into adrenal medullary granules (Kirshner, 1962; Carlsson, Hillarp and Waldeck, 1962, 1963) and adrenergic nerve granules [Euler and Lishajko, 1963 (1); Stjärne, 1964] suggests that these processes are metabolically dependent. This is particularly evident in partially depleted nerve granules which show only a small net uptake of noradrenaline (NA) in the absence of ATP but may be rapidly repleted to the original amount and even more with ATP. Medullary granules differ from nerve granules in this respect since they do not show net uptake of amines after depletion (Lishajko, 1969). Depletion and ATP-facilitated reuptake of NA can be repeated in vitro several times with nerve granules, suggesting that these, apparently in contrast to medullary granules, can be used for iterative loading and unloading with transmitter.

Amine uptake in nerve granules is facilitated in addition to ATP by UTP, ITP and CTP and also by ADP which lacks this action in adrenal medullary granules [Euler and Lishajko, 1963 (1), 1969; Carlsson et al., 1963]. AMP on the other hand has a weak action only and no effect was observed with cyclic AMP, even as dibutyryl compound, on nerve granules (Fig. 1). The latter finding is of interest since it tends to show that the formation of the "second messenger" (Suther-

LAND et al., 1968) is not part of the initial uptake process but is implicated in the secondary process by which the transmitter exerts its final effect.

These observations suggest that net uptake (and reuptake) of transmitter are processes which depend either on energy supply by granular ATP or by exogenous ATP, possibly provided by mitochondria (EULER and LISHAJKO, 1969) which seem to be a regular constituent of the axons and nerve terminals.

In view of the suggestive evidence for amine uptake in isolated nerve granules as an energy dependent process, we have tested a number of metabolic inhibitors and some substrates on the uptake and release of transmitter in granules. According to CARLSSON et al. (1963) some sulfhydryl reagents and the uncoupler pentachlorphenol exert a marked inhibitory effect on the incorporation of labeled amines

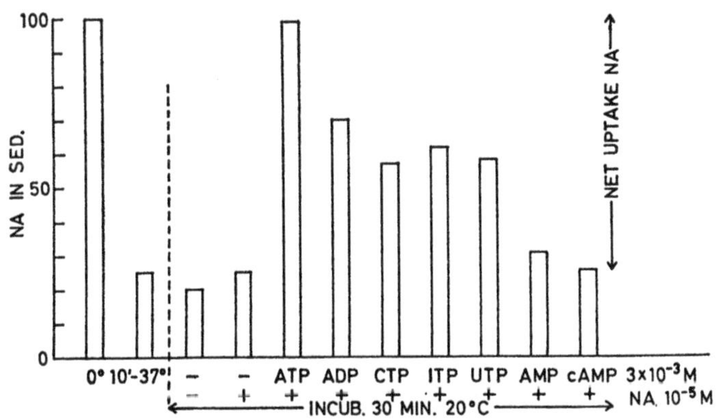

Fig. 1. Splenic nerve granules, incubated in 0.13 M K-phosphate pH 7.5. After preincubation 19 min, 37°, continued incubation 30 min 20° with addition of nucleotides and noradrenaline as indicated. Ordinate, per cent NA remaining in granules after incubation. (Net uptake experiment). [From EULER and LISHAJKO, Acta physiol. scand. (1969)]

in adrenal medullary granules while a number of other metabolic inhibitors seemed to influence the amine uptake only to a small extent or not at all.

As a result of our studies it became obvious that a number of known inhibitors at different sites in the respiratory chain inhibited both functions (EULER and LISHAJKO, 1969). NA uptake was thus inhibited by rotenone, chlorpromazine and antimycin A (Fig. 2). Cyanide and azide which act at the end of the chain had only a small and inconsistent action, however. Furthermore oligomycin, an inhibitor of oxidative phosphorylation, inhibited both release and uptake of NA in granules. Experiments with uncouplers (dinitrophenol, pentachlorphenol, desaspidin, carbonyl cyanide *m*-chlorophenylhydrazone) revealed that all of these enhanced the release and at the same time inhibited the ATP-dependent amine uptake. The results show several parallelisms between granule function and mitochondrial activity and suggest that transmitter uptake as well as release in nerve granules are controlled by a metabolic system, dependent on the function of at least a part of the electron transport chain and the phosphorylations linked to this chain. Further information has been adduced by the finding that neither atractylate, nor arsenate or fluoride had any consistent action on the two processes studied. While SH-reagents were active on adrenal medullary granules, they exercised relatively little action on the nerve granules. At high concentration, however, p-hydroxymercuribenzoic acid caused a complete and immediate release of all granule bound amines, presumably due to structural changes (D'Iorio, 1957; Hillarp, 1958). Pentachlorphenol at higher concentrations appears to act in the same way. Preliminary experiments have shown that isolated granule preparations have a low oxygen consumption (Giacobini, unpublished expts.) amounting to only a small fraction of that of mitochondria.

A number of substrates and different ionic media have also been tested as to their effect on the transmitter release and uptake in isolated nerve granules. Isotonic (0.13 M) potassium phosphate at pH 7.5 has been used as standard medium. In NaCl medium or sucrose there were only minor changes in release and

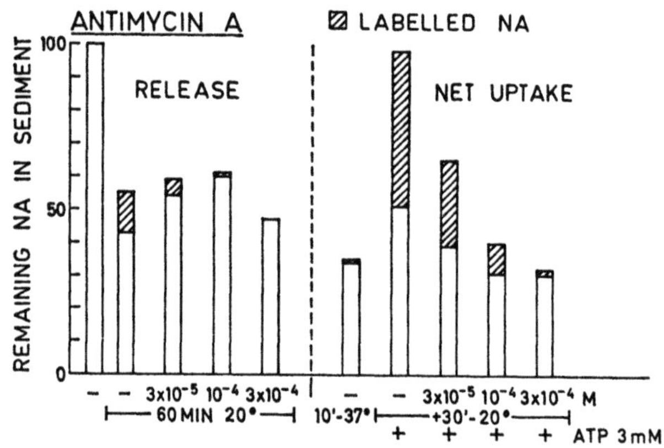

Fig. 2. Splenic nerve granules, incubated in 0.13 M K-phosphate pH 7.5. Left: Release expt. Ordinate per cent remaining NA in granules after incubation 60 min 20° with antimycin A in different concentrations. Right: Net uptake expt. as in Fig. 1. ATP facilitated NA uptake gradually diminished by addition of antimycin. [From Euler and Lishajko, Acta physiol. scand. (1969)]

uptake kinetics for NA as compared to potassium phosphate. Ca^{2+} above 5 mM caused inhibition of release. Of the various substrates added (glucose, glucose-1-phosphate, glucose-6-phosphate, succinate, lactate, citrate, glycerophosphate) none seemed to affect the ATP-dependent uptake or the release rate to any noticeable extent. 2-Phosphoglycerate 10^{-4} M moderately increased the release rate.

The observations with metabolic factors and inhibitors and uncouplers strongly suggest that the uptake is an energy-requiring process presumably involving both formation and utilization of ATP. The results with uncouplers suggest that the graded release is dependent on phosphorylation processes linked to the electron transport chain and that granular ATPase is actively involved (cf. Hillarp, 1958; Taugner and Hasselbach, 1966; Burger, Philippu and Schümann, 1968).

Uptake kinetics were studied on isolated nerve granules for (1) the spontaneous reuptake of NA without addition of exogenous ATP, (2) the net uptake of NA in partially depleted granules with addition of NA and ATP at 20 °C.

In the former case a relatively wide concentration range could be utilized, while the NA-uptake with added ATP only could be studied within a relatively narrow range. However, the results were uniform with ATP addition and the K_m value observed was about 1.5×10^{-6} M for l-NA (Fig. 3). In the reuptake experiments the variations were somewhat larger and the K_m value varied between 1.5 and 2×10^{-6} M (EULER and LISHAJKO, unpublished observations). For the d-NA the MICHAELIS constant was considerably higher.

In the isolated rat heart perfused with labeled amines in different concentrations IVERSEN (1963) found K_m values for l-NA of 0.27×10^{-6} M and for d-NA of 1.39×10^{-4} M. The K_m value for adrenaline (A) uptake in bovine adrenal medullary granules has been estimated to be 8×10^{-4} M (JONASSON, ROSENGREN and WALDECK, 1964).

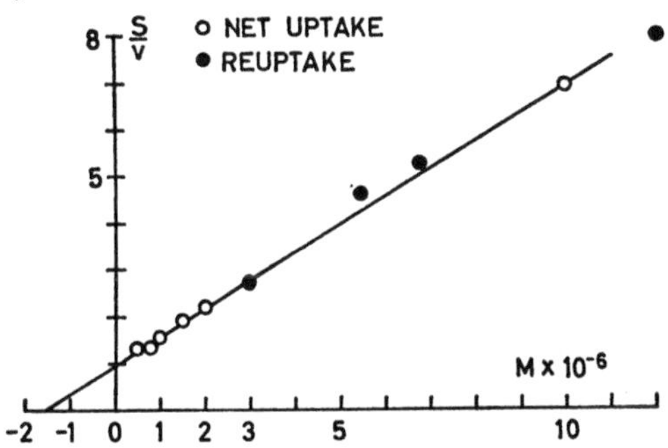

Fig. 3. Graphical determination of K_m-value for ATP-facilitated NA-uptake in isolated splenic nerve granules (circles). Dots: values from reuptake experiments. S, noradrenaline concentration in medium, V, initial uptake rate. Abscissa noradrenaline concentration ($\times 10^{-6}$)

2. Effect of Drugs on Adrenergic Nerve Transmission

Many attempts have been made to obtain further information about the mechanisms of adrenergic transmission at the granular and axon membrane level by the use of drugs. Since the drugs only exceptionally act on one cellular process only, the results must be judged with caution. On the other hand the members of certain groups of drugs sometimes seem to act in a similar way, thus giving some information as to the mechanisms involved. Several of these findings have been reviewed earlier and shall therefore only be briefly summarized.

a) Indirectly Acting Amines

As shown by SCHÜMANN and PHILIPPU (1962) tyramine liberates the transmitter from its stores in isolated granules by some kind of substitution. Further studies [EULER and LISHAJKO, 1968 (1)] have revealed that tyramine like several other indirectly acting amines, notably phenylethylamine, has a two-fold action on nerve granules, consisting in a direct release of transmitter by substitution, and inhibition of reuptake. In vivo the effect of indirectly acting amines may also involve release of transmitter from extragranular stores and inhibition of uptake at the axon membrane level (BURN and RAND, 1958; TRENDELENBURG, 1961; BURGEN and IVERSEN, 1965).

Low concentrations (10^{-6}— 10^{-5} M) of tyramine (TA) as well as amphetamine markedly enhance the contractile effect of transmural electrical stimulation of the isolated vas deferens of the guinea pig without causing any effect of their own

(Euler, 1969, unpublished observations). These effects may depend on reduced reuptake at the axonal level (Burgen and Iversen, 1965) or facilitated release from the granular or extragranular stores (cf. below). A weakly acting dose of TA (10 μg/kg) in the spinal adrenalectomized cat may be strongly enhanced in its cardiovascular action by a dose of reserpine which by itself causes no effect (Nasmyth, 1962; Bygdeman and Euler, unpublished observations). Since the blocking of reuptake at the granular level caused by reserpine has no visible effect of its own it appears likely that reserpine facilitates the action of TA on the extragranular stores possibly associated with the axon membranes. No effects of TA are seen after an efficient membrane entrance blocker like desmethylimipramine (DMI).

b) Adrenergic Blocking Agents

Several groups have noted the blocking effect on amine uptake in vivo and in vitro caused by adrenergic blocking agents [Axelrod, Hertting and Potter, 1962; Carlsson et al., 1963; Lindmar and Muscholl, 1964; Euler and Lishajko, 1965, 1968 (2)]. Carlsson et al. (1963) also expressed the view that the inhibitory effect observed on amine uptake in adrenal medullary granules may have a counterpart on the effector side, where amine uptake in some form must be assumed to initiate the action.

Although all blockers hitherto tested inhibit the amine uptake in isolated nerve granules, wide differences are observed in their efficiency as may be expected. Both α- and β-blockers are active in these respect. Among the most active α-blockers are dihydroergotamine, azapetine, plegicil, and chlorpromazine. Of the β-blockers pronethalol, propranolol and aptin were strongly active.

Inhibitory actions of the blockers have been observed also for the NA uptake in platelets (Bygdeman and Johnsen, 1969). In general the α-blockers as well as the β-blockers inhibit the release of NA from nerve granules on incubation. Of the blockers only two have been tested as isomers, propranolol and aptin. In both cases the l-forms had a stronger inhibitory action on the amine uptake than the d-form. There was no absolute stereospecificity, but it should be recalled that neither is this the case for the amine uptake in the absence of drugs.

In several cases the blocking of amine uptake and release in granules could be shown to be reversible and removed by washing. Thus no specific saturable binding could be demonstrated for DHE, aptin or the α-blocker N-ethoxycarbonyl-2-ethoxy-1,2-dihydroquinoline (Belleau et al., 1968).

c) Psychotropic Drugs

The strongly inhibitory effect of reserpine in low concentrations on the transmitter release from isolated nerve granules [Euler and Lishajko, 1961, 1963 (2)] has sometimes been regarded as paradoxical with regard to the depleting effect on the transmitter content in organs. Since reserpine also inhibits the ATP-dependent incorporation of amines in granules as shown for medullary granules by Kirshner (1962) and by Carlsson et al. (1963), the continuous release, although retarded, will still lead to a depletion of the stores. Similarly reserpine blocks the ATP-facilitated net uptake of NA in isolated nerve granules. Since there is a greatly retarded release, the concomitant reuptake is correspondingly inhibited. Until recently no correlate in vivo to the retardation of release has been demonstrated, however. This would also be difficult to show since control situations for comparison are hard to achieve experimentally. However, using the isolated vas deferens of the guinea pig it has been possible to show (Euler, 1969) that addition of reserpine in a concentration of about 10^{-6} M to the suspension medium may cause an acute inhibition of the response to hypogastric nerve stimulation (preganglionic)

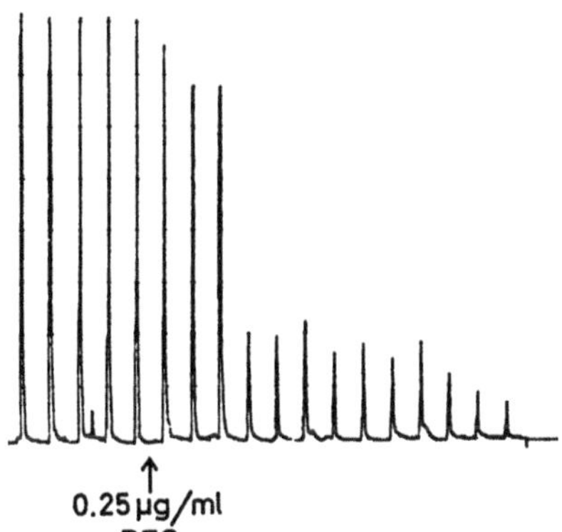

Fig. 4. Isolated vas deferens, guinea pig, 20 ml bath, 37°. Field stimulation 25/sec, 2 msec duration, 5 sec stimulation per minute. At arrow reserpine 0.25 µg/ml

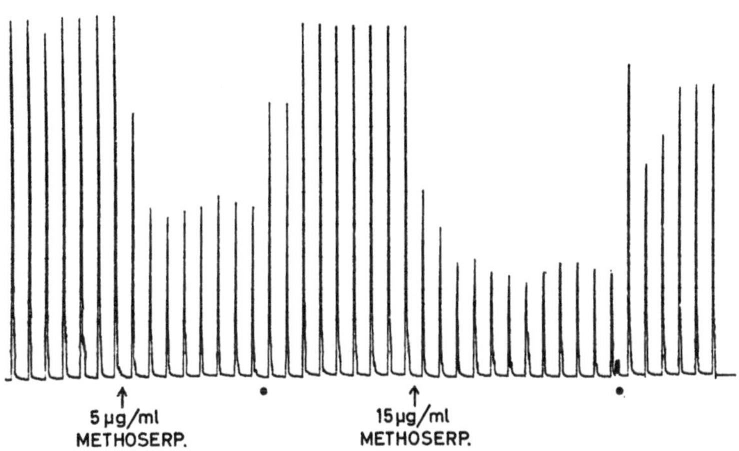

Fig. 5. Isolated vas deferens, guinea pig, 20 ml bath. Field stimulation 25/sec, 2 msec duration, 5 sec stimulation per minute. At arrows methoserpidine 5 and 15 µg/ml. Dots, washing

according to Hukovic (1961) or transmural stimulation (postganglionic) by the technique of Birmingham and Wilson (1963) (Fig. 4). The effect also occurs with methoserpidine in about 10 times higher concentration (Fig. 5). Since reserpine in the concentrations employed does not alter the response to NA added to the bath it is concluded that it inhibits the release of transmitter. The immediate effect rules out overall depletion of the stores and it seems therefore plausible that reserpine in some way prevents the filling of the stores drawn upon for immediate release of transmitter. These are almost certainly not the granules themselves but may consist of a smaller pool of extragranular stores supplied with transmitter from the larger pool of granules (cf. Trendelenburg, 1961). We are inclined to believe that this secondary "retail" pool is situated in close association with the axon membrane of the varicosity (cf. Fig. 10).

A tentative explanation of the acute inhibitory effect of reserpine described above would then be that the refilling mechanism by which the "retail" pool maintains its store becomes insufficient, due to the slow release of transmitter from the granules. In harmony with this concept is the fact that "neuronal rest" allows the effect of electrical stimulation to become temporarily restored. A blocking phenomen of apparently similar kind has been observed in the perfused rabbit ear by Day and Owen (1968) although no interpretation of the effect was offered. The acute inhibitory effect of reserpine was reversible when reserpine was allowed to act for only a few minutes, but became irreversible after longer exposure. In some instances the inhibitory effect was preceded by enhanced contractions and

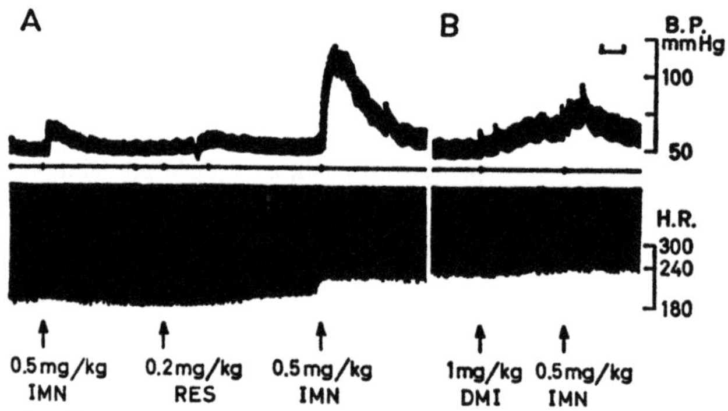

Fig. 6. Cat, spinal, adrenectomized, vagi cut. Upper curve blood pressure, lower curve heart rate. I.v. injections of iso-monomethylnicotinium bromide, reserpine and desmethylimipramine as indicated. Time mark 1 min. [From Euler and Persson, Acta physiol. scand. (1970)

reserpine occasionally produced contraction by itself. The cause of the "facilitating" effect is not known, but it may be explained by a temporarily increased supply of transmitter as a result of interaction of reserpine with the extragranular stores.

The lowered efficiency of adrenergic neuromuscular transmission after reserpine may contribute to the anti-hypertensive action of the drug observed clinically. Since the effect becomes overt only above a certain level of adrenergic nerve activity, it may not produce signs of adrenergic nerve insufficiency under normal conditions when low doses are employed, but act during increased sympathetic impulse flow.

Several observations point in the direction of an action of reserpine on the structural integrity of the cell. Thus Wilcken at al. (1967) have described structural changes in the mitochondria after repeated small doses of reserpine. An effect of this kind may also be involved in the amine releasing effect of reserpine in higher concentrations on isolated nerve granules (Euler and Lishajko, 1960).

Recently we have observed (Euler and Persson, 1970) that the quaternary nicotine analogue *iso*-monomethylnicotinium bromide, which has only a weak nicotinic effect on cardiovascular functions in the spinal cat, may become strongly active after a dose of reserpine which by itself causes no action (Fig. 6). It was assumed that the action of reserpine is to allow access of the nicotine analogues to otherwise unattainable transmitter stores. Since the quaternary nicotine compounds do not seem to affect the storage granules any more than nicotine itself [HEDQVIST, 1969 (1)] it appears reasonable to believe either that they act on the extragranular stores, presumably in or close to the axon membrane, or are enabled by reserpine to cause depolarization and transmitter release. Their action is prevented by DMI.

Several other psychotropic drugs have been tested on their effects on uptake and release in nerve granules. The majority of them inhibit release as well as reuptake and net uptake (DMI, protriptyline, nortriptyline, LU-3-010, LSD,

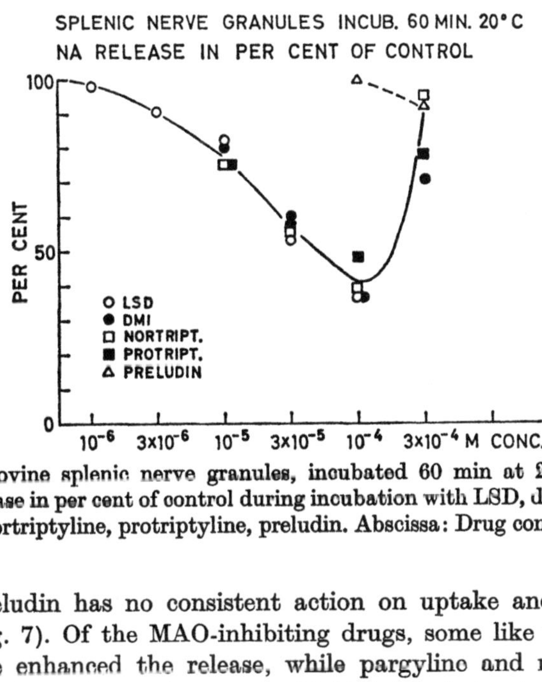

Fig. 7. Isolated bovine splenic nerve granules, incubated 60 min at 20° in K-phosphate. Noradrenaline release in per cent of control during incubation with LSD, desmethylimipramine (DMI) nortriptyline, protriptyline, preludin. Abscissa: Drug concentration

haliperidol). Preludin has no consistent action on uptake and release even at 3×10^{-4} M (Fig. 7). Of the MAO-inhibiting drugs, some like pheniprazine and tranylcypromine enhanced the release, while pargyline and nialamide had no consistent action in concentrations up to 10^{-3} M. Amphetamine enhances release like other indirectly acting amines.

There are thus considerable variations in the effects of the psychotropic drugs, but in general they influence the amine uptake and release at the granular level, which may contribute to the somatic and mental effects. So far there is no evidence that a psychotropic drug acts solely via the adrenergic nerve mechanisms, even though this may be true for a part of their effect.

3. Binding of Transmitter to Storage Granules

The binding mechanism for the adrenergic transmitter to nerve granules is still unknown in its details. It seems well established however, that ATP is a powerful mediator of the uptake, and since it occurs as a natural constituent of the granules it seems not unwarranted to assume that it operates also during the process of reuptake, occurring concomitantly with release on incubation of isolated granules. A reversible binding of ATP and transmitter may be mediated by some constituent of the granules (cf. Belleau, 1960; Hillarp, 1960). Drugs particularly of amine character may induce conformational changes in the proteins involved and thus influence the release and uptake process. A variety of conditions are known to affect the binding of transmitter to nerve granules. While hyposmotic lysis, freezing and thawing are only moderately active in this respect, mechanical factors such as high speed homogenization and sonication as well as surfactants rapidly cause a breakage of the binding. Various heavy metals have the same effect including the SH-reagent p-chloromercuribenzoate. Perhaps even more striking is the effect of prenylamine which causes a rapid loss of bound transmitter in a concentration of 3×10^{-4} M [Euler and Lishajko, 1968 (3)]. The uncouplers desaspidine and pentachlorphenol also have this effect.

It appears from these examples that the binding is either a very loose one or else that the postulated ATP-amine complex depends for its existence on the continuous and undisturbed activity of some metabolic process which can be put out of action by various factors. An ATP-amine-protein complex does not so far seem to have been demonstrated, however (Smith, 1968).

4. Actions of Acetylcholine and Prostaglandins in Adrenergic Neurotransmission

The mechanisms underlying the mobilization of Ca^{2+} ions, which are required for neurotransmission processes are still unknown (cf. Katz, 1969). While the effect of acetylcholine on postsynaptic neurons and on muscular endplates is not questioned, it is still under debate whether acetylcholine acts as an obligatory factor in the activation of the axon, as held by Nachmansohn (1966) and constitutes a link in adrenergic neurotransmission (Burn and Rand, 1959). Various observations in later years appear, however, to provide at least circumstantial evidence pointing in this direction which might warrant a discussion of this issue.

Cholinergic sympathetic nerves might preferably be excluded from this discussion since acetylcholine and not noradrenaline here serves as neuromuscular transmitter. The presence of ACh in splenic nerves (Euler, 1948), and of cholinesterase in adrenergic axon membranes (Burnstock and Robinson, 1967; Bell, 1969) suggests that ACh not only serves as a neurotransmitter at the cholinergic nerve endings but also is involved in axonal functions in adrenergic nerves (cf. Nachmansohn, 1966). The enhancing effect of low concentrations of ACh on the response of smooth muscle preparations to sympathetic nerve stimulation (Sjöstrand, 1961; Malik and Ling, 1969) are also suggestive of a participation of ACh in the neurotransmission process. Thus Sjöstrand showed that the response of the isolated vas deferens of the guinea pig, electrically stimulated by the hypogastric nerves (preganglionic), was strongly increased for ½ to 1 h after addition of ACh to about 10^{-7} M in the bath. In Malik and Ling's experiments which followed upon a study of Lindmar, Löffelholz and Muscholl (1968) on a muscarinic mechanism inhibiting the release of NA from peripheral adrenergic nerve fibres by nicotinic agents it was shown that ACh in the perfusing medium inhibits or blocks the vasoconstrictor response to stimulation of the mesenterial nerves but might also enhance it when applied for brief periods and in low concentrations. The nature of these effects are still obscure.

In recent own observations the enhancing effect of ACh on the response of the guinea pig vas deferens could be confirmed also with transmural stimulation of the postganglionic nerves (Fig. 8). The effect was readily blocked by atropine 10^{-7} g/ml which did not affect the muscle tone and only caused slight decrease in the stimulation response. The results suggested that if the ACh demonstrated in adrenergic nerves is a true component of the axons, this ACh may have important actions on the neuromuscular transmission process.

In these experiments the stimulation parameters were chosen so as to stimulate the postganglionic nerves while avoiding direct muscle stimulation as tested with neuronal blockers. Preganglionic stimulation did not occur as evidenced by the hexamethonium test.

As to the interpretation of the observed effect sensitization of the smooth muscle by ACh to the transmitter NA cannot be ruled out but did at least not appear with added NA. The possibility of an effect on the neuromuscular transmission process might therefore be considered. While ACh hardly penetrates the axon in its preterminal course it can gain access to the terminals, presumably as a consequence of their specialized structure (cf. Ferry, 1963; Armett and Ritchie, 1960). Because of the abundance of adrenergic nerve terminals in the vas deferens (Owman and Sjöstrand, 1966) the basis for an action of this kind is uniquely provided for in this organ. Since the response to pulses of 1 msec duration (25/sec, 5 sec stimulation) was clearly submaximal, it appears conceivable that the added ACh, in this preparation, activates a varying number of nerve terminals and thus causes increased transmitter release. Since ACh in low concentrations causes no obligatory direct contraction, it may be assumed that the added ACh causes a partial depolarization of the smooth muscle cells, unable to elicit contraction, but large enough to enable a train of subthreshold stimuli to become effective (cf. Burnstock and Holman, 1963). Effects of this kind should, by analogy also be possible to observe in the spleen.

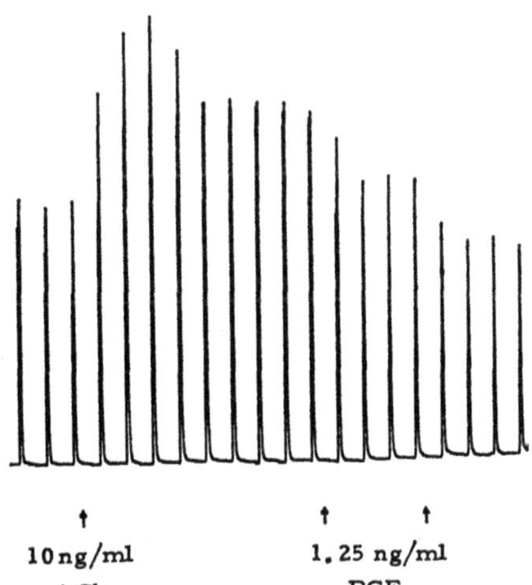

Fig. 8. Isolated vas deferens, guinea pig, 20 ml bath. Field stimulation 40/sec, 0.25 msec duration, 5 sec stimulation per minute. At arrows acetylcholine 10 μg/ml, prostaglandin E_1 1.25 μg/ml

Since prostaglandins occur widely in the organism and are released by nerve stimulation (Davies, Horton and Withrington, 1967), their possible role in neurotransmission may be considered. Some recent observations give certain support to this possibility.

Thus addition of prostaglandin E_1 (PGE_1) or PGE_2 to the medium perfusing the isolated spleen of the cat causes a marked reduction of the mechanical response to splenic nerve stimulation and a decreased action of injected NA and also a decrease in the release of labeled NA from the previously ^{3}H-NA-loaded spleen [Hedqvist and Brundin, 1969; Hedqvist, 1969 (2)]. The large increase in release of ^{3}H-NA regularly seen after addition of phenoxybenzamine to the perfusion fluid was similarly inhibited by PGE_2 [Hedqvist, 1969 (3)], which led to the conclusion that PGE_1 and PGE_2 interfere with the release of the transmitter in addition to an antagonistic effect on the effector organ, most marked with PGE_1.

More recently we have observed an effect of a similar kind on the isolated vas deferens of the guinea pig, stimulated transmurally by 5 sec trains of stimuli each

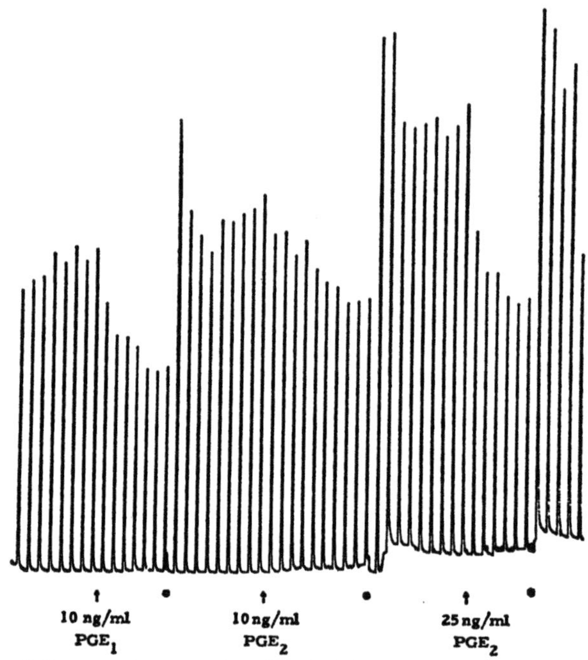

Fig. 9. Isolated vas deferens, guinea pig, 20 ml bath. Field stimulation 20/sec, 0.8 msec duration, 5 sec stimulation per minute. At arrows prostaglandin E_1 (PGE_1) 10 μg/ml, PGE_2 10 μg/ml, PGE_2 25 μg/ml. Dots, washing

half or whole minute, 25/sec, duration 0.2 to 1 msec (Euler and Hedqvist, 1969) (Fig. 9). Both PGE_1 and PGE_2 caused in concentrations from about 3×10^{-9} M in the bath an inhibition of the mechanic response. The effect set in rapidly and lasted until the PGE was washed out. A slight direct contractile effect was seen with the concentrations of PGE used and the sensitivity of the preparation towards added NA was moderately increased. Preliminary data indicate that prostaglandins in active concentrations may be released from the organ on nerve stimulation.

Higher concentrations of PGE_1 caused an increased response to transmural stimulation, as previously noted by Mantegazza and Naimzada (1965).

5. A Tentative Scheme for Adrenergic Neurotransmission

From the foregoing it appears that various recent observations tend to indicate involvement of acetylcholine in adrenergic neuromuscular transmission as originally suggested by BURN and RAND (1959) and BURN (1966) and of prostaglandins [HEDQVIST, 1969 (2)]. While the effect of prostaglandins shall not be discussed in the present communication beyond the comments made in connection with the presented data, the possible role of acetylcholine is schematically and tentatively illustrated in Fig. 10. The lack of precise knowledge does not allow a detailed scheme; the figure is intended to link together some facts and concepts. In brief the scheme adopts the idea of BURN and RAND (1959) of some kind of cholinergic link and assumes release of acetylcholine by the excitatory wave from axon membrane sites causing or amplifying presynaptic depolarization followed (1) by mobilization of calcium ions, (2) which by some unknown mechanism initiate the release

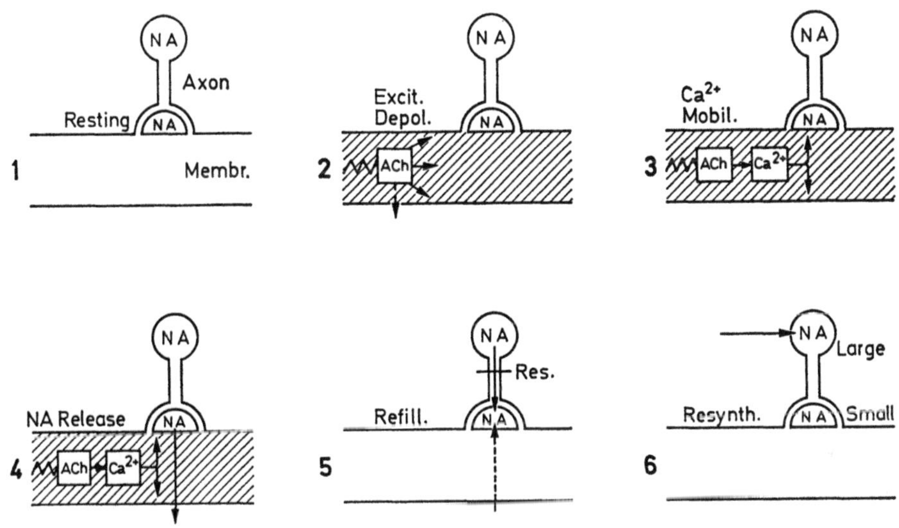

Fig. 10. Schematic representation of adrenergic neurotransmission. 1. Resting state. 2. Acetylcholine release coupled with depolarization. 3. Calcium mobilization. 4. Transmitter release. 5. Refilling of extragranular store (inhibited by reserpine). 6. Resynthesis of transmitter

of transmitter from membrane or juxtamembrane sites (3). As a result of the disturbed equilibrium between the granular and membrane stores, transmitter is transferred to the latter which may also receive contributions by recapture (4). Resynthesis is induced by negative feedback from the partially depleted store sites (5). The "quanta" released may be considerably smaller than the amount stored in one granule (cf. FOLKOW, HÄGGENDAL and LISANDER, 1968). The suggested system does not require the assumptions of a rapid consumption of granules by exocytosis or a large dependence on axonal reuptake, even though these phenomena may occur to varying extents. Even if no direct evidence for the existence of the membrane or juxtamembrane storage sites postulated in the scheme presented is available as far as the adrenergic nerve terminals are concerned, the network of "dense projections" in the presynaptic membrane recently reported by PFENNINGER et al. (1969) should be considered as possible transmitter sites and could have functional counterparts in other presynaptic areas.

Summary

The ATP facilitated uptake of transmitter in isolated adrenergic nerve granules, as well as the release, are inhibited or prevented by various metabolic inhibitors, such as rotenone, antimycin A and oligomycin. Uncouplers increase the release and inhibit uptake of transmitter. The results suggest that the uptake process is dependent on the functional integrity of the respiratory chain and electron transport linked phosphorylation.

Numerous observations suggest that the adrenergic neurotransmission process is initiated by a mechanism involving mobilization of acetylcholine and of calcium. The ensueing liberation of transmitter appears to occur from an extragranular site, localized in or close to the nerve terminal axon membrane, subsequently refilled from granular stores.

The acute neuromuscular block observed after reserpine may be due to retarded release of NA from the granules and insufficient refilling of the extragranular stores.

Prostaglandin E_1 and E_2 in oligodynamic concentrations inhibit adrenergic neurotransmission in the transmurally stimulated guinea pig vas deferens.

References

Armett, C. J., Ritchie, J. M.: The action of acetylcholine on conduction in mammalian non-myelinated fibres and its prevention by an anticholinesterase. J. Physiol. (Lond.) **152**, 141—158 (1960).

Axelrod, J., Hertting, G., Potter, L.: Effect of drugs on the uptake and release of ^{3}H-norepinephrine in the rat heart. Nature (Lond.) **194**, 297 (1962).

Bell, C.: Fine structural localization of acetylcholinesterase at a cholinergic vasodilator nerve-arterial smooth muscle synapse. Circulat. Res. **24**, 61—70 (1969).

Belleau, B.: Relationships between agonists, antagonists and receptor sites. Ciba Foundation Symposium "Adrenergic Mechanisms", p. 223—245. London: J. & A. Churchill Ltd. 1960.

—, Martel, R., Lacasse, G., Menard, M., Weinberg, N. L., Perron, Y. G.: N-Carboxylic acid esters of 1,2- and 1,4-dihydroquinolines. A new class of irreversible inactivators of the catecholamine α receptors and potent central nervous system depressants. J. Amer. Chem. Soc. **90**, 823 (1968).

Birmingham, A. T., Wilson, A. B.: Preganglionic and postganglionic stimulation of the guinea-pig isolated vas deferens preparation. Brit. J. Pharmacol. **21**, 569—580 (1963).

Burgen, A. S. V., Iversen, L. L.: The inhibition of noradrenaline uptake by sympathomimetic amines in the rat isolated heart. Brit. J. Pharmacol. **25**, 34—49 (1965).

Burger, A., Philippu, A., Schümann, H. J.: Untersuchungen zur Bedeutung einer ATPase aus Milznervengranula. Naunyn Schmiedebergs Arch. Pharmak. exp. Path. **260**, 101—102 (1968).

Burn, J. H.: Introductory remarks. In: Adrenergic transmission, Sec. V. Pharmacol. Rev. **18**, 459—470 (1966).

—, Rand, M. J.: The action of sympathomimetic amines in animals treated with reserpine. J. Physiol. (Lond.) **144**, 314—336 (1958).

— — Sympathetic postganglionic mechanism. Nature (Lond.) **184**, 163—165 (1959).

Burnstock, G., Holman, M. E.: Smooth muscle: Autonomic nerve transmission. Ann. Rev. Physiol. **25**, 61—85 (1963).

—, Robinson, P. M.: Localization of catecholamines and acetylcholinesterase in autonomic nerves. Circulat. Res. **21**, Suppl. III, 43—55 (1967).

Bygdeman, S., Johnsen, Ø.: Studies on the effect of adrenergic blocking drugs on catecholamine-induced platelet aggregation and uptake of noradrenaline and 5-hydroxytryptamine. Acta physiol. scand. **75**, 129—138 (1969).

Carlsson, A., Hillarp, N.-Å., Waldeck, B.: A Mg^{++}-ATP dependent storage mechanism in the amine granules of the adrenal medulla. Med. exp. **6**, 47—53 (1962).

CARLSSON, A., HILLARP, N.-Å., WALDECK, B.: Analysis of the Mg^{++}-ATP dependent storage mechanism in the amine granules of the adrenal medulla. Acta physiol. scand. **59**, Suppl. 215 (1963).
DAVIES, B. N., HORTON, E. W., WITHRINGTON, P. G.: The occurrence of prostaglandin E_2 in splenic venous blood of the dog following splenic nerve stimulation. J. Physiol. (Lond.) **188**, 38P—39P (1967).
DAY, M. D., OWEN, D. A. A.: The interaction between angiotensin and sympathetic vasoconstriction in the isolated artery of the rabbit ear. Brit. J. Pharmacol. **34**, 499—507 (1968).
D'IORIO, A.: The release of catecholamines from the isolated chromaffine granules of the adrenal medulla using sulphydryl inhibitors. Canad. J. Biochem. **35**, 395—400 (1957).
v. EULER, U. S.: Sympathin, histamine and acetylcholine in mammalian nerves. J. Physiol. (Lond.) **107**, 10P—11P (1948).
— Acute neuromuscular transmission failure in vas deferens after reserpine. Acta physiol. scand. **76**, 255—256 (1969).
—, HEDQVIST, P.: Inhibitory action of prostaglandins E_1 and E_2 on the neuromuscular transmission in the guinea pig vas deferens. Acta physiol. scand. **77**, 510—512 (1969).
—, LISHAJKO, F.: Effect of reserpine on the release of noradrenaline from transmitter granules in adrenergic nerves. Science **132**, 351—352 (1960).
— — Effect of reserpine on the release of catecholamines from isolated nerve and chromaffin cell granules. Acta physiol. scand. **52**, 137—145 (1961).
— — (1) Effect of adenine nucleotides on catecholamine release and uptake in isolated adrenergic nerve granules. Acta physiol. scand. **59**, 454—461 (1963).
— — (2) Effect of reserpine on the uptake of catecholamines in isolated nerve storage granules. Int. J. Neuropharmacol. **2**, 127—134 (1963).
— — Effect of drugs on the storage granules of adrenergic nerves. In: Pharmacology of cholinergic and adrenergic transmission, Proc. 2nd Internat. Pharmacol. Meeting, Prague, 1963, pp. 245—259, Praha: Czechoslovak Medical Press 1965.
— — (1) Effect of directly and indirectly acting sympathomimetic amines on adrenergic transmitter granules. Acta physiol. scand. **73**, 78—92 (1968).
— — (2) Inhibitory action of adrenergic blocking agents on reuptake and net uptake of noradrenaline in nerve granules. Acta physiol. scand. **74**, 501—506 (1968).
— — (3) Observations on the actions of prenylamine (Segontin) in vivo and on adrenergic transmitter granules. Biochim. appl. (Parma) **14**, Suppl. I, 17—32 (1968).
— — Effects of some metabolic co-factors and inhibitors on transmitter release and uptake in isolated adrenergic nerve granules. Acta physiol. scand. **77**, 298—307 (1969).
—, PERSSON. N.-Å.: Potentiation of adrenergic cardiovascular effects of two quaternary nicotine analogues by reserpine. Acta physiol. scand. **78**, 459—464 (1970).
FERRY, C. B.: The sympathomimetic effect of acetylcholine on the spleen of the cat. J. Physiol. (Lond.) **167**, 487—504 (1963).
FOLKOW, B., HÄGGENDAL, J., LISANDER, B.: Extent of release and elimination of noradrenaline at peripheral adrenergic nerve terminals. Acta physiol. scand. **72**, Suppl. 307 (1968).
HEDQVIST, P.: (1) On the mechanism of depletion of noradrenaline stores by *iso*-monomethylnicotinium bromide. Acta physiol. scand. **78**, 117—122 (1970).
— (2) Modulating effect of prostaglandin E_2 on noradrenaline release from the isolated cat spleen. Acta physiol. scand. **75**, 511—512 (1969).
— (3) Antagonism between prostaglandin E_2 and phenoxybenzamine on noradrenaline release from the cat spleen. Acta physiol. scand. **76**, 383—384 (1969).
—, BRUNDIN, J.: Inhibition of prostaglandin E_1 of noradrenaline release and of effector response to nerve stimulation in the cat spleen. Life Sci. **8**, 389—395 (1969).
HILLARP, N.-Å.: Enzymic systems involving adenosinephosphates in the adrenaline and noradrenaline containing granules of the adrenal medulla. Acta physiol. scand. **42**, 144—165 (1958).
— Some problems concerning the storage of catechol amines in the adrenal medulla. Ciba Foundation Symposium "Adrenergic Mechanisms", p. 481—501. London: J. & A. Churchill Ltd. 1960.
—, HÖGBERG, B., NILSON, B.: Adenosine triphosphate in the adrenal medulla of the cow. Nature (Lond.) **176**, 1032—1033 (1955).
HUKOVIC, S.: Responses of the isolated sympathetic nerve ductus deferens preparation of the guinea-pig. Brit. J. Pharmacol. **16**, 188—194 (1961).

Iversen, L. L.: The uptake of noradrenaline by the isolated perfused rat heart. Brit. J. Pharmacol. **21**, 523—537 (1963).

Jonasson, J., Rosengren, E., Waldeck, B.: Effects of some pharmacologically active amines on the uptake of arylalkylamines by adrenal medullary granules. Acta physiol. scand. **60**, 136—140 (1964).

Katz, B.: The release of neural transmitter substances. The Sherrington Lectures X. Liverpool: Univ. Press 1969.

Kirshner, N.: Uptake of catecholamines by a particulate fraction of the adrenal medulla. J. biol. Chem. **237**, 2311—2317 (1962).

Lindmar, R., Löffelholz, K., Muscholl, E.: A muscarinic mechanism inhibiting the release of noradrenaline from peripheral adrenergic nerve fibres by nicotinic agents. Brit. J. Pharmacol. **32**, 280—294 (1968).

—, Muscholl, E.: Die Wirkung von Pharmaka auf die Elimination von Noradrenalin aus der Perfusionsflüssigkeit und die Noradrenalinaufnahme in das isolierte Herz. Naunyn-Schmiedebergs Arch. exp. Path. Pharmak. **247**, 469—492 (1964).

Lishajko, F.: Release, reuptake and net uptake of dopamine, noradrenaline and adrenaline in isolated sheep adrenal medullary granules. Acta physiol. scand. **76**, 159—171 (1969).

Malik, K. U., Ling, G. M.: Modification by acetylcholine of the response of rat mesenteric arteries to sympathetic stimulation. Circulat. Res. **25**, 1—9 (1969).

Mantegazza, P., Naimzada, M. K.: Attivita della prostaglandina E_1 sul preparato nervo ipogastrico-deferente di varie specie animali. Atti Accad. med. lombarda **20**, 58—64 (1965).

Nachmansohn, D.: Role of acetylcholine in neuromuscular transmission. Ann. N. Y. Acad. Sci. **135**, 136—149 (1966).

Nasmyth, P. A.: An investigation of the action of tyramine and its interrelationship with the effects of other sympathomimetic amines. Brit. J. Pharmacol. **18**, 65—75 (1962).

Owman, Ch., Sjöstrand, N. O.: On short adrenergic neurons in the accessory male genital organs of the bull. Experientia (Basel) **22**, 759—761 (1966).

Pfenninger, K., Sandri, C., Akert, K., Eugster, C. H.: Contribution to the problem of structural organization of the presynaptic area. Brain Res. **12**, 10—18 (1969).

Schümann, H. J.: Über den Noradrenalin- und ATP-Gehalt sympathischer Nerven. Naunyn-Schmiedebergs Arch. exp. Path. Pharmak. **233**, 296—300 (1958).

—, Philippu, A.: The mechanism of catecholamine release by tyramine. Int. J. Neuropharmacol. **1**, 179—182 (1962).

Sjöstrand, N. O.: Effect of some smooth muscle stimulants on the motor response of the isolated guinea pig vas deferens to hypogastric nerve stimulation. Nature (Lond.) **192**, 1190—1191 (1961).

Smith, A. D.: Biochemistry of adrenal chromaffin granules. In: The interaction of drugs and subcellular components in animal cells, (Campbell, P. N., Ed.) London: J. A. Churchill Ltd. 1968.

Stjärne, L.: Studies of catecholamine uptake storage and release mechanisms. Acta physiol. scand. **62**, Suppl. 228 (1964).

Sutherland, E. W., Robison, G. A., Butcher, R. W.: Some aspects of the biological role of adenosine 3',5'-monophosphate (cyclic AMP). Circulation **37**, 279—306 (1968).

Taugner, G., Hasselbach, W.: Über den Mechanismus der Catecholamin-Speicherung in den „chromaffinen Granula" des Nebennierenmarks. Naunyn-Schmiedebergs Arch. Pharmak. exp. Path. **255**, 266—286 (1966).

Trendelenburg, U.: Modification of the effect of tyramine by various agents and procedures. J. Pharmacol. exp. Ther. **134**, 8—17 (1961).

Wilcken, D. E. L., Brender, D., Shorey, C. D., MacDonald, G. J.: Reserpine: Effect on structure of heart muscle. Science **157**, 1332—1334 (1967).

Prof. Dr. U. S. von Euler
Department of Physiology
Faculty of Medicine
Karolinska Institute
Stockholm/Sweden

Discussion

SCHÜMANN: I should like to ask Professor VON EULER about his stimulation experiments on the vas deferens. When you stimulated with postganglionic and preganglionic stimulation, was there a difference between the two kinds of stimulation with respect to the inhibition produced by reserpine?

VON EULER: No, with reserpine there was no clear cut difference between the two—so in the first series of experiments we used the preganglionic Hukovic's type but in later experiments we have been using postganglionic field stimulation and the effect comes with both types of stimulation.

SCHÜMANN: An additional question: There exist experiments on the vas deferens preparation of SJÖSTRAND in which he tried to inhibit the effect of the electrical stimulation of the hypogastric nerve by ganglionic blocking agents, for instance, by hexamethonium. The effect of stimulation with a frequency of 12 imp/sec could be completely abolished, whereas that of 50 imp/sec was only slightly diminished. Have you observed such dependence on frequency?

VON EULER: The inhibitory effect of reserpine is, to some extent, dependent on the stimulation frequency also when post-ganglionic stimulation is used, which we usually check with hexamethonium.

KIRSHNER: Have you looked at the vas deferens from animals shortly after treating them with reserpine before the stores were depleted?

VON EULER: I think that such experiments have been made earlier by many and we have not repeated these because we wanted to see whether there was any acute effect when there is no question of any depletion, i.e. overall depletion. We think that there is local depletion of the 'available' store.

EFRON: I would like to ask Professor VON EULER if he has data on some psychotropic drugs, e.g. members of the phenothiazine group, such as chlorpromazine, stelazine, etc?

VON EULER: Would you like to specify any of them?

EFRON: Any representative of them, say chlorpromazine, or some butyrophenones, or tricyclic compounds.

VON EULER: Chlorpromazine has about the same type of effect as rotenone and inhibits the respiratory chain at the same site but since in addition it is an adrenergic blocker, it is not perhaps a very pure type. Promethazine in higher concentrations has a strong depleting effect and releases amines rapidly, but that may be due to structural changes in the membrane.

CARLSSON: I should like to continue along Dr. KIRSHNER's line about in vitro versus in vivo problems. Professor VON EULER, you had this scheme. Reserpine blocks the transfer from the big to the small pool. I think the in vivo evidence after injection of reserpine does not support the idea of any block here because you can go on stimulating and have responses for hours, as long as you have reasonable amounts in the store. That means, that if you inject reserpine into the animal in a reasonable dose you have the block only at the site of uptake by the storage granules; whereas, interestingly enough, in your new experiments where you add reserpine, maybe in ten times higher concentrations in vitro in your vas deferens preparation, it looks as if you get an immediate effect.

VON EULER: The extragranular store does not seem to be inhibited, while the granular store is, in spite of its content of transmitter.

CARLSSON: Well, it is a transmission failure that you will finally end up with when the store is empty.

VON EULER: When it is empty, of course!

THOENEN: The blocking effect of reserpine on the sympathetic transmission in the vas deferens preparation in vitro is rapidly reversible. I should like to ask you whether you have tried to prevent the restoration of transmission by inhibiting the tyrosine hydroxylase, since it could be assumed that the rapid recovery is due to newly synthesized norepinephrine.

VON EULER: No, we have not tried that but we did notice that if reserpine in the concentration of say 0.5 to 1 μg per ml in the bath is there for more than 10 to 15 min or so, there is an irreversible inhibition of the transmission or response. Therefore, one cannot restore the response by washing the neuronal rest. Thus, if one leaves the organ for 15 min and then stimulates, one gets an excellent first contraction but after that, the effect rapidly decreases.

PHILIPPU: Is the presence of magnesium ions necessary for the net uptake of catecholamines by the nerve granules?

VON EULER: Yes, for the ATP dependent uptake, magnesium is absolutely necessary.

Bayer-Symposium II, 161—165 (1970)

Saturation of the Uptake of Noradrenaline as a Determinant of Pharmacological Effects

U. Trendelenburg, P. R. Draskóczy, K.-H. Graefe, and H. Hennemann

With 2 Figures

Intraneuronal uptake is responsible for the removal of l-noradrenaline from the fluid perfusing isolated rabbit hearts (Lindmar and Muscholl, 1964). In this species, the uptake mechanism is characterized by lacking stereospecificity and by being saturable; however, in spite of the lack of stereospecificity, the rate of uptake of d-noradrenaline is lower than that of the l-isomer when the heart is perfused with a concentration of the amine (200 ng/ml) that causes partial saturation of uptake (Draskóczy and Trendelenburg, 1968).

From this observation a working hypothesis was developed to account for the phenomenon of saturation of uptake. It is proposed that the rate of net uptake of noradrenaline across the neuronal cell membrane is a function not only of the extracellular but also of the cytoplasmic concentration of this amine, and that saturation occurs whenever the cytoplasmic concentration of noradrenaline rises substantially. The intraneuronal cytoplasmic concentration of noradrenaline may be assumed to be determined by uptake into the storage granules as well as by deamination by monoamine oxidase (MAO). Since granular uptake is stereospecific (Stjärne and von Euler, 1965), the intraneuronal fate of d-noradrenaline must differ from that of the l-isomer; for equal rates of uptake through the neuronal membrane, d-noradrenaline might reach higher cytoplasmic concentrations than those achieved by the l-isomer. This difference might be accentuated by the fact that MAO exhibits a slight degree of stereospecificity in favor of the l-isomer (Giachetti and Shore, 1966). It is conceivable that the postulated difference in cytoplasmic concentration of the isomers is negligible when the extracellular concentration is low; this assumption has to be made to account for the fact that rates of net uptake were equal for both isomers, when the perfusion concentration was 20 ng/ml, while a significant difference in rates of net uptake was observed with a concentration of 200 ng/ml (Draskóczy and Trendelenburg, 1968).

In order to test the hypothesis, rabbit hearts were perfused with 200 ng/ml of either isomer after the intraneuronal mechanisms of inactivation had been impaired by pretreatment with reserpine (1 mg/kg s.c. 24 h prior to the experiment, to block granular uptake) or pargyline (100 mg/kg s.c. 16 h prior to the experiment, to block MAO) or both. As in the earlier study, arterial and venous concentrations of noradrenaline were measured 10 to 15 min after the beginning of the perfusion, and net uptake was calculated from the arterio-venous difference (Draskóczy and Trendelenburg, 1968).

According to the hypothesis, pretreatment with reserpine should reduce the net uptake of l-noradrenaline from a perfusion with 200 ng/ml to values corresponding to those observed for d-noradrenaline in untreated hearts (Fig. 1). Such a reduction in the net uptake of l-noradrenaline should be observed, although pretreatment with reserpine was found to have no influence on the net uptake of this isomer, when the perfusion concentration is very low (40 ng/ml, LINDMAR and MUSCHOLL, 1964). Moreover, pretreatment with reserpine should not reduce the net uptake of d-noradrenaline, even at the high perfusion concentration of 200 ng/ml, since block of the stereospecific granular uptake mechanism should not

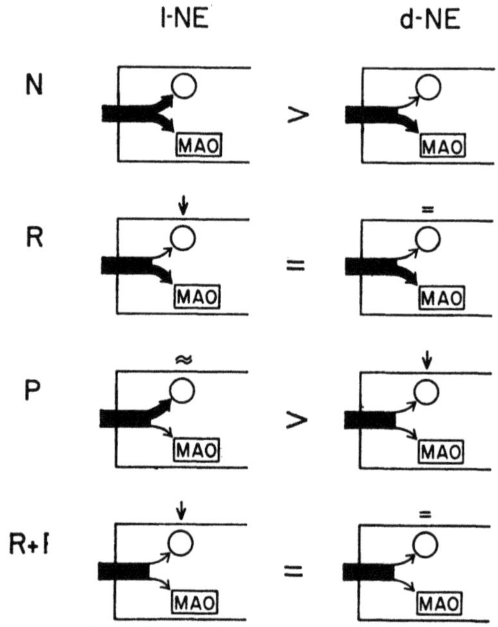

Fig. 1. Schematic representation of postulated changes in the cytoplasmatic concentration of noradrenaline in adrenergic nerve endings of organs perfused with 200 ng/ml of noradrenaline. Shown are models of the neuron with the neuronal membrane, the membranal uptake of noradrenaline (arrows), intraneuronal MAO (MAO) and granular storage sites (circles). Pretreatment: N = normal heart; R = 24 h after pretreatment with reserpine to block granular storage sites; P = 16 h after pretratment with pargyline to block intraneuronal MAO; R + P = combined pretreatment. Discontinuity of the heavy arrow indicates situations in which, according to the model, the cytoplasmatic noradrenaline concentration should increase For details see text. The symbols compare expected rates of net uptake of neighbouring experimental situations (> greater than; = equal to; ≈ similar to; ↓ smaller than in experimental situation depicted to the left or above): Left column refers to perfusions with l-noradrenaline, right column to perfusions with the d-isomer

significantly affect the intraneuronal disposition of this isomer. Pretreatment with pargyline, on the other hand, should reduce rates of net uptake for both isomers, since MAO may be assumed to play a role in the intraneuronal degradation of both isomers. Finally, the combined pretreatment should reduce rates of net uptake of both isomers to very low levels (Fig. 1).

The results of this series of experiments are presented in Table 1. They show very good agreement between experimental observations and the postulates derived from the working hypothesis.

From the working hypothesis a second postulate may be derived and tested. When MAO has been inhibited by pretreatment with pargyline, only one mechanism (i.e., granular uptake) is left to keep the cytoplasmic concentration of l-noradrenaline at low levels. Since granular storage capacity is limited, it is to be expected that, in pargyline-treated hearts, the storage capacity be exhausted in the course of a prolonged perfusion with l-noradrenaline. Hence, the cytoplasmic concentration should increase, saturation should develop, and the net uptake from the perfusion fluid should decline with time.

In order to test this postulate, hearts were perfused with 60 ng/ml of l-noradrenaline for 60 min. The concentration was reduced from 200 to 60 ng/ml in order to delay the expected exhaustion of the storage capacity; second, the lower concentration was chosen in order to achieve a higher percentage value for removal of noradrenaline from the perfusion fluid. Under these conditions, 5 normal hearts

Table 1. *Removal of the isomers of noradrenaline from a perfusion (200 ng/ml) through isolated rabbit hearts*

	l-noradre-naline	d-noradre-naline	P
controls	14.3 ± 1.54	7.9 ± 1.42	<0.01
after pretreatment with reserpine (1 mg/kg, 24 h)	8.4 ± 1.45	9.2 ± 2.88	N. S.
after pretreatment with pargyline (100 mg/kg, 16 h)	6.6 ± 2.50	2.6 ± 1.24	N. S.
after pretreatment with reserpine and pargyline	2.0 ± 0.78	3.1 ± 1.45	N. S.

Shown are means (± S. E.) of 100 (A-V)/A, where A = inflow concentration and V = outflow concentration. Measurements were made 10 to 15 min after beginning of perfusion. Each group consisted of 6 to 14 hearts. P refers to significance of difference between the isomers. N. S. = not significant.

removed about 25% of the perfused l-noradrenaline, and the rate of uptake remained constant throughout the perfusion period. In 5 pargyline-pretreated hearts, on the other hand, uptake declined gradually during the 60-min perfusion and reached very low levels. It is concluded that the observed time course of changes in uptake by pargyline-pretreated rabbit hearts supports the working hypothesis.

Although the evidence is by necessity indirect, it supports the view that the net movement of noradrenaline across the neuronal membrane is slowed when the cytoplasmic noradrenaline concentration rises, and that the latter is determined by the combined effects of granular uptake and MAO. According to this interpretation of the results, the effect of reserpine and pargyline on the net movement of noradrenaline across the neuronal membrane can be explained in terms of their well known intraneuronal effects; it is not necessary to postulate any effects of these agents on the membranal uptake mechanism.

If the described results can be applied to other adrenergically innervated organs, pretreatment with pargyline should increase the sensitivity of effector organs to those sympathomimetic amines which are taken up by adrenergic nerve endings and degradated by MAO. However, this increased sensitivity should be characterized by requiring time, i.e., responses to the amine should increase gradually as the capacity of the granular stores becomes exhausted. Such a slowly developing increase in the response to noradrenaline of isolated guinea-pig atria whose MAO had been inhibited has been observed by FURCHGOTT and SANCHEZ-GARCIA (1968). They coined the term „secondary sensitization", since the quick initial response to noradrenaline was followed by a secondary slow increase in response. The same phenomenon is observed in the isolated nictitating membrane after pretreatment with pargyline (TRENDELENBURG, unpublished observations).

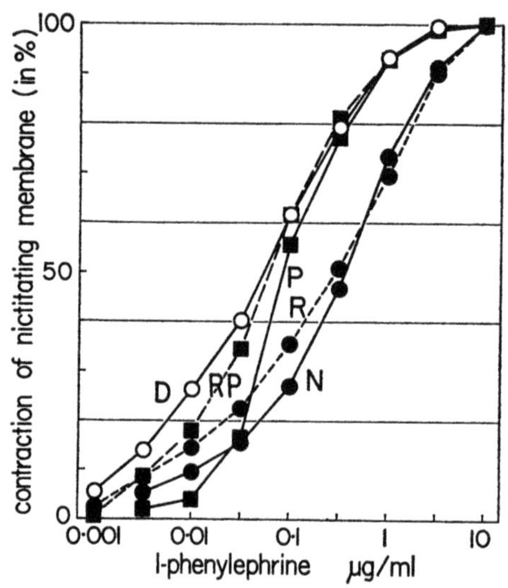

Fig. 2. Response of the isolated nictitating membrane to l-phenylephrine after various types of pretreatment. Ordinates: contraction of inferior or medial muscle in percent of maximum contraction. Abscissae: concentration of l-phenylephrine in the bath (µg/ml, log scale). Shown are means of 5 to 6 experiments per group. N = untreated innervated membranes; R = 24 h after pretreatment with 1 mg/kg of reserpine s.c.; P = after a 30 min incubation of the isolated muscle with 100 µg/ml of pargyline (to block MAO) and subsequent removal of the agent from the bath; R + P = combined pretreatment; D = denervated nictitating membranes of all experimental groups (N, R, P and R + P). The results for denervated membranes were pooled, since the various pretreatments did not cause significant changes in the response of the denervated membrane to l-phenylephrine. Note that the magnitude of changes in sensitivity (i.e., the horizontal shift of dose-response curves to the left) were qualitatively and quantitatively similar to changes in rate of uptake presented in Table 1

There are no reports in the literature indicating that block of MAO (or the combination of block of MAO and pretreatment with reserpine) causes a pronounced increase in sensitivity to directly acting sympathomimetic amines. However, this may well be due to the fact that the increase in response is observed only when care is taken to obtain steady state responses. Since the initial quick response of such a tissue to noradrenaline is not increased (FURCHGOTT and SANCHEZ-GARCIA, 1968), the increase in response may be missed when the secondary slow further increase is not recorded.

For a study of this problem, l-phenylephrine was chosen, since it is not a substrate of catechol O-methyl transferase (to avoid complicating factors), since it is a substrate of MAO, and since it seems to be taken up by adrenergic nerves

(LANGER and TRENDELENBURG, 1969). Dose-response curves were determined on normal and denervated isolated nictitating membranes, and care was taken to ensure that steady state response were obtained (Fig. 2). Pretreatment with reserpine caused a small increase in the sensitivity of the innervated muscle to phenylephrine; pretreatment with pargyline was more effective; the combined pretreatment was most effective and increased sensitivity to phenylephrine to that observed for denervated muscles. On the other hand, the various pretreatments did not cause any significant changes in the sensitivity of the denervated muscle. However, the response of the denervated muscle differed from that of innervated ones (after the combined pretreatment) in that it was quick. A second difference lies in the fact that the combined pretreatment failed to increase the response of innervated muscles to low concentrations of phenylephrine to that of denervated muscles. On the basis of the working hypothesis this is not surprising, since very low concentrations of phenylephrine cannot be expected to produce high cytoplasmic levels within a reasonable period. Hence, the combined pretreatment should be more effective against higher concentrations of the amine than against lower concentrations. This was observed.

Conclusions. When the intraneuronal mechanisms of inactivation (i.e., granular uptake and enzymic degradation) fail to keep the intraneuronal cytoplasmic concentration of sympathomimetic amines low, the net movement of the amine across the neuronal membrane is decreased. As a result, the concentration of the amine at the receptors increases, and an increase in sensitivity ensues. This proposal provides an explanation for changes in sensitivity after pretreatment with reserpine and/or pargyline; it is not necessary to postulate a direct effect of these two agents on membranal uptake.

References

DRASKÓCZY, P. R., TRENDELENBURG, U.: The uptake of l- and d-norepinephrine by the isolated perfused rabbit heart in relation to the stereospecificity of the sensitizing action of cocaine. J. Pharmacol. exp. Ther. **159**, 66—73 (1968)

FURCHGOTT, R. F., SANCHEZ-GARCIA, P.: Effects of inhibition of monoamine oxidase on the actions and interactions of norepinephrine, tyramine and other drugs on guinea-pig left atrium. J. Pharmacol. exp. Ther. **163**, 98—122 (1968).

GIACHETTI, A., SHORE, P. A.: Optical specificity of monoamine oxidase. Life Sci. **5**, 1373—1378 (1966).

LANGER, S. Z., TRENDELENBURG, U.: The effect of a saturable uptake mechanism on the slopes of dose-response curves for sympathomimetic amines and on the shifts of dose-response curves produced by a competitive antagonist. J. Pharmacol. exp. Ther. **167**, 117—142 (1969).

LINDMAR, R., MUSCHOLL, E.: Die Wirkung von Pharmaka auf die Elimination von Noradrenalin aus der Perfusionsflüssigkeit und die Noradrenalinaufnahme in das isolierte Herz. Naunyn-Schmiedebergs Arch. exp. Path. Pharmak. **247**, 469—492 (1964).

STJÄRNE, L., EULER, U. S. v.: Stereospecificity of amine uptake mechanism in nerve granules. J. Pharmacol. exp. Ther. **150**, 335—340, 1965.

Prof. Dr. U. TRENDELENBURG
Institut für Pharmakologie und
Toxikologie der Universität
D-8700 Würzburg
Koellikerstraße 2

Discussion

GEFFEN: I would like to ask whether you have taken into account in your explanations the possibility of uptake of noradrenaline into the cardiac muscle cells. LIGHTMAN and IVERSEN have recently redefined uptake which was previously thought to occur only at high concentrations. There is considerable uptake into cardiac muscle cells, even at low concentrations, but because this is normally metabolized one does not see it appearing as noradrenaline in the cardiac muscle. Now most of your explanations, as I understand them, referred to neuronal uptake. How would muscle uptake complicate this situation?

TRENDELENBURG: No doubt, the participation of extraneuronal uptake would complicate things. I should like to keep the model as simple as possible until solid evidence forces me to add complications to the model. This should not be interpreted as meaning that nature invariably keeps things simple. We may well have to adjust the model to forthcoming evidence. With regard to extraneuronal uptake, the main difficulty lies in the fact that we are not yet able to affect it pharmacologically in a strictly predictable way. On the other hand, I don't doubt the existence of extraneuronal uptake. We demonstrated its presence in the isolated nictitating membrane of the cat.

In that preparation it is not reduced by pretreatment with reserpine. Hence, I would regard it as most unlikely that extraneuronal uptake influenced the results obtained after pretreatment with reserpine (see preceding report). After pretreatment with paragyline, the situation is different in so far as we found a small increase in extraneuronal uptake in the isolated nictitating membrane after this pretreatment. Hence there may be an influence of pargyline on extraneuronal uptake. However, this is unlikely to affect net uptake since such an effect should change the sensitivity of the denervated nictitating membrane, and this was not observed. Thus, the effects described in our communication seem to be due to neuronal uptake. If further experiments lead to a different conclusion, the model will have to be adjusted.

VON EULER: We have an old observation which I wonder whether Dr. TRENDELENBURG will comment upon—in perfusion experiments on the human placenta which were made long ago in London, we found that adrenaline increased the pressure of the perfusing system using a constant output pump. We then added cocaine to the preparation and found that there was a greatly increased response to that.

TRENDELENBURG: I have no experience with the perfused placenta. However, it may be relevant that there is an increasing body of evidence indicating that cocaine can have postsynaptic effects, in addition to its well known presynaptic action (block of neuronal uptake). For instance, KALSNER and NICKERSON have reported that on isolated rabbit aortic strips, cocaine potentiates the effects of methoxamine (which is not taken up by adrenergic nerve endings, and which is a substrate of neither COMT nor MAO). However, there must be species and/or organ differences, since we did not find any potentiation by cocaine of the effects of methoxamine on the cat's nictitating membrane. On the whole, I would guess that any tissue which has no adrenergic innervation, might be a promising preparation for the observation of the weak postsynaptic effects of cocaine. No doubt, it is no longer tenable to maintain that *all* effects of cocaine must be presynaptic (i.e. related to its ability to block neuronal uptake).

Bayer-Symposium II, 168—186 (1970)

Cholinomimetic Drugs and Release of the Adrenergic Transmitter

E. MUSCHOLL

With 8 Figures

Introduction

In this paper an attempt will be made to describe the actions of cholinomimetic drugs on the peripheral adrenergic nerve fibres as far as its capacity to retain, or to release, the transmitter is concerned. The conventional classification of cholinomimetic responses in nicotinic and muscarinic ones will be adhered to. As to be shown by the results, the classification can also be extended to describe the responses of the peripheral adrenergic nerve fibre to acetylcholine, nicotine, methacholine, and to several of their congeners. Hexamethonium was used to antagonize the activation of nicotine receptors (in the following the more frequently applied term "nicotinic" will be used); atropine, under more careful control of dose-response relationships, was employed to antagonize activation of muscarine (in the following "muscarinic") receptors.

Whereas the effect of drugs acting on nicotinic receptors mediating a release of the adrenergic transmitter has been known for some time, the effect of drugs acting on muscarinic receptors of the peripheral adrenergic fibre was discovered only recently. The occurrence of muscarinic receptors on the terminal part of the adrenergic neuron is less surprising than their long-dated neglect since there is a great body of evidence for muscarinic receptor sites in sympathetic ganglion cells (for a comprehensive review see TRENDELENBURG, 1967). However, muscarinic actions on peripheral adrenergic nerves may have been overlooked because of the proximity of receptor sites situated in nerves and in end organs. Since in the majority of experiments the response of the end organ is taken as a measure of the quantity of the transmitter released, the effects of drugs having similar affinities for receptors in both nerve and indicator organ cannot be analyzed properly. These difficulties are circumvented if a suitable organ is perfused and the transmitter output rather than the ensuing response of the organ to the released transmitter is studied. In our laboratory the rabbit isolated heart has been extensively used for investigations of uptake, storage and release of noradrenaline, and the details of the methods employed have been reported elsewhere (HUKOVIC and MUSCHOLL, 1962; LINDMAR and MUSCHOLL, 1964; MUSCHOLL, 1965; LÖFFELHOLZ and MUSCHOLL, 1969).

As implicated by the restriction to phenomena occurring at peripheral adrenergic nerves, the mechanisms of catecholamine release from adrenal medulla or other chromaffin tissue will not be dealt with.

Nicotinic Mechanisms of Noradrenaline Release

Nicotinic Excitation. Excitatory responses of the heart and several smooth muscular organs to nicotine have been known for some time. DIXON (1924) who reviewed the pharmacological actions of nicotine, ascribed its stimulatory effects on the isolated heart to stimulation of intracardiac sympathetic ganglia. The work of HOFFMANN et al. (1945) has shown that the positive inotropic and chronotropic effects of nicotine or acetylcholine on atropinized heart preparations of different animal species were produced by the release of a substance resembling adrenaline in its actions. These findings were soon confirmed (McNAMARA et al., 1948) and from indirect evidence it was concluded that intracardiac ganglion cells were the point of attack of the nicotinic agents (KOTTEGODA, 1953).

Since stimulatory effects were also observed in preparations devoid of intrinsic sympathetic ganglia it was suggested that nicotinic drugs liberate catecholamines from peripheral adrenergic fibres (MIDDLETON et al., 1956; LEE and SHIDEMAN, 1959; FURCHGOTT, 1960). The now overwhelming evidence for this proposal will not be mentioned in the present paper because it was extensively discussed in earlier publications (LEE and SHIDEMAN, 1959; COOPER, 1966; LINDMAR et al., 1968).

Fluorimetric methods revealed that nordrenaline was the principal amine liberated from perfused rabbit or guinea-pig hearts in response to acetylcholine (RICHARDSON and WOODS, 1959; ANGELAKOS and BLOOMQUIST, 1966; LÖFFELHOLZ, 1967), nicotine (LÖFFELHOLZ, 1970) or 1,1-dimethyl-4-phenyl piperazine (DMPP) (LINDMAR and MUSCHOLL, 1961; LINDMAR et al., 1968). There is also ample evidence for the mediation by noradrenaline of sympathomimetic effects produced with nicotinic agents. Depending on the type of response to noradrenaline on the particular organs the responses after nicotinic agents may be stimulatory or inhibitory; they are blocked by procedures like treatment with reserpine or postganglionic sympathetic denervation which cause a loss of the adrenergic transmitter (for a review see BURN and RAND, 1962). Since hexamethonium as well as various agents with local anaesthetic activity blocked the noradrenaline release from the rabbit heart produced by DMPP (LINDMAR and MUSCHOLL, 1961) it was concluded that DMPP released the transmitter by a depolarization of the adrenergic nerve fibre as suggested by FURCHGOTT (1960). Atropine did not affect the noradrenaline release following DMPP unless concentrations above 10^{-6} g/ml were administered (LINDMAR et al., 1968). In contrast, atropine (10^{-9} to 10^{-6} g/ml) greatly facilitated the noradrenaline release following acetylcholine. This observation will be discussed further below.

Autoinhibition of Nicotinic Excitation. When a constant infusion of a nicotinic drug was applied, the noradrenaline output of the heart reached a maximum within the first minutes of infusion. Subsequently, the output decreased very rapidly and reached the preinfusion level 3 to 6 min after start of the infusion (LINDMAR et al., 1967; LÖFFELHOLZ, 1967; see also Fig. 6, lower panel, effect of acetylcholine in the presence of atropine).

In order to study the time course of the decay in noradrenaline output more closely, LÖFFELHOLZ (1970) infused nicotinic agents and collected the venous effluent of the hearts in periods of 15 or 30 sec (Fig. 1). Since the log noradrenaline output plotted against time gave straight lines the decay was exponential. The slopes of the three regression lines in Fig. 1 were similar. Thus, the half-time of decay of the noradrenaline output was not related to the total amount of transmitter released into the perfusate. The total outputs of noradrenaline were: 1836 ng after acetylcholine in the presence of atropine, 888 ng after DMPP and 378 ng after acetylcholine. The noradrenaline output during the first collection period was omitted from the calculation of the regression lines. It was always lower than that of the second period. This was partly due to the dead space of the perfusion apparatus and partly to the intravascular volume, both of which together caused a delay of approximately 7 sec until the drug reached the capilla-

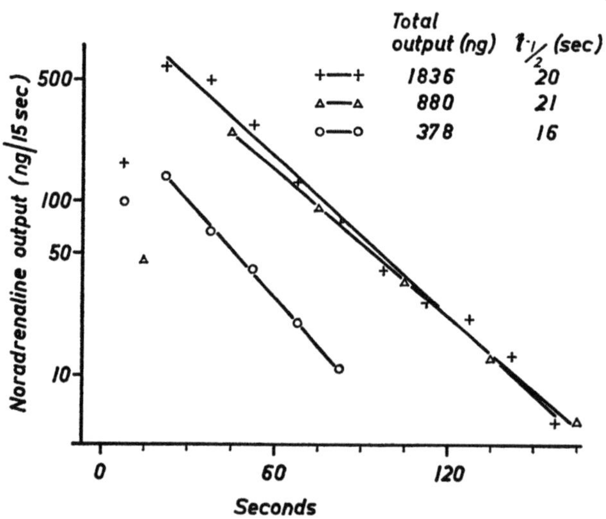

Fig. 1. Time course of the output of noradrenaline during continuous infusion of nicotinic drugs into the rabbit heart. Ordinate, log. noradrenaline output (ng/15 sec). Abscissa, time (sec) after start of drug infusion. o Acetylcholine 3.8×10^{-5} g/ml; + acetylcholine 3.8×10^{-5} g/ml, 15 min after and during atropine 10^{-6} g/ml; Δ DMPP 10^{-5} g/ml. Each point represents a noradrenaline estimation. $t_{1/2}$, half-time of exponential decay

ries. Diffusion of the drug onto the receptors and the process of stimulation—secretion coupling are some other events which may have contributed to the result that the noradrenaline output measured in the first collection period was not representative for the whole time of sampling (15 or 30 sec) but for a much shorter interval.

The question now arises whether the results shown in Fig. 1 represent a simple washout curve for noradrenaline suddenly released from the nerve fibres or whether they reflect a process of release maintained for 1 to 2 min though at a steadily declining rate. The washout of noradrenaline was tested in the following way. Concentrations of 100, 200 and 500 ng/ml were perfused for 4 min. The perfusion with noradrenaline was then terminated and the effluent of the heart was collected in periods of 15 sec for about 2 min. As shown by previous experiments (LINDMAR and MUSCHOLL, 1964) the concentration of noradrenaline in the effluent reaches a constant level 2 min after beginning of a noradrenaline infusion. The plot of the washout of noradrenaline (log. ng against time) yielded values indicating a biphasic process (Fig. 2). The fast component ($t_{1/2}$ = 5 to 8 sec) probably represents washout from the aortic cannula and the intravascular space, the slower component ($t_{1/2}$ = 14 to 21 sec) washout from a compartment containing that noradrenaline which had not been removed by uptake into adrenergic nerves. The half-times of the latter component agree with those calculated from the experiments of Fig. 1. This would favour the first alternative presented above, namely a sudden termination of the release of the transmitter by nicotinic agents.

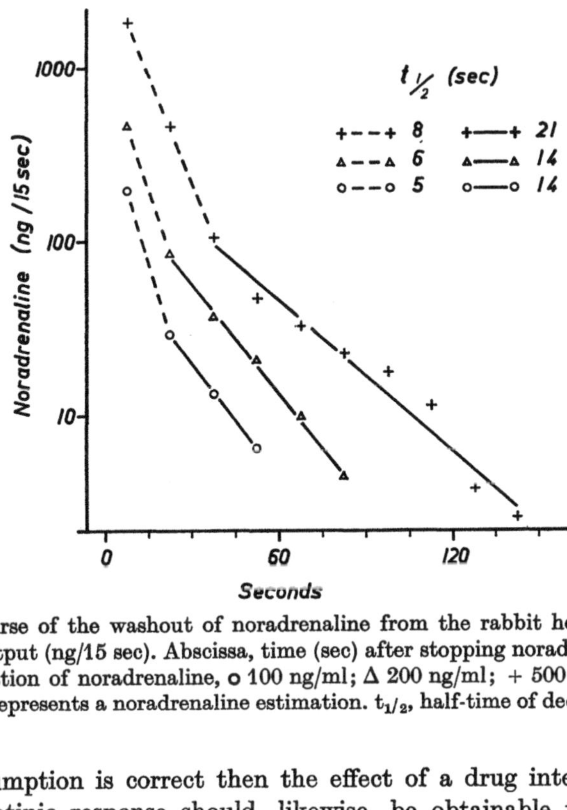

Fig. 2. Time course of the washout of noradrenaline from the rabbit heart. Ordinate, log. noradrenaline output (ng/15 sec). Abscissa, time (sec) after stopping noradrenaline infused for 4 min. Concentration of noradrenaline, o 100 ng/ml; Δ 200 ng/ml; + 500 ng/ml. Each point represents a noradrenaline estimation. $t_{1/2}$, half-time of decay

If this assumption is correct then the effect of a drug interfering with the excitatory nicotinic response should, likewise, be obtainable for only a short period of time. The rabbit heart was perfused with a standard concentration of acetylcholine in the presence of atropine. Hexamethonium was added at different time intervals (Löffelholz, 1970). Fig. 3 shows that simultaneous infusion of hexamethonium abolished the noradrenaline releasing effect of acetylcholine (plus atropine). However, if hexamethonium was added only 10 or 20 sec after the beginning of the acetylcholine perfusion, the noradrenaline output did not differ from that of the controls. Hexamethonium added 5 sec after the beginning of the acetylcholine perfusion had partly lost its blocking action because the noradrenaline output was about 10% of the control output. Thus, the period of time during which the process of nicotinic release could be modified was restricted to 5 to 10 sec.

The output of noradrenaline in response to nicotinic drugs was always short-lived irrespective of the amount of transmitter released. Therefore, the decay in

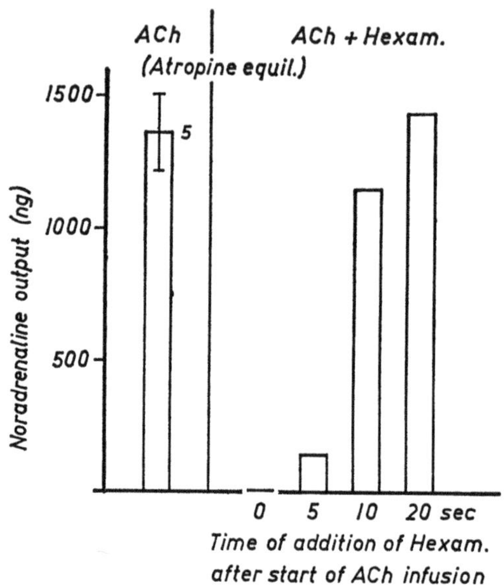

Fig. 3. Block of acetylcholine-induced noradrenaline output by hexamethonium. Height of columns represents total noradrenaline output (ng) from rabbit hearts after infusion of acetylcholine (3.8×10^{-5} g/ml), atropine (10^{-6} g/ml) present throughout experiment. Left panel, output of 5 hearts (mean $\pm$ S.E.). Right panel, single experiments in which hexamethonium (3×10^{-6} g/ml) was added 0, 5, 10 or 20 sec after beginning of perfusion with acetylcholine

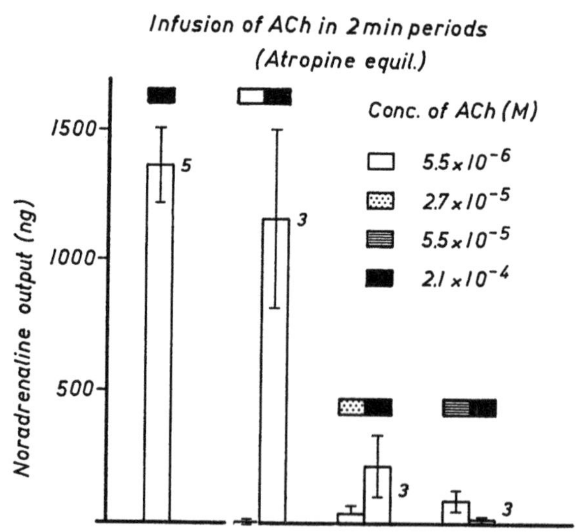

Fig. 4. Inhibitory effect of preinfusion with a low concentration of acetylcholine on noradrenaline output evoked by a high concentration of acetylcholine. Height of columns indicates total noradrenaline output (ng). Vertical bars indicate S.E. and figures the number of experiments. Infusion of acetylcholine for 2 min with 10^{-6} g/ml (5.5×10^{-6} M), 5×10^{-6} g/ml (2.7×10^{-5} M) or 10^{-5} g/ml (5.5×10^{-5} M) immediately followed by a 2 min infusion of the standard concentration of 3.8×10^{-5} g/ml (2.1×10^{-4} M). Atropine (10^{-6} g/ml) present throughout the experiments

output was not caused by a momentary exhaustion of the transmitter available for release. These results with continuous perfusions of nicotinic drugs can be explained by a short-lived excitation and subsequent inhibition of nicotinic receptors. This is reminiscent of the phenomenon of autoinhibition which was observed when the effects of nicotinic agents on ganglionic synapses were studied (VAN ROSSUM, 1962). Since high doses of nicotine cut short the stimulatory action, a

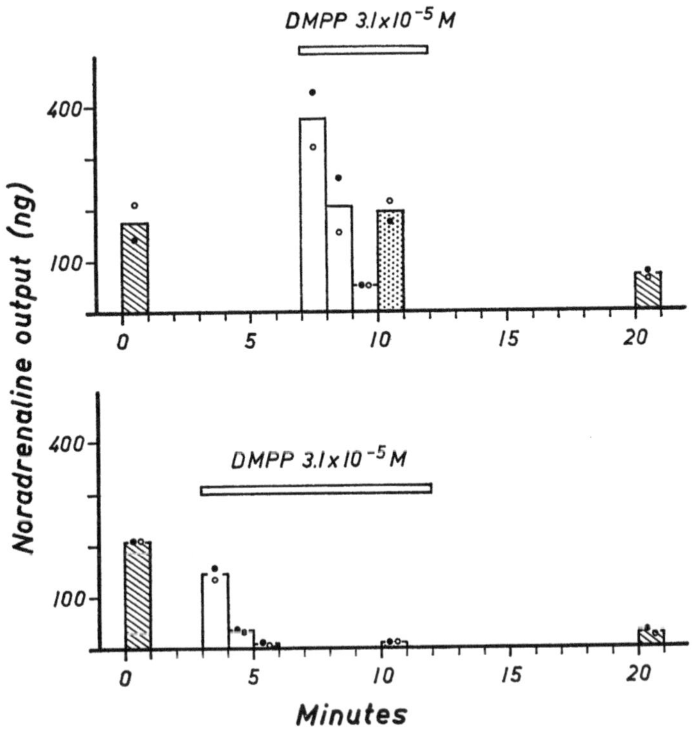

Fig. 5. Effect of DMPP on the noradrenaline output of the rabbit heart evoked by sympathetic nerve stimulation. Ordinate, total output of noradrenaline (ng). Abscissa, time course of experiment (min). Columns indicate mean outputs found in the individual experiments marked by the symbols. Hatched columns, noradrenaline output evoked by 1 min stimulation of postganglionic sympathetic nerves (right and left nerves stimulated separately 30 sec each, pulses of 1 msec at 10/sec) in the absence of DMPP (0 to 1; 20 to 21 min). Stippled columns, noradrenaline output during nerve stimulation (10 to 11 min) in the presence of DMPP (10^{-5} g/ml = 3.1×10^{-5} M). Open columns, noradrenaline output evoked by DMPP. The horizontal bars indicate infusion of DMPP

bell-shaped dose-response curve will result (VAN ROSSUM, 1962; TRENDELENBURG, 1967). In accordance with this reasoning the dose-response curve of the noradrenaline releasing action of nicotine showed an ascending and a descending limb.

In another series of experiments it was investigated how the two properties of acetylcholine, to cause a noradrenaline release and to produce an autoinhibition, are related. In the presence of atropine a standard concentration of acetylcholine causing a maximum noradrenaline output was administered for 2 min. If this was preceded by an infusion of a low concentration (10^{-6} g/ml) of acetylcholine which did not evoke a detectable noradrenaline output, the standard dose was still effective (Fig. 4). Preinfusion with 5×10^{-6} g/ml acetylcholine caused a small but noticeable noradrenaline output, whereas the effect of the standard dose was greatly inhibited. Furthermore, preinfusion with 10^{-5} g/ml acetylcholine produced an output little higher than after 5×10^{-6} g/ml and depressed the effect of the

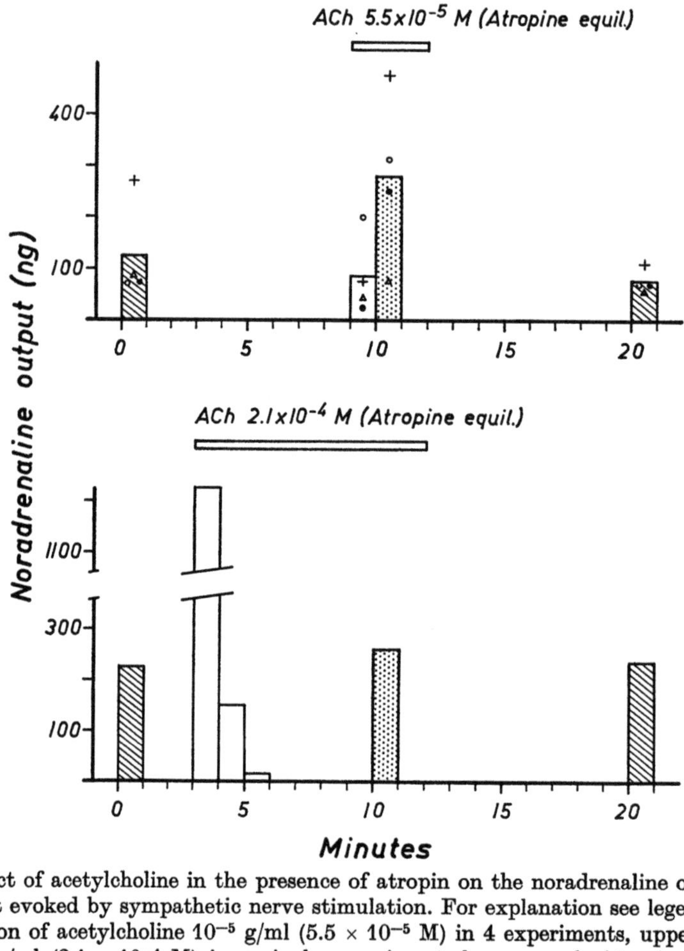

Fig. 6. Effect of acetylcholine in the presence of atropin on the noradrenaline output of the rabbit heart evoked by sympathetic nerve stimulation. For explanation see legend of Fig. 5. Concentration of acetylcholine 10^{-5} g/ml (5.5×10^{-5} M) in 4 experiments, upper panel, and 3.8×10^{-5} g/ml (2.1×10^{-4} M) in a single experiment, lower panel. Atropine (10^{-6} g/ml) present throughout the whole experiment

standard dose completely. Apparently, the nicotinic blocking action can be observed only if there was a preceding nicotinic excitation.

As one would expect from the foregoing, the occupation of the nicotinic receptors produces a block only towards nicotinic agents. In fact, when nicotinic autoinhibition had been achieved a few minutes after administration of DMPP (Fig. 5, upper panel) or acetylcholine in the presence of atropine (Fig. 6) the noradrenaline release evoked by sympathetic nerve stimulation was not decreased. The increase in noradrenaline output observed in some experiments (Fig. 6, upper panel) has not been studied sufficiently for a satisfactory explanation.

Muscarinic Inhibition of Noradrenaline Release Evoked by Nicotinic Drugs

It has generally been assumed that addition of atropine to incubation or perfusion solutions of adrenergically innervated organs merely blocks muscarinic receptors of the end organs and thus unmasks the proper nicotinic effects of acetylcholine. However, atropine was recently found to facilitate the acetylcholine-induced release of noradrenaline from the rabbit heart and muscarinic drugs were shown to decrease the noradrenaline release caused by DMPP (LÖFFELHOLZ et al., 1967). These observations were followed up in a more detailed study (LINDMAR et al., 1968) from which the existence of a muscarinic mechanism inhibiting the noradrenaline release caused by nicotinic drugs was inferred. In short, the evidence for this is based on the following findings. Acetylcholine (3.8×10^{-5} g/ml) became increasingly more effective in raising the noradrenaline output as its muscarinic action was antagonized by atropine (10^{9-} to 10^{-6} g/ml). However, these concentrations of atropine did not alter the noradrenaline output following DMPP (10^{-5} g/ml) which on the rabbit heart releases the transmitter by a purely nicotinic action (cf. paragraph on non-nicotinic actions of DMPP). The noradrenaline output evoked by DMPP was gradually diminished and finally abolished when increasing concentrations of muscarinic drugs (methacholine, pilocarpine, acetylcholine) were added.

The possibility was considered that atropine, by inhibiting the removal of acetylcholine from the perfusion fluid during its passage through the heart, increased the effective concentration of acetylcholine and thus facilitated noradrenaline release. However, atropine altered neither the removal of acetylcholine nor that of noradrenaline.

Maximum inhibition of nicotinic noradrenaline release occurred with concentrations of acetylcholine, methacholine and pilocarpine that did not themselves cause a noradrenaline release. Hence, on the isolated rabbit heart a substantial noradrenaline release can be brought about by acetylcholine only if its muscarinic action is blocked.

The muscarinic inhibition of noradrenaline release during infusion of acetylcholine was recently confirmed by HAEUSLER et al. (1968), who used the isolated cat heart. In addition, electrophysiological evidence for muscarinic receptors on terminal adrenergic nerve fibres was obtained by the finding that the amplitude of antidromically conducted discharges of cardiac sympathetic fibres produced by acetylcholine was decreased by pilocarpine which, on its own, did not cause antidromic firing. From some indirect evidence HAEUSLER et al. (1968) suggested that stimulation of muscarinic receptors causes a hyperpolarization of the axonal membrane that counteracts the following sequence of events: depolarization by nicotinic drugs, increased calcium influx, and release of transmitter.

Muscarinic Inhibition of Noradrenaline Release Evoked by Nerve Stimulation

Excitation of muscarinic receptors does not only inhibit noradrenaline release by nicotinic agents but also noradrenaline release evoked by electrical stimulation of sympathetic nerves. In Fig. 7 (taken from LÖFFELHOLZ and MUSCHOLL, 1969) the effect of various concentrations of acetylcholine on the noradrenaline output of the perfused rabbit heart produced by a standard stimulation procedure is shown. The sympathetic fibres leaving the right and the left stellate ganglion were each stimulated at 10/sec for 30 sec. The venous effluent of the heart was collected during the 1 min period of stimulation and the following minute. In 17 hearts electrical nerve stimulation caused increases in heart rate (mean: 75 beats/min, range 24 to 132) and amplitude of contraction (mean: 38%, range 11 to 85%) above the control value. The stimulatory effects did not significantly differ between

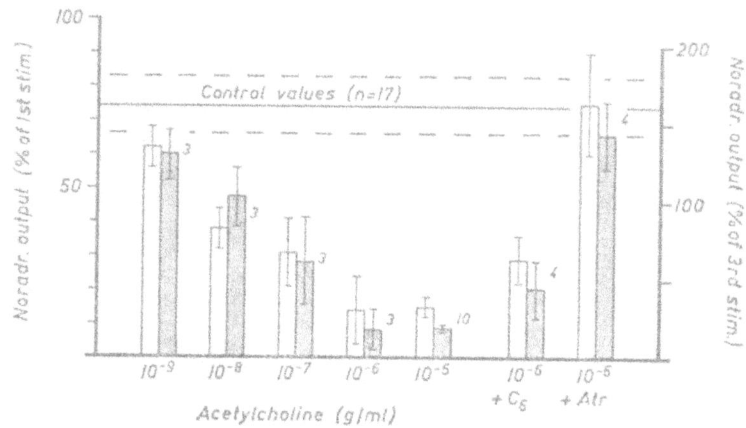

Fig. 7. Effect of acetylcholine on the noradrenaline output of the rabbit heart evoked by sympathetic nerve stimulation. The columns indicate noradrenaline output evoked during the second period of stimulation (in the presence of acetylcholine) in percent of either one or the other control value obtained in the first period (ordinate at left margin, open columns) and in the third period (ordinate at right margin, shaded columns). The scale of the ordinate at the right margin was adapted in such a way that the two control values lay on one horizontal line at the top of the graph. The lower broken line indicates the S.E. of the control value corresponding to the left ordinate, the upper broken line shows the S.E. of the control value corresponding to the right ordinate. Vertical bars at the tops of the columns indicate S.E. of the mean and the figures besides the columns the number of experiments. C_6, hexamethonium 3×10^{-6} g/ml; Atr, atropine 10^{-6} g/ml present throughout the experiment. Fig. taken from LÖFFELHOLZ and MUSCHOLL, 1969)

each of three periods of stimulation carried out at 10 min intervals. However, the noradrenaline output decreased slightly from the first to the third stimulation. Acetylcholine was infused 1 min before and during the second period of stimulation. Then the noradrenaline output measured at the second stimulation during acetylcholine was compared with the output to be expected from the first stimulation and the known rate of decay. Since the action of acetylcholine was easily reversed by washing, the noradrenaline output of the third period of stimulation could also be used as a reference value for estimating the deviation of the output during the second period. The results with acetylcholine expressed in both ways agreed well (Fig. 7). The lowest concentration of acetylcholine tested (10^{-9} g/ml) did not significantly alter the noradrenaline output during nerve stimulation, but concentrations of 10^{-8} to 10^{-6} g/ml decreased it in a dose-dependent manner. The maximum inhibition was obtained with 10^{-6} g/ml, 10^{-5} g/ml being no more effective. Low concentrations of acetylcholine (10^{-9} to 10^{-8} g/ml) did not alter spontaneous rate and amplitude of contraction, higher ones (10^{-7} g/ml) decreased both parameters by 36 ± 14beats/min and by 11 ± 5%, respectively, or caused arrest of the heart (10^{-6} to 10^{-5} g/ml). Possibly because of the direct effects of acetylcholine on the pacemaker the alterations of rate and amplitude of the Langendorff heart induced by sympathetic nerve stimulation could not simply be related to the decrease in the output of noradrenaline shown in Fig. 7. The coronary flow tended to decrease during infusion of acetylcholine and periods of sympathetic stimulation, but the difference from the flow rates observed during the first period of stimulation (without acetylcholine) never reached the 5% level of significance.

The inhibitory effect of acetylcholine (10^{-6} g/ml) on the noradrenaline output elicited by electrical nerve stimulation can be assigned to the muscarinic rather than to the nicotinic action of acetylcholine since atropine (10^{-6} g/ml) but not hexamethonium (3×10^{-6} g/ml) completely reversed the inhibition (Fig. 7). The concentration of hexamethonium employed blocked the noradrenaline release in response to DMPP (2.7×10^{-5} g/ml) but did not alter the noradrenaline output evoked by electrical nerve stimulation (HUKOVIC and MUSCHOLL, 1962; LÖFFELHOLZ and MUSCHOLL, 1969).

If the inhibition of noradrenaline output caused by acetylcholine is not related to its nicotinic action, a nicotinic drug such as DMPP should not decrease noradrenaline output under conditions which permitted acetylcholine to interfere with the release of the adrenergic transmitter. A dose of 10^{-5} g/ml was used which is known to cause a nearly maximum release of noradrenaline that occurs mainly during the first 2 min of infusion of the drug into the rabbit heart (LINDMAR et al., 1967, 1968). For this reason the infusion of DMPP was started 3 min before the second period of stimulation and continued for another 2 min during the stimulation of the nerves and the collection of the venous effluent. Two experiments are shown in Fig. 5, upper panel. The mean output in response to DMPP was 580 ng noradrenaline during the first 2 min of infusion and only 51 ng in the minute preceding nerve stimulation. The noradrenaline output evoked by nerve stimulation was 113% of that observed in the first stimulation period. Hence, the effect of nerve stimulation on noradrenaline release was not decreased when the nicotinic inhibitory action had attained its maximum. However, electrical nerve stimulation was ineffective to release noradrenaline during the 8th min of an infusion of DMPP (10^{-5} g/ml) and had not regained its noradrenaline releasing ability 9 to 10 min after stop of the infusion (Fig. 5, lower panel). It is noteworthy that a decrease in noradrenaline output occurred when the time interval between the onset of infusion of DMPP and the nerve stimulation was greater than 4 min. This "late" effect of DMPP is obviously related to its peculiar "bretylium-like" action which will be discussed later on in the paragraph dealing with non-nicotinic actions of DMPP. In order to evade an interference of neurone blockade with the assessment of nicotinic effects on noradrenaline release evoked by electrical nerve stimulation, the set-up of the DMPP experiments was repeated with acetylcholine in the presence of atropine as a nicotinic drug. Fig. 6 (upper panel) shows that acetylcholine (10^{-5} g/ml) immediately after the nicotinic noradrenaline release greatly facilitated rather than inhibited the transmitter output evoked by nerve stimulation. A dose of acetylcholine (3.8×10^{-5} g/ml) which caused a maximum output of noradrenaline decreased the transmitter output neither in the 8th min of infusion nor 8 to 9 min after stop of the infusion (Fig. 6, lower panel).

Significance of Muscarinic Inhibition of Noradrenaline Release

The interactions of acetylcholine with the peripheral adrenergic nerve fibre are generally intrepreted in the sense that acetylcholine, injected or liberated locally from cholinergic fibres, releases noradrenaline. A great number of investigations pertaining to this subject have been carried out on heart preparations; some of these papers were quoted in the paragraph on nicotinic mechanisms of noradrenaline release. However, on the perfused heart of the rabbit and guinea pig (LINDMAR et al., 1968) as well as the cat (HAEUSLER et al., 1968) acetylcholine does not cause a noticeable noradrenaline release unless its muscarinic action is blocked. Moreover, acetylcholine interferes with noradrenaline release evoked by adrenergic nerve stimulation even if its nicotinic action is excluded (Fig. 7).

A hypothesis dealing with cholinergic-adrenergic interactions was put forward by LEADERS and LONG (1962) in order to explain the action of nicotine in non-atropinized cat atria. There was a cardioinhibition followed by a positiv chronotropic response. Mainly from experiments utilizing sympathectomy, pretreatment with reserpine, or addition of hemicholinium, the authors suggested that the parasympathetic nervous system is involved in both the negative and the positive phases of the response to nicotine and that, after parasympathetic activity, catecholamine liberation occurs from sympathetic nervous stores. Subsequently, LEADERS (1963) suggested that acetylcholine released by activity of parasympathetic fibres may directly liberate noradrenaline from sympathetic nerve stores. Observations of two-phasic actions of nicotine on rat atria appeared to be consistent with this idea (CHIANG and LEADERS, 1965). However, experiments on guinea-pig atria by KHAN et al. (1965) made it unlikely that acetylcholine was involved in the liberation of noradrenaline by nicotine. A fall in the bath temperature abolished the negative chrono- and inotropic effects of both vagal stimulation and nicotine but left the cardiostimulatory action of the latter unchanged. More recently, CHIANG and LEADERS (1968) confirmed that the cardiostimulation in response to nicotine was not necessarily due to a mechanism involving the parasympathetic nervous system since they, likewise, achieved a dissociation of the negative and positive actions of nicotine on rat atria by administration of certain drugs. In the light of present knowledge, an excitatory effect of acetylcholine on adrenergic nerve fibres as proposed in the early paper by LEADERS (1963) is unlikely to occur. Since the muscarinic inhibitory response to acetylcholine is elicited by much lower concentrations of acetylcholine than the excitatory response (Fig. 8) there will be an inhibition of the acetylcholine-induced noradrenaline release unless a drug blocking muscarinic receptors is administered.

A facilitation of noradrenaline release by atropine (cf. Fig. 8) may be brought about when acetylcholine derived from stimulation of cholinergic nerves diffuses, under favourable circumstances, onto adrenergic nerve terminals. In a previous paper (LINDMAR et al., 1968) the following example was mentioned. A summary of the evidence for the "cholinergic link" hypothesis (BURN, 1967) includes reports by various authors who used 5 different tissues on which the responses to postganglionic sympathetic nerve stimulation at low frequencies were potentiated by cholinesterase inhibitors. Interestingly enough, in all cases hyoscine or atropine was administered with the aim to exclude the actions of acetylcholine on the end organs. If the nerves had contained only a small proportion of cholinergic fibres, the acetylcholine released and protected from being hydrolyzed by cholinesterase might have reached adrenergic terminals. The anticholinergic drugs may have facilitated, rather than only unmasked, the effect of nerve stimulation under the conditions mentioned. However, it has been pointed out by BURN (1968) that administration of atropine or hyoscine is no prerequisite for obtaining positive

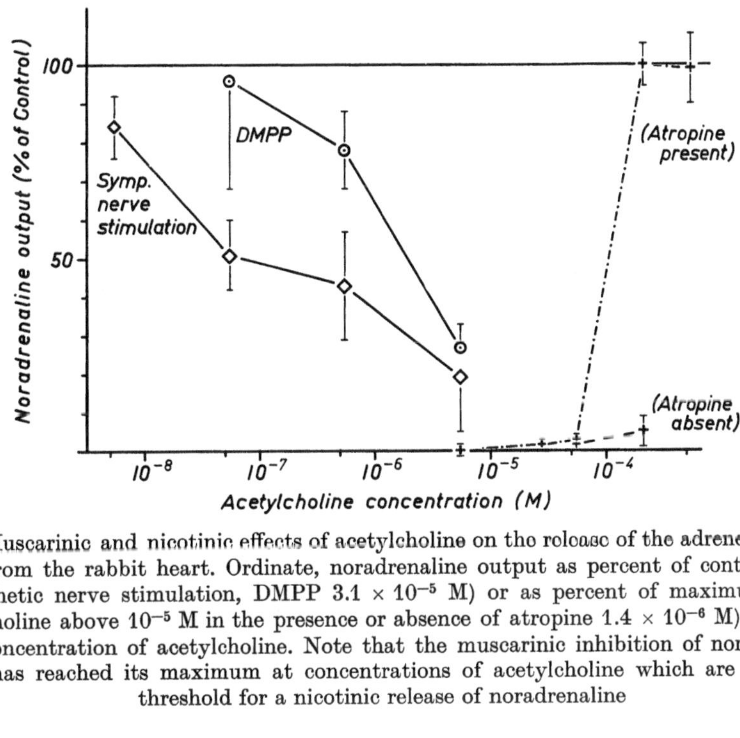

Fig. 8. Muscarinic and nicotinic effects of acetylcholine on the release of the adrenergic transmitter from the rabbit heart. Ordinate, noradrenaline output as percent of control output (sympathetic nerve stimulation, DMPP 3.1×10^{-5} M) or as percent of maximum output (acetylcholine above 10^{-5} M in the presence or absence of atropine 1.4×10^{-6} M). Abscissa, molar concentration of acetylcholine. Note that the muscarinic inhibition of noradrenaline release has reached its maximum at concentrations of acetylcholine which are below the threshold for a nicotinic release of noradrenaline

evidence for the "cholinergic link" hypothesis because in two investigations (BURN and RAND, 1960; McGIFF et al., 1967) anticholinesterases enhanced the responses to sympathetic nerve stimulation in the absence of anticholinergic drugs. Yet in the experiments on rabbit ear vessels BURN and RAND "did not find it easy to abolish", as they expressed it, the vasoconstrictor effect of stimulation of the mixed nerve at the base of the ear by pretreatment of the animals with three doses of reserpine (1.5 mg/kg each). This casts serious doubts on the supposition that the vasoconstriction was mediated only by adrenergic nerves. In the experiments on dog renal vessels (McGIFF et al., 1967) it was noted that atropine did not enhance the vasoconstrictor response to sympathetic nerve stimulation as the "cholinergic link" hypothesis would predict. In contrast, atropine greatly decreased the responses to nerve stimulation although it converted the vasodilation following intraarterial acetylcholine into a vasoconstriction. It seems, therefore, that the role of atropine or hyoscine in the experiments providing evidence for the "cholinergic link" hypothesis still needs to be clarified.

In the experiments shown in Fig. 7 acetylcholine caused a 50% inhibition of adrenergic transmitter output evoked by nervous impulses if applied at concen-

Table. *Effect of hexamethonium or pretreatment with reserpine on positive chronotropic and inotropic responses of isolated guinea-pig atria elicited by nicotine and DMPP*

	BARNETT et al. (1966)	BHAGAT (1966)	LINDMAR (1962)	TRENDELENBURG (1960)
Autonomic nerves to atria	attached	dissected	dissected	dissected
Atropine present throughout experiment, g/ml	1×10^{-6}	5×10^{-8}	—	6.7×10^{-7}
Nicotine g/ml	$6 \times 10^{-6} - 1 \times 10^{-5}$	1×10^{-5}	5×10^{-6}	1.3×10^{-5}
DMPP g/ml	—	1×10^{-5}	5×10^{-6}	—
Hexamethonium g/ml	1×10^{-6}	2×10^{-5}	$5 \times 10^{-6} - 1 \times 10^{-3}$	6.7×10^{-6}
Reserpine i.p. mg/kg; h before exp.	3; 24	2.5; 24	3; 20 – 18	5 – 10; 48 – 24
Response to nic. after hex.				
rate	2[a] s	11.1 s	13.6 s	—
amplitude	2[a] s	4.2 s	3.4 s	8[b]
Response to nic. after res.				
rate	97 ns	16.7 s	7.8 s	—
amplitude	—	35.9 s	18.8 s	19.0 s
Response to DMPP after hex.				
rate	—	75.6 ns	57.5 ns	—
amplitude	—	81.0 ns	53.2 ns	—
Response to DMPP after res.				
rate	—	4.9 s	6.4 s	—
amplitude	—	24.3 s	17.8 s	—

[a] Figure extrapolated from a graph.
[b] Significance not stated or not evident from the data given.

Results are given as percent of control response in absence of hexamethonium or without reserpine pretreatment. Absolute values given in original publications were transformed into percentages. The data taken from BARNETT and BENFORADO (1966) were selected in order to match concentrations used by the other workers as closely as possible. Significance of difference from control response (100%) indicated by *s*, $P < 0.01$; *ns*, $P > 0.05$.

trations as low as 10^{-8} to 10^{-7} g/ml. Since there is morphological evidence for the occurrence of juxtaposed cholinergic and adrenergic terminal axons in the autonomic ground plexus (HILLARP, 1959; EHINGER and FALCK, 1966) discovered in many organs, a peripheral muscarinic inhibition of sympathetic responses caused by a release of acetylcholine onto adrenergic nerve fibres may be of importance physiologically. This has been discussed in detail in a recent paper (Löffelholz and Muscholl, 1969).

Whereas all investigators agree that pretreatment with reserpine abolishes the stimulating effect of tyramine on heart tissue (for a review see Muscholl, 1966) one group of workers found no alteration of the effect of nicotine or acetylcholine after reserpine (Barnett and Benforado, 1966) while other authors observed a significant although incomplete inhibition of the stimulatory action of nicotine (Bhagat, 1966; Lindmar, 1962; Trendelenburg, 1960). As can be seen from the table, the doses of nicotine used in these experiments differed by a factor of 2.6 at the most. However, in the single paper showing an unaltered chronotropic response to nicotine (Barnett and Benforado) a concentration of atropine 20 times that used by Bhagat was employed. Lindmar did not add atropine prior to nicotine. In the detailed paper by Barnett and Benforado it was demonstrated that the concentration of atropine was critical in determining the nicotinic effects of acetylcholine. For instance, higher concentrations were required to unmask the positive inotropic effects of acetylcholine than were needed for chronotropic actions. The concentration of atropine (10^{-6} g/ml) used by Barnett and Benforado in the nicotine experiments produced a maximum increase in heart rate after acetylcholine.

Under conditions in which noradrenaline release by acetylcholine alone is insufficient to produce sympathomimetic effects (for example, after partial depletion of the transmitter by reserpine) the combined action of acetylcholine and atropine may enhance the noradrenaline release to a level resulting in sympathomimetic responses. The discrepancies of the results concerning the action of nicotine (Table) may be explained by assuming that nicotine liberated acetylcholine from cholinergic fibres. The acetylcholine, in turn, released noradrenaline from adrenergic fibres. The concentration of atropine used by Barnett and Benforado facilitated the chronotropic responses but in the other experiments the concentrations were insufficient to facilitate inotropic responses. In spite of the treatment with reserpine enough transmitter seems to have been available for release by the mechanism discussed. This agrees with the observations by Trendelenburg (1965) and Barnett and Benforado (1966) that pretreatment with 3 or 5 mg/kg reserpine decreased but did not abolish the effect of sympathetic nerve stimulation on guinea-pig atria exposed to atropine.

Non-nicotinic Actions of DMPP

Since in most of our experiments DMPP rather than nicotine was used as a nicotinic drug the question arises as to whether the action of DMPP differs in some respects from that of nicotine. In fact, there are several reports on effects of DMPP that are not obtained with nicotine.

"Tyramine-like" Action of DMPP. Lindmar (1962) found that DMPP caused a positive ino- and chronotropic action on isolated guinea-pig and rat atria which was antagonized by dichloroisoproterenol or pretreatment of the animals with reserpine. However, the stimulatory action of DMPP was unaltered by hexamethonium on rat atria and not significantly inhibited on guinea-pig atria. In contrast, positive ino- and chronotropic actions of nicotine on guinea-pig atria were fully antagonized

by hexamethonium. This led LINDMAR (1962) to suggest that DMPP exerted a "tyramine-like" effect on the preparations mentioned. The same conclusion was reached by BHAGAT (1966) who used guinea-pig atria and obtained essentially the same results as LINDMAR (see Table). Consistent with these observations are the findings by BHAGAT et al. [1967 (2)] that hexamethonium interfered with the release of tritiated noradrenaline from isolated guinea-pig atria by nicotine but failed to alter the release by DMPP or tyramine. In another study on guinea-pig atria extraneuronally stored H^3-noradrenaline was partially released by DMPP or tyramine but not by nicotine [BHAGAT et al., 1967 (1)]. Similarly, a "tyramine-like" action of low doses of DMPP was observed on the cat nictitating membrane (HAEFELY et al., 1966).

While the evidence for a "tyramine-like" effect of DMPP on guinea-pig atria is strong there is no indication for such an action on the perfused rabbit heart. For instance, hexamethonium completely antagonized the stimulation of the heart and the noradrenaline release produced by DMPP but did not alter the cardio-stimulation and the noradrenaline release caused by tyramine (LINDMAR and MUSCHOLL, 1961). Furthermore, the noradrenaline release from the rabbit heart by tyramine was largely unaffected by doses of amethocaine or d-cocaine which abolished the noradrenaline release evoked by DMPP (LINDMAR and MUSCHOLL, 1961). The differential effect of the local anaesthetic drugs on noradrenaline release supports the proposal that DMPP lacks a "tyramine-like" effect on the rabbit heart.

Another procedure which allows complete separation of the effects of DMPP on the one hand, and tyramine, on the other, is perfusion of the rabbit heart with a calcium-deficient solution. When the normal calcium-concentration of the Tyrode solution (3.6 meq/l) was lowered to 0.2 meq/l the noradrenaline release by DMPP was abolished whereas that produced by tyramine remained unaltered (LINDMAR et al., 1967). Finally, on the perfused rabbit heart the time courses of the actions of DMPP and tyramine applied as continuous infusions are entirely different. DMPP, like acetylcholine, caused a transient release of noradrenaline which wore off with a half-time of about 20 sec (see Fig. 1). This agrees with the previous findings that the noradrenaline output after DMPP (LINDMAR et al., 1967), like that after acetylcholine in the presence of atropine (LÖFFELHOLZ, 1967), occurred in the first few minutes of an infusion only and had ceased after the 4th min. In contrast, the tyramine-induced noradrenaline output remained practically constant during a perfusion for 15 min (LINDMAR et al., 1967). If DMPP, in addition to its nicotinic action, had a "tyramine-like" effect on the rabbit heart, the noradrenaline output during a continuous infusion of DMPP should have attained a constant level after the nicotinic block had abolished the initial nicotinic release of noradrenaline. This was not the case although tyramine, infused during the phase of nicotinic block produced by acetylcholine, was still effective to release noradrenaline into the perfusate of the rabbit heart (LÖFFELHOLZ, unpublished). Thus, the experimental evidence available rules out the possibility that the noradrenaline release caused by DMPP is, at any time, due to a "tyramine-like" effect.

"Bretylium-like" Action of DMPP. On several preparations DMPP has been shown to exert a "bretylium-like" adrenergic neurone blocking action. Thus, the inhibitory effects of stimulation of the periarterial nerves to the small intestine of the guinea pig (WILSON, 1962; BIRMINGHAM and WILSON, 1965) and rabbit

(BENTLEY, 1962; BIRMINGHAM and WILSON, 1965) or the vasoconstrictor effect of stimulation of the periarterial adrenergic nerves to the isolated mesenteric arteries of the rat [MALIK and LING, 1969 (1, 2)] was blocked by DMPP but not by hexamethonium. The block was not due to a decreased responsiveness of the smooth muscle to noradrenaline. The neurone blocking action of DMPP, like that of bretylium or guanethidine (DAY, 1962), was antagonized by dexamphetamine [BIRMINGHAM and WILSON, 1965; MALIK and LING, 1969 (1, 2)] or temporarily overcome by raising the calcium concentration of the perfusion fluid [MALIK and LING, 1969 (1, 2)].

BIRMINGHAM and WILSON described the neurone blockade caused by DMPP as a "bretylium-like" action. In spite of the failure of hexamethonium to antagonize the DMPP-induced neurone blockade MALIK and LING [1969 (2)] supposed that the block was due to the property of DMPP to occupy nicotinic receptors. However, from our results (Figs. 5 and 6) it follows that at the time when the nicotinic block was fully developed (3 to 4 min after start of DMPP, 3.1×10^{-5} M; 7 to 8 min after start of acetylcholine $2.1 \times 10^{4-}$ M, equilibration with atropine) the noradrenaline release evoked by nerve stimulation was, by no means, decreased. Only with DMPP and provided the infusion was continued for more than 4 min (for example, 8 min in Fig. 5, lower panel) a neurone blocking effect was obtained on the rabbit heart. Similarly, WILSON (1962) and BIRMINGHAM and WILSON (1965) emphasized that the ganglion stimulating and blocking effects of DMPP which occur rapidly are temporally dissociated from the neurone blocking action which occurs later.

The evidence that the nicotinic and the "bretylium-like" effects of DMPP can be separated is based on two kinds of observations. 1. Hexamethonium abolishes the nicotinic actions but does not antagonize the neurone block. 2. The nicotinic actions precede the neurone block.

Summary

The effect of cholinomimetic drugs was investigated on the isolated perfused rabbit heart. The noradrenaline output was measured fluorimetrically.

Activation of nicotinic receptors by acetylcholine, nicotine or dimethylphenyl piperazine (DMPP) resulted in a dose-dependent release of noradrenaline which was terminated after 5 to 10 sec. Then the noradrenaline output measured in the venous effluent of the heart followed the washout curve obtained with infused noradrenaline. The transitory character of the noradrenaline release after nicotinic drugs was not due to a momentary exhaustion of the transmitter available for release. Rather, it was the result of autoinhibition of nicotinic receptors by the stimulating agent. When the nicotinic block was established the noradrenaline release evoked by electrical nerve stimulation was not inhibited.

Muscarinic drugs (acetylcholine, methacholine, pilocarpine) inhibited the noradrenaline release caused by nicotinic drugs or by nerve stimulation. The significance of this phenomenon for the interpretation of results obtained with anticholinergic drugs is discussed.

My thanks are due to the Deutsche Forschungsgemeinschaft for supporting the work and to Dr. LÖFFELHOLZ for allowing me to present some of his recent results.

References

Angelakos, E. T., Bloomquist, E.: Release of norepinephrine from isolated hearts by acetylcholine. Arch. int. Physiol. Biochem. **73**, 397—402 (1965).

Barnett, A., Benforado, J. M.: The nicotinic effects of choline esters and of nicotine in guinea pig atria. J. Pharmacol. exp. Ther. **152**, 29—36 (1966).

Bentley, G. A.: Studies on sympathetic mechanisms in isolated intestinal and vas deferens preparations. Brit. J. Pharmacol. **19**, 85—98 (1962).

Bhagat, B.: Response of isolated guinea-pig atria to various ganglion-stimulating agents. J. Pharmacol. exp. Ther. **154**, 264—270 (1966).

—, Bovell, G., Robinson, I. M.: (1) Influence of cocaine on the uptake on H^3-norepinephrine and on the responses of isolated guinea-pig atria to sympathomimetic amines. J. Pharmacol. exp. Ther. **155**, 472—478 (1967).

—, Robinson, I. M., West, W. L.: (2) Mechanism of sympathomimetic responses of isolated guinea-pig atria to nicotine and dimethylphenylpiperazinium iodide. Brit. J. Pharmacol. **30**, 470—477 (1967).

Birmingham, A. T., Wilson, A. B.: An analysis of the blocking action of DMPP iodide on the inhibition of isolated small intestine produced by stimulation of the sympathetic nerves. Brit. J. Pharmacol. **24**, 375—386 (1965).

Burn, J. H.: Release of noradrenaline from the sympathetic postganglionic fibre. Brit. med. J. **1967/II**, 197—201.

— Discussion. In: Adrenergic neurotransmission, p. 39. (Wolstenholme, G. E. W., O'Connor, M., Eds.). London: Churchill 1968.

—, Rand, M. J.: Sympathetic postganglionic cholinergic fibres. Brit. J. Pharmacol. **15**, 56—66 (1960).

— — A new interpretation of the adrenergic nerve fibre. Advanc. Pharmacol. **1**, 1—30 (1962).

Chiang, T. S., Leaders, F. E.: Mechanism for nicotine and DMPP on the isolated rat atria-vagus nerve preparation. J. Pharmacol. exp. Ther. **149**, 225—232 (1965).

— — Dissociation between the initial negative and the secondary positive chronotropic and inotropic effects of nicotine in rat atria. Arch. int. Pharm. **172**, 347—355 (1968).

Cooper, T.: Surgical sympathectomy and adrenergic function. Pharmacol. Rev. **18**, 611—618 (1966).

Day, M. D.: Effect of sympathomimetic amines on the blocking action of guanethidine, bretylium and xylocholine. Brit. J. Pharmacol. **18**, 421—439 (1962).

Dixon, W. E.: Nicotin, Coniin, Piperidin, Lupetidin, Cytisin, Lobelin, Spartein, Gelsemin. Mittel, welche auf bestimmte Nervenzellen wirken. Heffter-Heubners Handb. exp. Pharmak. **2**, 656—736 (1924).

Ehinger, B., Falck, B.: Concomitant adrenergic and parasympathetic fibres in the rat iris. Acta physiol. scand. **67**, 201—207 (1966).

Furchgott, R. W.: Discussion. In: Adrenergic mechanism, p. 511—512. (Vane, J. R., Wolstenholme, G. E. W., O'Connor, M., Eds.) London: J. & A. Churchill Ltd. 1960.

Haefely, W., Hürlimann, A., Thoenen, H.: The effect of stimulation of sympathetic nerves in the cat treated with reserpine, α-methyldopa and α-methylmetatyrosine. Brit. J. Pharmacol. **26**, 172—185 (1966).

Haeusler, G., Thoenen, H., Haefely, W., Huerlimann, A.: Electrical events in cardiac adrenergic nerves and noradrenaline release from the heart induced by acetylcholine and KCl. Naunyn-Schmiedebergs Arch. Pharmak. exp. Path. **261**, 389—411 (1968).

Hillarp, N.-A.: The construction and functional organization of the autonomic innervation apparatus. Acta physiol. scand. **46**, Suppl. 157, 1—68 (1959).

Hoffmann, F., Hoffmann, E. J., Middleton, S., Talesnik, J.: The stimulating effect of acetylcholine on the mammalian heart and the liberation of an epinephrine-like substance by the isolated heart. Amer. J. Physiol. **144**, 189—198 (1945).

Hukovic, S., Muscholl, E.: Die Noradrenalin-Abgabe aus dem isolierten Kaninchenherzen bei sympathischer Nervenreizung und ihre pharmakologische Beeinflussung. Naunyn-Schmiedebergs Arch. exp. Path. Pharmak. **244**, 81—96 (1962).

Khan, M., Mantegazza, P., Piccinini, F.: Effect of low temperatures on the responses of guinea-pig isolated atria to nicotine and to sympathetic and parasympathetic stimulation. Brit. J. Pharmacol. **25**, 119—125 (1965).

KOTTEGODA, S. R.: Stimulation of isolated rabbit auricles by substances which stimulate ganglia. Brit. J. Pharmacol. 8, 83—86 (1953).

LEADERS, F. E.: Local cholinergic-adrenergic interaction: Mechanism for the biphasic chronotropic response to nerve stimulation. J. Pharmacol. exp. Ther. **142**, 31—38 (1963).

— LONG, J. P.: Mechanism of the positive chronotropic response to nicotine. J. Pharmacol. exp. Ther. **137**, 206—212 (1962).

LEE, W. C., SHIDEMAN, F. E.: Mechanism of the positive inotropic response to certain ganglionic stimulants. J. Pharmacol. exp. Ther. **126**, 239—249 (1959).

LINDMAR, R.: Die Wirkung von 1,1-Dimethyl-4-Phenyl-Piperazinium-Jodid am isolierten Vorhof im Vergleich zur Tyramin- und Nicotinwirkung. Naunyn-Schmiedebergs Arch. exp. Path. Pharmak. **242**, 458—466 (1962).

— LÖFFELHOLZ, K., MUSCHOLL, E.: Unterschiede zwischen Tyramin und Dimethylphenylpiperazin in der Ca^{++}-Abhängigkeit und im zeitlichen Verlauf der Noradrenalin-Freisetzung am isolierten Kaninchenherzen. Experientia (Basel) **23**, 933—934 (1967).

— — — A muscarinic mechanism inhibiting the release of noradrenaline from peripheral adrenergic nerve fibres by nicotinic agents. Brit. J. Pharmacol. **32**, 280—294 (1968).

— MUSCHOLL, E.: Die Wirkung von Cocain, Guanethidin, Reserpin, Hexamethonium, Tetracain und Psicain auf die Noradrenalin-Freisetzung aus dem Herzen. Naunyn-Schmiedebergs Arch. exp. Path. Pharmak. **242**, 214—227 (1961).

— — Die Wirkung von Pharmaka auf die Elimination von Noradrenalin aus der Perfusionsflüssigkeit und die Noradrenalinaufnahme in das isolierte Herz. Naunyn-Schmiedebergs Arch. exp. Path. Pharmak. **247**, 469—492 (1964).

LÖFFELHOLZ, K.: Untersuchungen über die Noradrenalin-Freisetzung durch Acetylcholin am perfundierten Kaninchenherzen. Naunyn-Schmiedebergs Arch. Pharmak. exp. Path. **258**, 108—122 (1967).

— Autoinhibition of nicotinic release of noradrenaline from postganglionic sympathetic fibres. Naunyn-Schmiedebergs Arch. Pharmak., in press (1970).

— LINDMAR, R., MUSCHOLL, E.: Der Einfluß von Atropin auf die Noradrenalin-Freisetzung durch Acetylcholin. Naunyn-Schmiedebergs Arch. Pharmak. exp. Path. **257**, 308 (1967).

— MUSCHOLL, E.: A muscarinic inhibition of the noradrenaline release evoked by postganglionic sympathetic nerve stimulation. Naunyn-Schmiedebergs Arch. Pharmak. **265**, 1—15 (1969).

MALIK, K. U., LING, G. M.: (1) Modification by acetylcholine of the response of rat mesenteric arteries to sympathetic stimulation. Circulat. Res. **25**, 1—9 (1969).

— — (2) The effect of 1,1-dimethyl-4-phenyl-piperazinium on the response of mesenteric arteries to sympathetic nerve stimulation. J. Pharm. Pharmacol. **21**, 514—519 (1969).

MCGIFF, J. C., BURNS, R. B. P., BLUMENTHAL, M. R.: Role of acetylcholine in the renal vasoconstrictor response to sympathetic nerve stimulation in the dog. Circulat. Res. **20**, 616—629 (1967).

MCNAMARA, B., KROP, S., MCKAY, E. A.: The effect of calcium on the cardiovascular stimulation produced by acetylcholine. J. Pharmacol. exp. Ther. **92**, 153—161 (1948).

MIDDLETON, S., OBERTI, C., PRAGER, R., MIDDLETON, H. H.: Stimulating effect of acetylcholine on the papillary myocardium. Acta physiol. lat.-amer. **6**, 82—89 (1956).

MUSCHOLL, E.: Drugs interfering with the storage and release of adrenergic transmitters. In: Pharmacology of cholinergic and adrenergic transmission, p. 291—302. (KOELLE, G. B., DOUGLAS, W. W., CARLSSON, A., Eds.). Oxford: Pergamon Press 1965.

— Indirectly acting sympathomimetic amines. Pharmacol. Rev. **18**, 551—559 (1966).

RICHARDSON, J. A., WOODS, E. F.: Release of norepinephrine from the isolated heart. Proc. Soc. exp. Biol. (N. Y.) **100**, 149—151 (1959).

ROSSUM, J. M., VAN: Classification and molecular pharmacology of ganglionic blocking agents. Part I. Mechanism of ganglionic synaptic transmission and mode of action of ganglionic stimulants. Int. J. Neuropharmacol. **1**, 97—110 (1962).

TRENDELENBURG, U.: The action of histamine and 5-hydroxytryptamine on isolated mammalian atria. J. Pharmacol. exp. Ther. **130**, 450—460 (1960).

— The effect of sympathetic nerve stimulation on isolated atria of guinea pigs and rabbits pretreated with reserpine. J. Pharmacol. exp. Ther. **147**, 313—318 (1965).

TRENDELENBURG, U.: Some aspects of the pharmacology of autonomic ganglion cells. Ergebn. Physiol. **59**, 1—85 (1967).

WILSON, A. B.: An adrenergic neurone blocking action of DMPP. J. Pharm. Pharmacol. **14**, 700 (1962).

Prof. Dr. E. MUSCHOLL
Pharmakologisches Institut der Universität
D-6500 Mainz
Langenbeckstraße 1

Discussion

VON EULER: I realise, Dr. MUSCHOLL, that you have studied the output but I wonder how you would account for the stimulating effect of acetylcholine in low concentrations.

MUSCHOLL: You mean its effect, for instance, on noradrenaline release evoked by nerve stimulation. We found a stimulating effect of acetylcholine only immediately after acetylcholine was added. There are at least three possibilities. Firstly, there might be a small depolarization left and we know from other drugs causing depolarization that the output of transmitter is increased—one of these drugs is amphetamine. Secondly, in the period immediately after the nicotinic excitation, the small depolarization may cause inhibition of re-uptake of transmitter and, therefore, perhaps the increase in noradrenaline output. There is also a third possibility that the calcium influx is increased immediately after a period when the sodium concentration in the fibre was elevated. I did not want to comment on this in the paper because of shortness of time. We have not done any experiments to test these possibilities but as I said, this can be investigated experimentally—we should like to do it. The effect of very low concentrations of acetylcholine observed by MALIK, we could not explain because what we have seen here is certainly a nicotinic effect of acetylcholine. We have not given such small concentrations as MALIK did, we stopped at 10^{-9} g/ml but may be if we had taken 10^{-12} or so, we would have seen an increase in output—we would certainly like to try this.

TRENDELENBURG: In experiments with the superior cervical ganglion one can see a parallel to the increased responses which you observed in the presence of nicotinic drugs: after a long lasting block of ganglionic transmission has been produced by repeated injections of nicotine, the response of the ganglion to various non-nicotinic agents (histamine, muscarinic drugs, 5-hydroxytryptamine, various polypeptides) is greatly enhanced.

MUSCHOLL: I should like to add that recently NEDERGAARD and BEVAN also found an increased response of their pulmonary artery preparation to sympathetic nerve stimulation if nicotinic drugs had been applied before. But I think that's no explanation, it's only another experimental observation. It is clear that one should do additional experiments.

von Euler: In this case you mention the possibility of two axons lying close together, which has been repeatedly observed and which we saw yesterday. Do you now consider it a possibility that the effect is not only one of suppression but also one of excitation?

Muscholl: Well, as I showed you, the action which acetylcholine exerts is to depress the output of transmitter and only if very high concentrations are applied or when a muscarinic blocking drug is applied, will there then be an increase in output of transmitter. I think it is perhaps not bad that the adrenergic nerve fibre is in some case protected from contact with acetylcholine (which may decrease its output) because of the presence of cholinesterase which you have mentioned. This could protect the nerve fibre from diffusing acetylcholine and decreasing its response too much. Perhaps one should consider this possibility.

Philippu: Did you find any effect on the release of catecholamines by atropine alone?

Muscholl: No, I must correct myself: the effects we have observed were small and not statistically significant but there was a slight increase in noradrenaline output after atropine. We want to follow up this observation more closely. It could be that the stimulation parameters we have used also cause some stimulation of cholinergic fibres or that cholinergic fibres may be present in the sympathetic trunk. If acetylcholine is usually released along with excitation of the sympathetic nerves, there may be a small inhibition of the output, but if we give atropine this inhibition is abolished and we can elevate the output a little bit—however the elevation was not greater than about 15%.

Schümann: Did you perform experiments with tyramine in which you enhanced noradrenaline output and then tried to see whether acetylcholine would diminish it.

Muscholl: We have investigated tyramine as well. It is impossible to show an inhibiting effect of muscarinic drugs on the output caused by tyramine and it is not possible to alter the output by the nicotinic action of acetylcholine or other nicotinic drugs. So the nicotinic and the muscarinic effects do not affect the tyramine induced noradrenaline release.

Bayer-Symposium II, 188—198 (1970)

Session 4b

Effects of Drugs on Uptake and Release of Catecholamines (II)

Chairman: A. Carlsson

Membrane Effects of Catecholamine Releasing Drugs

D. Palm, H. Grobecker, and I. J. Bak

With 8 Figures

Drugs like reserpine, prenylamine, chlorpromazine, imipramine etc. exert well defined effects on the uptake and storage mechanisms of catecholamines in vitro as well as in vivo. It is assumed that the site of action of these drugs is localised at cellular and subcellular membranes (for review see Carlsson et al., 1963; von Euler and Lishajko, 1965; Holtz and Palm, 1966; Seeman, 1966). Partly from methodological and also from pharmacological reasons the interest of most investigators was directed mainly to those membranes, which are involved in the transport of biogenic amines, particularly catecholamines.

The aim of our investigations was to demonstrate similar effects of the drugs mentioned above on *membranes* of subcellular particles *not associated directly with catecholamine transport and storage.*

It was postulated by several authors, that the main effects of drugs interfering with catecholamine transport mechanisms at the membrane of storage particles depend on the inhibition of an ATP-Mg^{++} catalysed transport process through these membranes (Kirshner, 1962; Carlsson et al., 1962; von Euler and Lishajko, 1965). In addition, it could be shown, for instance by Taugner and Hasselbach (1966), and Burger et al. (1969), that an ATPase, probably involved in this transport, was inhibited by drugs like reserpine and prenylamine.

We have investigated the action of a number of drugs known to inhibit catecholamine transport, and the respective ATPase as well, on a similar enzyme, namely a *Mg^{++}-activated ATPase* localised in *heart mitochondria* of the rat.

The *highly lipophilic drugs reserpine, prenylamine*, and also *chlorpromazine* and *propranolol* proved to be inhibitors of the Mg^{++}-stimulated enzyme activity. With respect to reserpine and prenylamine, this inhibition was achieved by concentrations which also inhibited the ATPase of catecholamine storage particles. As can be seen from Fig. 1, a 50% inhibition was obtained with all drugs in concentrations of about 5×10^{-5} M. Similar results were also obtained by Schöne (1968). Therefore it must be assumed, that the action of drugs affecting catecholamine transport are *not specifically* directed to enzymes which are localised only at specific storage sites.

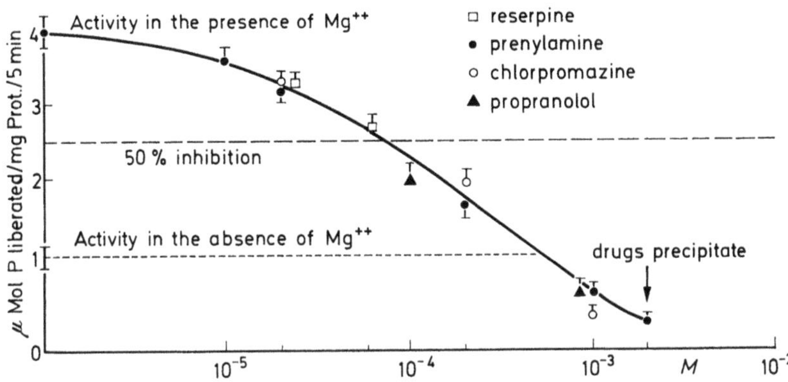

Fig. 1. Inhibition of Mg^{++}-activated ATPase of rat heart mitochondria by reserpine, prenylamine, chlorpromazine and D,L-propranolol. (For details of the method see Kielly and Kielly, 1953)

In low concentrations drugs like reserpine, prenylamine, chlorpromazine, propranolol etc. inhibit not only the uptake (Carlsson et al., 1963; Jonasson et al., 1964), but also the spontaneous and the drug induced release of catecholamines in isolated granules (von Euler and Lishajko, 1961; von Euler et al., 1964; von Euler and Lishajko, 1966).

In high concentrations, they cause a rapid release of catecholamines from the storage particles (von Euler and Lishajko, 1961; Schöne and Lindner, 1962; Weil-Malherbe and Posner, 1963; Philippu et al., 1965). In preparations from bovine medullary granules, this release caused by prenylamine was temperature independent and the dose-effect curves were very steep, indicating an all-or-non phenomenon (Fig. 2).

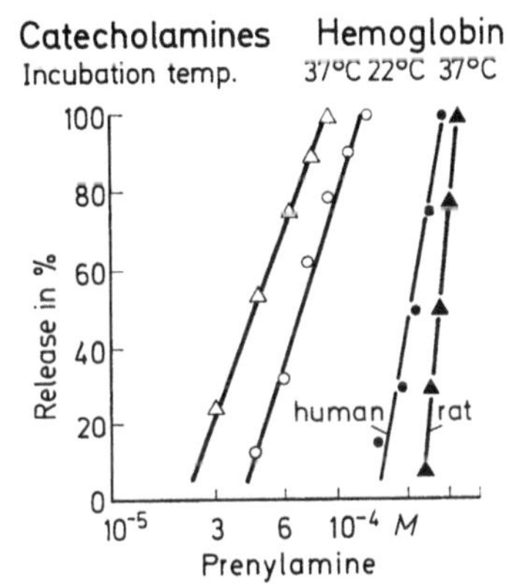

Fig. 2. Comparison between release of catecholamines from bovine medullary granules and hemoglobin from human and rat erythrocytes (normotonic saline) caused by prenylamine. (For detail see Grobecker et al., 1968)

Comparing the release of catecholamines with the uptake of the releasing agent, i.e. ^{14}C-labelled prenylamine, the following results were obtained (Fig. 3): In the concentration of about 4×10^{-5}M, two to three moles of catecholamines were released per mol prenylamine taken up by the granules. This ratio, however, increased with higher concentrations of the d-ug, reaching a *constant maximum of 15 to 17 moles* amine released per mole prenylamine fixed by the granules, which is in accord with previous data (Philippu et al., 1965). The phase of maximal release is characterised by the straight part of the curve and occurred at concentrations of about 7×10^{-5} M. These experimental results, i.e. the steep dose-effect curve (cf. Fig. 2) and the straight line, the calculated part of which leads directly to the point of the original amine content of the preparation (Fig. 3, dashed line), led us to assume a severe impairment of the granular membrane.

This assumption was confirmed by *electron-microscopic investigations*. After incubation of isolated bovine medullary granules with concentrations from 6×10^{-5} to 2×10^{-4} M the electronmicrographs (Fig. 4) showed *swelling* of the granules and also *severe damage of the granular membranes*. Only small islets of electron dense material remained inside the particles.

Similar effects of the drug could be shown also on other membranes, e.g. the *cytoplasmic membrane of erythrocytes*. After incubation of human or rat erythro-

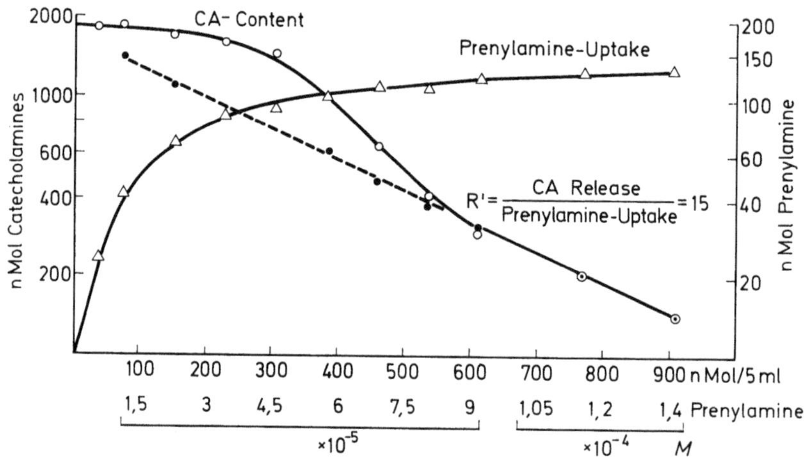

Fig. 3. Comparison between uptake of ^{14}C-prenylamine and release of catecholamines in isolated bovine medullary granules. Experimental conditions: 0.3 M sucrose pH 7.4, 37 °C, 30 min. Δ—Δ: uptake of ^{14}C-prenylamine; o—o: release of catecholamines; ●—●: calculated values of catecholamine release (nM = R'x prenylamine uptake)

cytes in *normotonic* saline with prenylamine in concentrations of about 2×10^{-4} M a rapid and complete hemolysis could be observed. The dose-effect curves of the release of hemoglobin was as steep as that of the catecholamine release from adrenal medullary granules (cf. Fig. 2).

However, the hemolysis induced by *hypotonic* saline could be prevented by prenylamine in concentrations from 10^{-6} to 3×10^{-4} M (Fig. 5). The latter concentration, which nearly completely *prevented* the *hypotonically* induced hemolysis, had caused a *complete release* of hemoglobin in *normotonic* saline (cf. Fig. 2).

The drugs used in our experiments are known to possess a high *lipid solubility* which may explain their affinity to all kinds of membranes (SEEMAN, 1966). Moreover, these compounds efficiently *lower surface tension* (SEEMAN and BIALY, 1963; GROBECKER et al., 1968). Therefore, it is conceivable, that *high concentrations* lead to a "*labilisation*" of membranes, causing a release of catecholamines from granule preparations and of hemoglobin from erythrocytes suspensions. *Low concentrations* however, obviously lead to a "*stabilisation*" of the respective membrans. Therefore they prevent, for instance, the hypotonically induced release of hemo-

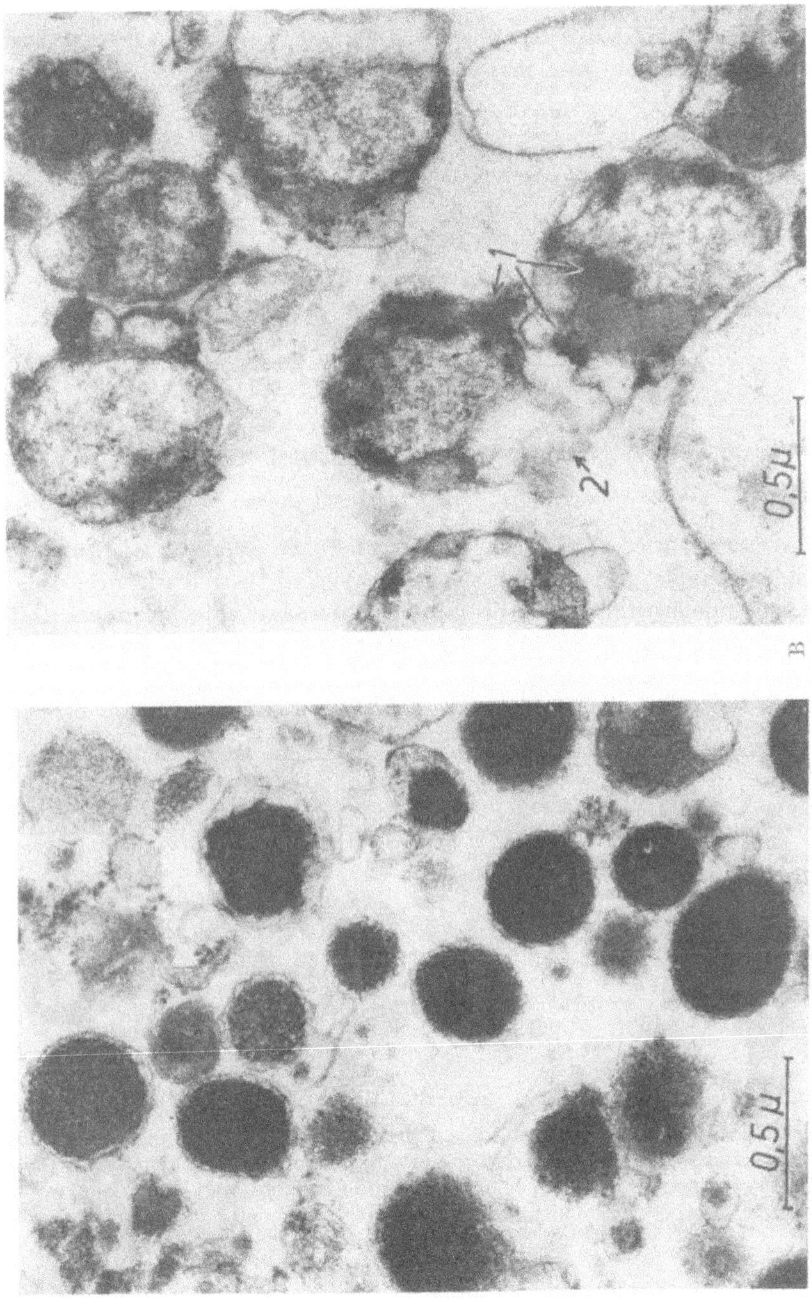

Fig. 4. Electron micrographs of chromaffine granules before (A) and after (B) incubation with prenylamine (2.4×10^{-4} M, 37 °C, 30 min). (From GROBECKER et al., 1968)

globin. This stabilising effect in erythrocytes is due to an increase of the surface area/volume ratio, i.e. an extension of the membrane by a certain physico-chemical processes (SEEMAN, 1966; METCALFE et al., 1968). It is well known, that beside prenylamine reserpine and also many other drugs, possess these *labilising and*

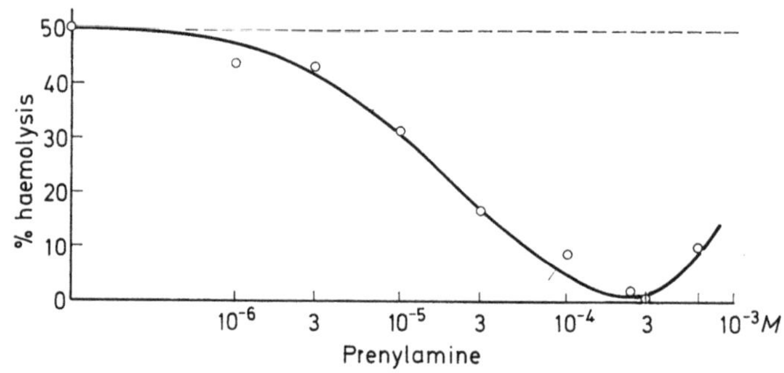

Fig. 5. Protection by prenylamine of human erythrocytes against hypotonic hemolysis. (From GROBECKER et al., 1968)

stabilising properties in erythrocytes (SEEMAN, 1966). In isolated granules, particularly in splenic nerve granules, low concentrations of these drugs were able to inhibit the spontaneous and also the drug-induced release of catecholamines. In the same range of concentrations the ATP-Mg^{++}-stimulated uptake of catecholamines

Table 1. *Surface tension lowering effects of drugs in relation to inhibition of uptake and release of catecholamines*

	surface tension − 10 dyn/cm (a)	medullary granules uptake (b)	nerve uptake	granules release (c-f)
Reserpine	10^{-6} M	1.6×10^{-6} M → 89	10^{-6} M → 80	10^{-6} M → 67
Chlorpromazine	3×10^{-5} M	2.8×10^{-5} M → 55	2×10^{-5} M → 54	2×10^{-5} M → 53
Imipramine	1.1×10^{-4} M	3.2×10^{-5}M → 39		
Propranolol	→ 10^{-3} M		3×10^{-4} M → 57	3×10^{-4} M → 62

Column 1: equieffective concentrations of drugs. Column 2, 3, 4: concentrations, which inhibit uptake and release by → %. Data, given in the columns were published by (a) SEEMAN and BIALY, 1963; (b) JONASSON et al., 1964; (c) v. EULER and LISHAJKO, 1961; (d) v. EULER et al., 1964; (e) STJÄRNE, 1964; (f) v. EULER and LISHAJKO, 1966.

into the granules was inhibited. These effects produced by *psychotropic drugs* with *completely different chemical configuration* seem to be well correlated to their *potency, to lower surface tension.* Table 1 shows this relationship. Obviously, the same is true for *α-and-β-adrenergic blocking drugs.* As illustrated in Table 2, with a decreasing potency to lower surface tension, also the inhibiting effects on uptake and release of catecholamines are diminished.

One may speculate, that the so-called *stabilising effect* offers a *common mechanism*, by which inhibition of spontaneous release, active transport mechanisms at the granular membrane—and possibly also the inhibition of ATPases at granular, mitochondrial and sarcoplasmic membranes [cf. BALZER et al., 1968 (1)] can be explained.

Table 2. *Surface tension lowering effects of β-adrenergic blocking agents in relation to inhibition of catecholamine uptake and release in isolated splenic nerve granules. (Action of propranolol = 1)*

	rel. surface tension lowering (a)	rel. inhibition of uptake (3×10^{-5} M) (b, c)	rel. inhibition of release (3×10^{-4} M) (b, c)
Propranolol	1	1	1
Pronethalol	0.7	1.1	0.82
Kö 592	0.59	0.61	—
INPEA	0.27	0.24	—
MJ 1999	0	0.19	0

The values are calculated from data published by (a) LEVY (1968); (b, c) v. EULER and LISHAJKO (1966, 1968).

Because of their high *lipid solubility* an *accumulation* of these compounds in *cellular and subcellular membranes* [cf. BALZER et al., 1968 (2)] should be expected also after administration in vivo.

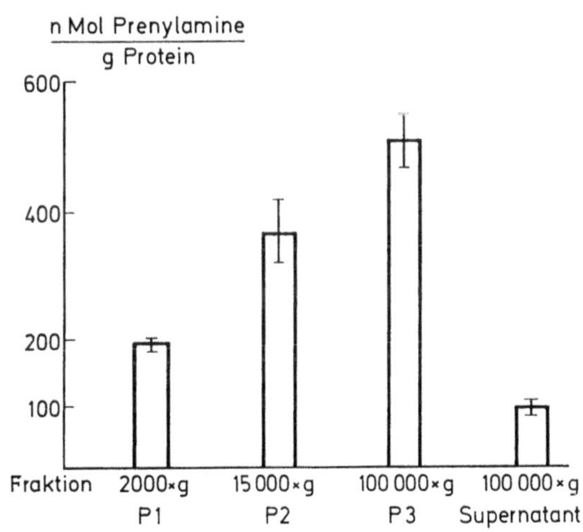

Fig. 6. Subcellular distribution of radioactivity in the guinea pig heart 4 h after injection of 30 mg/kg ^{14}C-prenylamine i.p. Note the high accumulation of the drug in fraction P_2 (mitochondria) and P_3 (microsomes)

Previous experiments from our laboratory with ^{14}C-prenylamine (Fig. 6; GROBECKER et al., 1965) and similar experiments reported by STJÄRNE et al. (1969) using reserpine have shown, that these drugs do not accumulate in catecholamine storing particles specifically, but to nearly the same extent also in mitochondria.

If the degree of accumulation of these drugs is high enough one should expect similar morphological membrane alterations as seen in our in vitro experiments with chromaffine granules (cf. Fig. 4).

We have studied—by means of *electron-microscopy*—the effect of reserpine (1 mg/kg), chlorpromazine (33 mg/kg) and prenylamine (33 mg/kg) on rat heart mitochondria after prolonged oral administration of these drugs over a period of 10 to 14 days. Controls received saline (cf. GROBECKER et al., 1969).

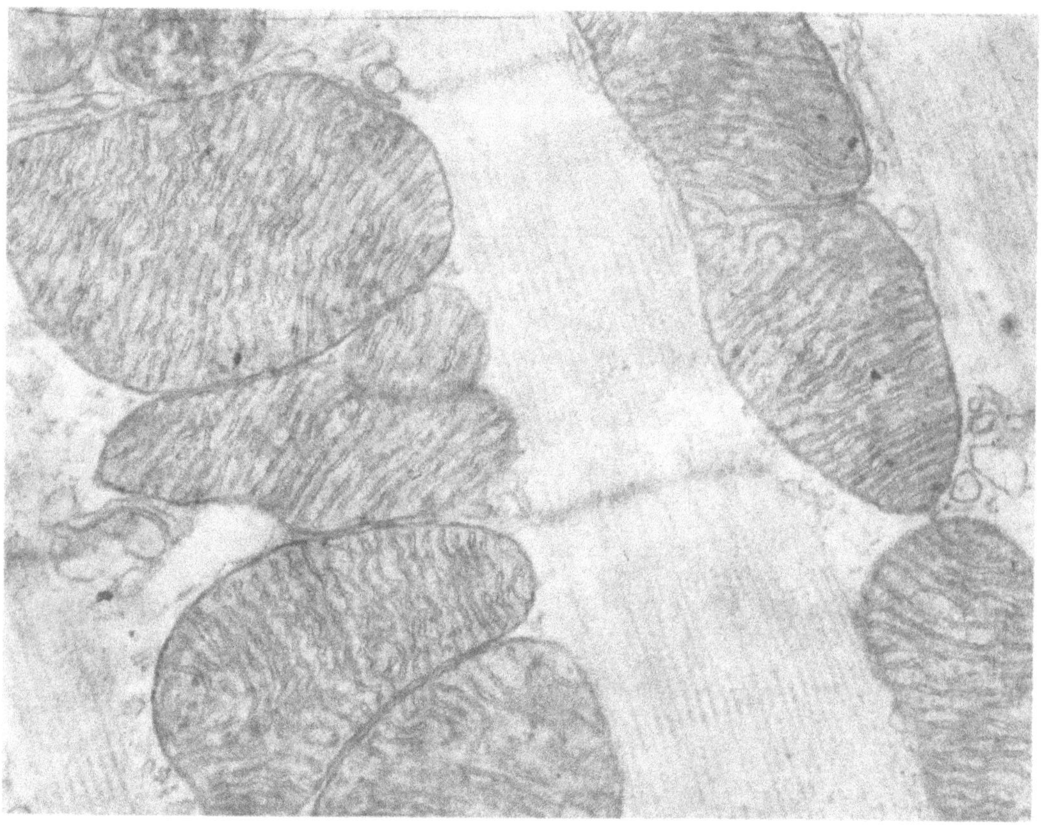

Fig. 7. A high powered electron micrograph taken from control group depicts healthy looking mitochondria with regulary lamellated cristae and electron dense matrix. The mitochondrial cristae are further characterized by showing the uniform size of internal space (× 32000)

The electron-micrographs of sections from heart muscle of untreated animals (Fig. 7) show numerous mitochondria situated between muscle fibres. The lamellated mitochondrial cristae are regularly arranged and the mitochondrial matrices are electron dense. *In treated animals however (e.g. prenylamine 33 mg/kg), the mitochondria appeared in a markedly swollen state, the mitochondrial matrices showed little electron density and the mitochondrial cristae are dearranged and partially damaged* (Fig. 8). In addition *vacuolisation and mitochondrial* degeneration processes could be observed with the same dose of the drug. *Similar alterations were observed after treatment with chlorpromazine and reserpine.* There occurred also intramitochondrial *accumulation of electron dense material,* as it was already described for the action of reserpine on heart mitochondria in dogs and mice (WILCKEN et al., 1967; SUN et al., 1968). Furthermore the *sarcoplasmic reticulum was in an extremely swollen state after* administration of 100 mg/kg prenylamine over a period of 5 days.

At the present time it would be a speculation only, to see a *connection between the ultrastructural alterations* described in our experiments and the *myocardial*

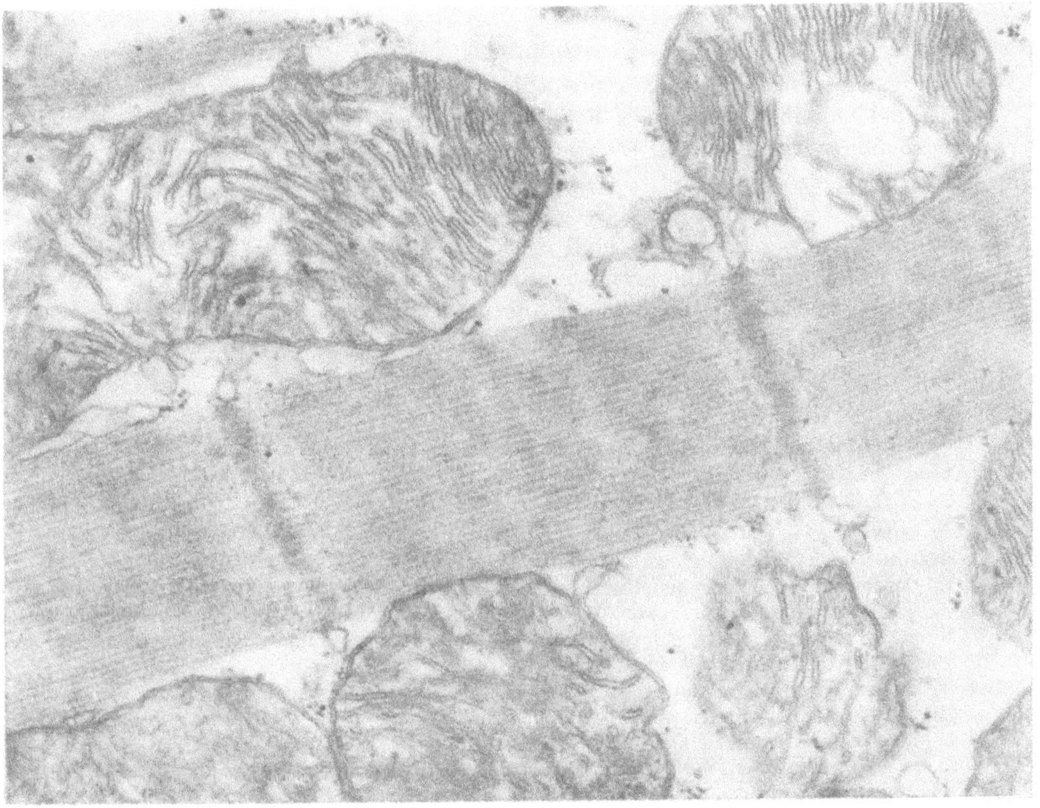

Fig. 8. A high powered electron micrograph prepared from the group treated with prenylamine 33 mg/kg) depicts are markedly swollen state of mitochondria. Due to the loss of mitochondrial matrix, it shows a pale appearance in the back ground. In addition, dearrangement of mitochondrial cristae and presence of vacuolization are further evident (× 32000)

lesions seen *in man after a long term treatment with psychotropic drugs* [ALEXANDER, 1968 (1), (2); RICHARDSON et al., 1966].

These mitochondrial alterations should lead to an impairment of biochemical functions as already suggested by ZAIMIS (1961). In this context it is of interest that, for instance, reserpine inhibits oxidative phosphorylation in heart mitochondria, as shown by SCHWARTZ and LEE (1960). Furthermore, an enhancement of mitochondrial as well as microsomal MAO activity in several organs was described by IZUMI et al. (1967, 1969) and also by YOUDIM and SANDLER (1968).

From our results it should be assumed that certain pharmacological actions of drugs like reserpine, prenylamine, chlorpromazine, and others are mainly based on their *physico-chemical properties*, i.e., their high *lipid solubility* and especially their *high surface activity*.

Therefore one may speculate that by these drugs not only the *membranes of catecholamine storing particles*, but also those of *other subcellular particles* are impaired in their functional and morphological state. Presumably this is brought about by a common, rather unspecific mechanism, due to physico-chemical properties of the drugs causing *alterations of the substructure of all kinds of membranes*.

Summary

Drugs like reserpine, prenylamine, chlorpromazine, and propranolol because of their lipid solubility and surface activity not only affect membranes involved in catecholamine transport and storage processes but also other cellular and subcellular membranes.

In *low concentrations*, these drugs seem to impair biological processes in a reversible manner, whereas in high concentrations membranes are affected irreversibly, e.g. chromaffine granules and erythrocytes are lysed in vitro, mitochondria of the rat heart are strongly affected in vivo.

It is assumed, that irrespectively of the chemical structure, the surface activity of these psychotropic drugs mainly determines their potency to affect all kinds of membranes, especially that of catecholamine storing particles.

Supported by a grant from the Deutsche Forschungsgemeinschaft.

References

ALEXANDER, C. S.: (1) Cardiotoxic effects of certain psychotropic drugs including electron microscopic changes on myocardial biopsy. Circulation **37/38**, Suppl. VI, 32 (1968).

— (2) Cardiotoxic effects of phenothiazine and related drug. Circulation **38**, 1014 (1968).

BALZER, H., MAKINOSE, M., HASSELBACH, W.: (1) The inhibition of sarcoplasmic calcium pump by prenylamine, reserpine, chlorpromazine and imipramine. Naunyn-Schmiedebergs Arch. Pharmak. exp. Path. **260**, 444—455 (1968).

— — FIEHN, W., HASSELBACH, W.: (2) The binding of calcium transport inhibitors reserpine, chlorpromazine and prenylamine to the lipid of the membranes of the sarcoplasmic reticulum. Naunyn-Schmiedebergs Arch. Pharmak. exp. Path. **260**, 456—473 (1968).

BURGER, A., PHILIPPU, A., SCHÜMANN, H. J.: ATP-Spaltung und Aminaufnahme durch Milznervengranula. Naunyn-Schmiedebergs Arch. Pharmak. exp. Path. **262**, 208—220 (1969).

CARLSSON, A., HILLARP, N.-Å., WALDECK, B.: A Mg^{++}-ATP dependent storage mechanism in the amine granules of the adrenal medulla. Med. exp. **6**, 47—53 (1962).

— — — Analysis of the Mg^{++}-ATP dependent storage mechanism in the amine granules of adrenal medulla. Acta physiol. scand. **59**, Suppl. 215, 5—36 (1963).

EULER, U. S. v., LISHAJKO, F.: Effect of reserpine on the release of catecholamines from isolated nerve and chromaffine cell granules. Acta physiol. scand. **52**, 137—145 (1961).

— — Effect of drugs on the storage granules of adrenergic nerves. In: Pharmacology of cholinergic and adrenergic transmission. Proc. 2nd int. Pharm. Meet. Prague 20—23 Aug. 1963. Oxford: Pergamon Press **3**, 245—259 (1965).

— — Inhibitory action of adrenergic blocking agents on catecholamine release and uptake in isolated nerve granules. Acta physiol. scand. **68**, 257—262 (1966).

— — Inhibitory action of adrenergic blocking agents on reuptake and net uptake of noradrenaline in nerve granules. Acta physiol. scand. **74**, 501—506 (1968).

— STJÄRNE, L., LISHAJKO, F.: Effects of reserpine, segontin and phenoxybenzamine on the catecholamines and ATP of isolated nerve and adreno-medullary storage granules. Life Sci. **3**, 35—40 (1964).

GROBECKER, H, BAK, I. J., SCHMID, B., PALM, D.: Drug induced changes in ultrastructure and biochemical function of rat heart mitochondria. Fourth International Congress on Pharmacology, Abstracts, p. 187. Basel 1969.

— HOLTZ, P., PALM, D., BAK, I. J., HASSLER, R.: In vitro lysis of erythrocytes and chromaffine granules by prenylamine. Experientia (Basel) **24**, 701—703 (1968).

— PALM, D., SCHÜMANN, H. J.: Über den Mechanismus der Freisetzung von Brenzcatechinaminen durch Segontin. Naunyn-Schmiedebergs Arch. exp. Path. Pharmak. **251**, 158—159 (1965).

HOLTZ, P., PALM, D.: Brenzcatechinamine und andere sympathicomimetische Amine. Biosynthese und Inaktivierung, Freisetzung und Wirkung. Ergebn. Physiol. **58**, 1—580 (1966).

IZUMI, F., OKA, M., YOSHIDA, H., IMAIZUMI, R.: Effect of reserpine on monoamine oxidase activity in guinea pig heart. Life Sci. **6**, 2333—2343 (1967).

— — — — Stimulatory effect of reserpine on monoamine oxidase in guinea pig heart. Biochem. Pharmacol. **18**, 1739—1748 (1969).

JONASSON, J., ROSENGREN, E., WALDECK, B.: Effects of some pharmacologically active amines on the uptake of arylalkylamines by adrenal medullary granules. Acta physiol. scand. **60**, 136—140 (1964).

KIELLY, W. W., KIELLY, R. K.: A specific adenosintriphosphatase of liver mitochondria. J. biol. Chem. **200**, 213—222 (1953).

KIRSHNER, N.: Uptake of catecholamines by a particulate fraction of the adrenal medulla. J. biol. Chem. **237**, 2311—2317 (1962).

LEVY, J. V.: Surface tension effects of β-adrenergic blocking drugs. J. Pharm. Pharmacol. **20**, 813—815 (1968).

METCALFE, J. C., SEEMAN, P., BURGEN, A. S. V.: The proton relaxation of benzyl alcohol in erythrocyte membranes. Molec. Pharmacol. **4**, 87—95 (1968).

PHILIPPU, A., PALM, D., SCHÜMANN, H. J.: Effect of segontin and reserpine on isolated medullary granules. Nature (Lond.) **205**, 183 (1965).

RICHARDSON, H. L., GRAUPNER, K. I., RICHARDSON, M. E.: Intramyocardial lesions in patients dying suddenly and unexpectedly. J. Amer. med. Ass. **195**, 254—260 (1966).

SCHÖNE, H.: Untersuchungen zum Wirkungsmechanismus von Prenylamin. Abstr. 2nd Int. Symp. Pharm. Chem., p. 24. Münster, July 22—26, 1968.

— LINDNER, E.: Über die Wirkung von N-[(3′-Phenylpropyl-(2′)]-1,1-diphenylpropyl-(3)-amin auf den Katecholamin-Stoffwechsel. Klin. Wschr. **40**, 1196—1200 (1962).

SCHWARTZ, A., LEE, K. S.: Effect of reserpine on heart mitochondria. Nature (Lond.) **188**, 948—949 (1960).

SEEMAN, P. M.: Membrane stabilization by drugs: tranquilizers, steroids, and anesthetics. Int. Rev. Neurobiol. **9**, 145—221 (1966).

— BIALY, H. S.: The surface activity of tranquilizers. Biochem. Pharmacol. **12**, 1181—1191 (1963).

STJÄRNE, L.: Studies of catecholamine uptake storage and release mechanisms. Acta physiol. scand. **62**, Suppl. 228, 1—97 (1964).

— LAGERCRANTZ, H., HEDQVIST, P.: Subcellular distribution of reserpine in sympathetic nerves and in the adrenal medulla. Abstr. 4th Int. Congr. on Pharmacol., p. 495. Basel, July 1969.

SUN, S.-C., SOHAL, R. S., COLCOLOUGH, H. L., BURCH, G. E.: Histochemical and electronmicroscopic studies of the effects of reserpine on the heart muscle of mice. J. Pharmacol. exp. Ther. **161**, 210—221 (1968).

TAUGNER, G., HASSELBACH, W.: Über den Mechanismus der Catecholamin-Speicherung in den chromaffinen Granula des Nebennierenmarks. Naunyn-Schmiedebergs Arch. Pharmak. exp. Path. **255**, 266—286 (1966).

WEIL-MALHERBE, H., POSNER, H.: The effect of drugs on the release of epinephrine from adreno-medullary particles in vitro. J. Pharmacol. exp. Ther. **140**, 93—102 (1963).

WILCKEN, P. E. L., BRENDER, D., MACDONALD, G. J., SHOREY, C. D., HINTERBERGER, A.: Effect of reserpine on the structure of heart mitochondria and the relation to catecholamine depletion. Circulat. Res. **20/21** Suppl. III, 203—210 (1967).

YOUDIM, M. B. H., SANDLER, M.: Activation of monoamine oxidase by reserpine. Europ. J. Pharmacol. 4, 105—108 (1968).
ZAIMIS, E.: Reserpine-induced circulatory failure. Nature (Lond.) 192, 521—523 (1961).

Prof. Dr. D. PALM
Pharmakologisches Institut der Universität
D-6000 Frankfurt a. M.
Ludwig Rehn-Straße 14

Discussion

STJÄRNE: I was very interested in Dr. PALM's talk because I feel that there has been a tendency to be rather dogmatic about the action of drugs. For instance, the action of reserpine has been interpreted frequently as being exclusively related to inhibition of uptake of amines into the granules. This led us to try to look for the affinity of reserpine for different intracellular organelles. This is what we found in the adrenal medulla after incubation with labelled and cold reserpine and centrifugation of the adrenal medullary large granule fraction on a sucrose density gradient. On the top panel of this slide you see the distribution of reserpine in the gradient. On the bottom panel you can see that we get a very adequate separation of mitochondria, marked by cytochrome oxidase, from the catecholamine granules, marked by fluorimetric amine determination. We also have the protein distribution here. There can be no doubt that there is a tremendously much higher affinity of reserpine for mitochondria than for the amine granules. Of course, it is interesting to speculate on the implications of this, since the amine granules need ATP for the uptake of catecholamines, and this ATP has somewhere to be generated, and you would rather assume that the mitochondria are the donors of ATP. I would not dare to pretend that the exclusive action of reserpine should now instead be dogmatically stated to be the one on the mitochondria. But I would suggest that at least during the recovery phase after reserpine induced amine depletion, damage to the mitochondria may very well explain the tremendous lag before you have complete restoration of amine storage function.

PHILIPPU: You have mentioned that you also examined the effect of α-blockers. Did you find any relationship between the surface tension activity of these compounds and their ability to release catecholamines?

PALM: We did not make any experiments; I mentioned only experiments done by Professor VON EULER's group and especially with β-blocking agents. I have some figures also for α-blocking drugs; if one compares, for instance, the surface activity of γ-adrenergic blocking drugs with their potency to inhibit spontaneous release, as well as the active uptake of noradrenaline, there is a very close relationship and, as you mentioned, there is no specific difference, for instance, for D- and L-isomers because the latter have nearly the same surface tension. This holds true not only for β-blockers; if one compares the surface tension lowering potency of reserpine, it is the most potent drug and the most potent drug which also causes haemolysis. These are data from SEEMAN. I have calculated in this slide the relative potency to lower surface tension from the β-blockers taken from data of LEVY and the data on amine release and uptake are from Professor VON EULER's group. If surface activity decreases also inhibition of uptake and release is diminished, i.e. drugs with the lowest surface activity are nearly without influence on the granules.

SCHÜMANN: You mentioned just in the discussion that reserpine is the most active drug with respect to lowering the surface tension. What is the relative potency of prenylamine in comparison to that of reserpine?

PALM: I am sorry, we cannot directly compare reserpine and prenylamine because we do not have the respective data. But if one compares the surface activity, for instance, in the concentration of 10^{-6} M of reserpine it is about 10 dyn/cm, with propanenol the most active β-sympatholytic drug, one needs to lower surface tension by —10 dyn/cm more than 20×10^{-3} Mol per litre. Therefore, the surface tension lowering is well correlated with all these effects.

SCHÜMANN: It might be interesting in this connection that according to our experiments on isolated sympathetic nerve granules, the ATP and Mg^{++} stimulated uptake mechanism for noradrenaline can be inhibited more effectively by reserpine than by prenylamine.

CARLSSON: Professor PALM said that the activity of a drug like reserpine is mainly based on physical properties. My first question is—how do you explain then the tremendous difference in activity between reserpine and iso-reserpine? I think iso-reserpine is about 100 times less active. My second question is—how did you manage to make your rats survive on a dose of 1 mg/kg per day; how was their body weight, water and food intake and body temperature?

PALM: May I answer the first question: I am not aware of the chemical difference between iso-reserpine and reserpine. There is, for instance, as far as I can remember one example given by LEVY in which he compared the lipid solubility of D- and L-isomers of β-blockers and found no difference. However, there is a difference between the ($\pm$) racemate and the (+)-isomer of about 2 to 3 in lowering the surface tension. Therefore only small steric changes may alter the surface activity.

CARLSSON: Would it not be interesting to study the surface activity of iso-reserpine, that would strengthen your case considerably.

PALM: I think so.

CARLSSON: The other question about the rats.

PALM: Reserpine was administered orally. It is true that loss in weight occurred and body temperature decreased. However, this did not happen in the case of prenylamine and not in the case of phenothiazine. I know, this is a factor which has to be taken into account.

DE ROBERTIS: The action of reserpine is very long lasting. After only one injection it takes about one week for the nerve ending to recover. By that time reserpine is probably very low. How can you explain this effect?

Palm: I did not get your argument?

de Robertis: My argument is that the effect of reserpine is a very long lasting one, so that after a single injection of reserpine, in order to have the nerve endings back to normal, you have to wait about one week. In the meantime nerve endings show no alterations, only the lack of granulated vesicles. The effects observed by you I think can be explained by the very high doses used in which surface tension plays a very important role. I do not think this action could explain the more specific effect on the binding of catecholamines in the granule.

Palm: Professor de Robertis, the presence of reserpine in the tissues lasts for days too, therefore one cannot say, as Dr. Brodie did, "it's a hit and run drug", as recently mentioned, for instance, by Shore. Shore found after days still radioactive reserpine in the organs. Furthermore, one must take into account that if one incubates isolated particles (erythrocytes) with a concentration of 10^{-5} M of chlorpromazine, one finds an enrichment of the drug, which I think is about three or four orders higher (Seeman) than in the outer medium and therefore one can well explain these effects by enrichment of the drug inside the membrane.

Muscholl: I wonder whether you can find morphological effects on mitochondria with smaller doses of reserpine. It has been shown that doses as low as 0.1 mg/kg or 0.05 mg/kg, given for about one week in rabbits, rats or dogs lead to an almost 100% depletion of the transmitter. Have you correlated this with morphological effects on mitochondria?

Palm: I must say that we have worked with high doses in rats. There were some results published in 1967 by Australian authors, they gave 25 μg/kg reserpine in dogs and found nearly the same as we found with high doses in rats.

Carlsson: I should also like to comment a little bit on the problem of ATP deficiency in the cell. I think that this is an interesting point which was brought up by Dr. Stjärne. We have, in fact, evidence that the ATP production of the adrenergic neurons is still going on after reserpine treatment. Thus, the mitochondria are functioning, because we know that the membrane pump is functioning very well and this is energy dependent. Apparently the mitochondria can fulfil that part of the ATP production and thus it seems likely that it can also provide the granules with some ATP.

Kirshner: We have studied glucose utilization by the adrenal medulla using radioactive glucose and found very little of the glucoses metabolized by the Krebs-cycle, most of it goes by glycolyses to lactate. We have also studied the energy requirements for secretion from the adrenal medulla and find that if you block either glycolysis or oxidate phosphorylation, it has no effect on the response to acetylcholine, but if you block both of them by giving antimycin or oligomycin, as well as iodioacetic acid, then you get no response. Neither glycolysis alone nor oxidate phosphorylation alone is sufficient enough to sustain the energy requirements for secretion.

Corrodi: May I add something about this property of drugs to stick unspecifically to membranes seen in metabolic and distribution studies with many experimental drugs. We have learned that it is quite a common feature of amines of a certain molecular weight over quite a wide range. They disappear from the blood very quickly and they are stored in the organs and every tissue unspecifically. In the β-receptor field, substances of similar structure have the same distribution pattern whether they are β-blockers or not. Phenothiazines, indoles or tricyclic antidepressants of different structures have exactly the same behaviour, so this might be an unspecific feature of all medium molecular weight amines.

PHILIPPU: Dr. STJÄRNE, your suggestion that reserpine may damage the mitochondria so that no more ATP is available for the active transport of the amines into the granules is very interesting indeed. But as you know, very small amounts of reserpine inhibit also in vitro the ATP-magnesium-dependent uptake of the amines by the granules, although no mitochondria are present.

STJÄRNE: Well, just a short comment—I think it was interesting to learn from Dr. DAHLSTRÖM that the rate of axonal transport of mitochondria is considerably less than that of the amine granules—if I understood her correctly. I wonder if the very great difference between the time required during the recovery phase after reserpine treatment, to regain normal amine uptake capacity and also normal neuro-effector transmission, and that required to regain normal amine storage capacity, could, to some extent, be related to differences in the rate of axonal transport of granules and mitochondria. Reserpine damages mitochondria as well as amine granules. It would take a longer time to replace the old mitochondria with new ones so that enough ATP can be generated to build up the amine stores, than to replace the damaged granules. This is just speculation.

Bayer-Symposium II, 202—209 (1970)

Effect of Angiotensin on Noradrenaline Release of the Isolated Rabbit Heart

Hans-Joachim Schümann

With 8 Figures

A review of the literature about the mechanism of action of angiotensin on the heart shows conflicting results. Some authors (Fowler and Holmes; Koch-Weser) conclude that angiotensin acts exclusively by a direct action, whereas

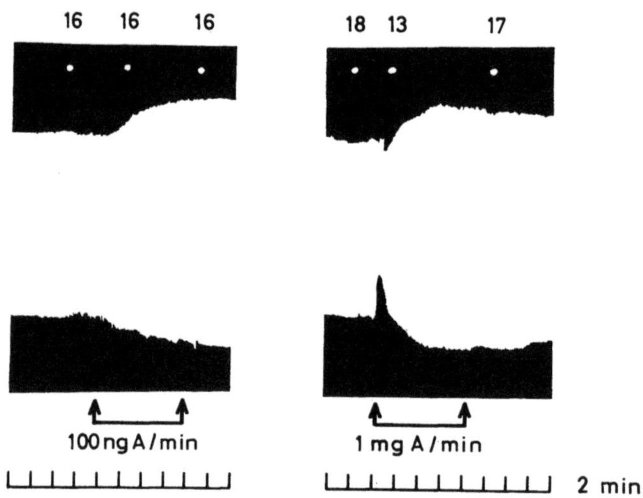

Fig. 1. Effect of angiotensin (A) infusion on conctractility and frequency/10 sec of the isolated rabbit heart

others as for instance Krasney et al. (1965, 1966) postulate that it acts in addition by an indirect adrenergic mechanism. Since the argumentation is mostly based on experiments with cardiac denervation and antagonists without determination of the adrenergic transmitter, it was our aim to contribute to the elucidation of the mechanism of action of angiotensin by measuring the noradrenaline output of the isolated perfused heart at different experimental conditions.

First of all I should like to mention that the following experiments have been performed together with my colleagues Dr. Starke and Dr. Werner.

In a first series of experiments we analysed the effects of angiotensin on the isolated with constant pressure perfused rabbit heart with sympathetic nerve supply. The method used is described by Hucović and Muscholl. — Fig. 1 shows the effects of an angiotensin infusion on contractility and frequency on the spontaneously beating heart. 100 ng angiotensin/min caused a positive inotropic effect but did not influence the heart rate. The high concentration of 1 mg angiotensin/min exerted a biphasic inotropic effect: a transient inhibition was followed by an increase of contractility. The rate of beat decreased and than turned to normal, but did not increase. — The positive inotropic effect of angiotensin in concentrations from 10 ng up to 1 mg/min was not accompanied by the appearence of increased amounts of noradrenaline in the venous effluent.

This is illustrated in Fig. 2. Infusion of 1 mg angiotensin/min did not release any noradrenaline during four collection periods of 2 min, although the positive inotropic effect was of the same order of magnitude as that of tyramine which released about 40 ng of noradrenaline per collection period.

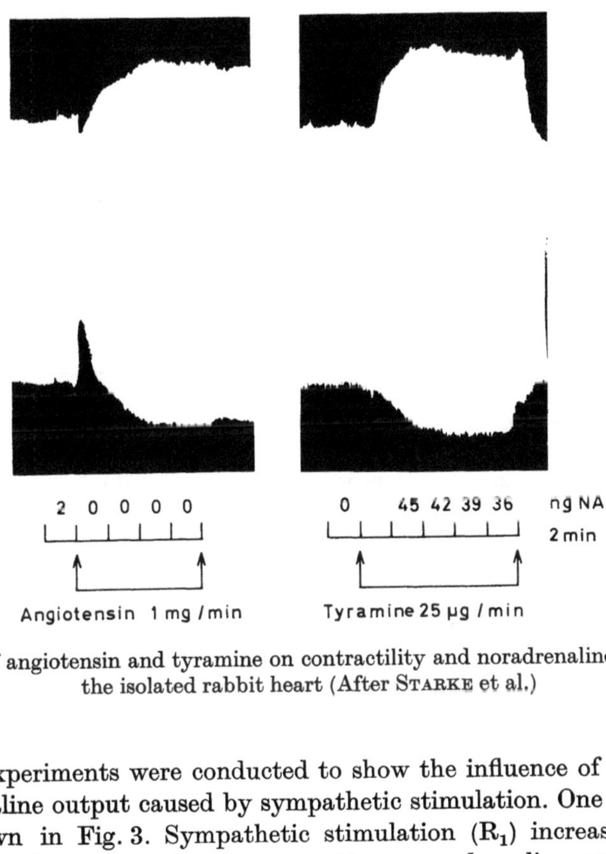

Fig. 2. Effect of angiotensin and tyramine on contractility and noradrenaline (NA) output on the isolated rabbit heart (After Starke et al.)

Further experiments were conducted to show the influence of angiotensin on the noradrenaline output caused by sympathetic stimulation. One typical experiment is shown in Fig. 3. Sympathetic stimulation (R_1) increased heart rate, contractility and noradrenaline output. 22 ng noradrenaline appeared in the effluent of the 2 min collection period. Three successive sympathetic stimulations during a continous infusion of 100 ng angiotensin caused an increased output of noradrenaline of 38, 34, and 30 ng noradrenaline, respectively, whereas angiotensin had no influence on the heart rate and was ineffective to release noradrenaline on the spontaneously beating heart as mentioned before.

A significant increase of noradrenaline output during stimulation could already be obtained after 32 ng angiotensin (Fig. 4). It amounted to about 50% and lasted during three consecutive stimulation periods.

The control experiments without angiotensin infusion showed, in contrast, a decline of the noradrenaline output from stimulation to stimulation by about 8%. — Since all our calculations are related to the effect of the stimulations

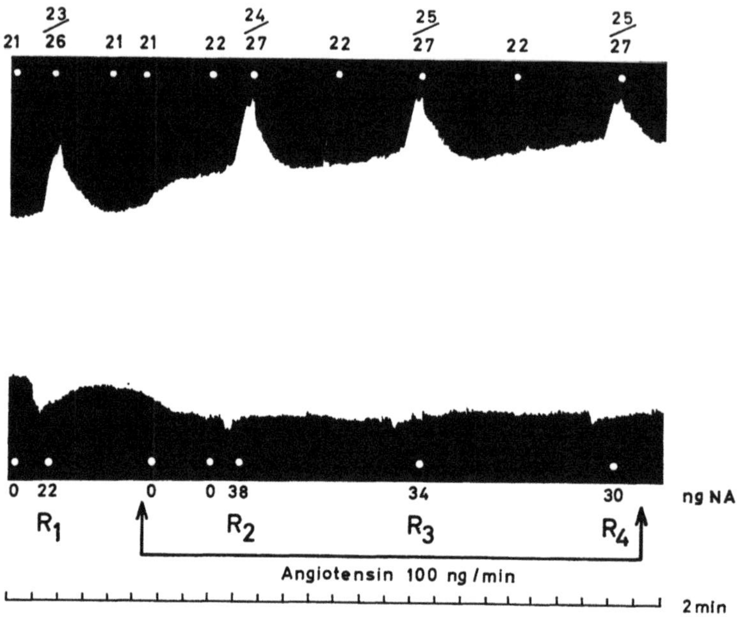

Fig. 3. Influence of angiotensin on the effects of sympathetic stimulation (R). Heart rate/10 sec as counted 20 to 30 and 35 to 45 sec after the onset of stimulation (5 imp/sec, 3 msec, 8 mA alternatively 15 sec left and right sympathetic nerve for 1 min) noradrenaline (NA) output per 2 min collection period. (After Starke et al.)

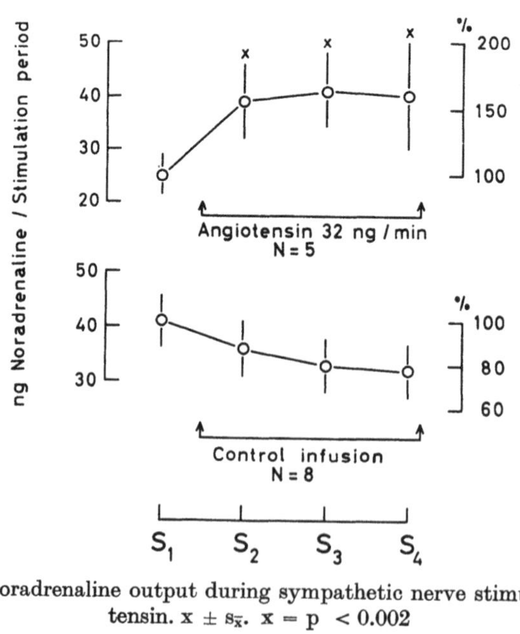

Fig. 4. Increase of noradrenaline output during sympathetic nerve stimulation (S) by angiotensin. $x \pm s_{\bar{x}}$. $x = p < 0.002$

preceding those during angiotensin infusion the real increases of noradrenaline output should be even greater.

It is a well known fact that angiotensin decreases coronary flow. We were therefore interested to see whether there exists a relationship between changes of coronary flow and noradrenaline output during sympathetic stimulation (Fig. 5). Flow and output of the preceding control stimulation was set at 100%. — It can be seen that the infusion of increasing amounts of angiotensin from 10 to 320 ng/min caused an increased and dose-dependent noradrenaline output and furthermore a diminution of the coronary flow. This reduction of flow, however, reached already a maximum after 32 ng angiotensin. It amounted to about 22%.

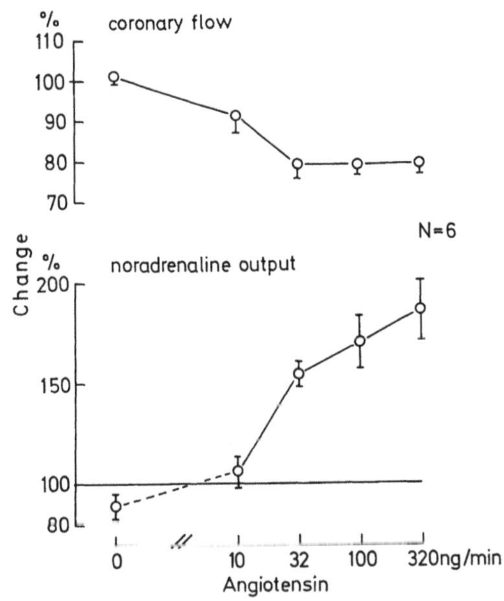

Fig. 5. Changes of coronary flow and noradrenaline output during sympathetic nerve stimulation by angiotensin. $x \pm s_{\bar{x}}$. N = 6. Flow and output of preceding control stimulation = 100%

In order to examine the importance of the flow reduction more in detail we perfused hearts with *constant flow* by means of a roller pump. This procedure allowed us to reduce mechanically the perfusion volume and thereby the coronary flow. The results are given in Fig. 6. The noradrenaline output induced by sympathetic nerve stimulation decreased proportionally to the degree of flow reduction. A flow reduction of 75% for instance decreased the noradrenaline output by more than 60%. After readjustment of the perfusion volume to the original value, also the amine output returned to the control level.

The described relationship between the size of perfusion volume and the height of noradrenaline output could also be demonstrated in experiments with the vasoconstrictor hormone vasopressin. Increasing amounts of arg^8– vasopressin from 1 to 25 mU/min decreased proportionally to the degree of flow reduction the output of noradrenaline. The conformity of the reactions elicited by vasopressin and mechanical reduction of flow imply that vasopressin very likely interferes with cardiac sympathetic nerves only by causing vasoconstriction.

Fig. 7 shows a comparison between the effects of vasopressin and angiotensin on the isolated heart perfused with constant pressure. In a: infusion of 100 ng angiotensin caused about the same reduction of flow as did 5 mU of vasopressin. However, these—with respect to the reduction of flow—equieffective doses of angiotensin and vasopressin had opposite effects on the noradrenaline output: vasopressin decreased, whereas angiotensin increased the output of noradrenaline. This observation is in contrast to the results of Hertting and Schiefthaler obtained with angiotensin and vasopressin on the isolated perfused cat spleen. In this organ both peptides reduced noradrenaline output.

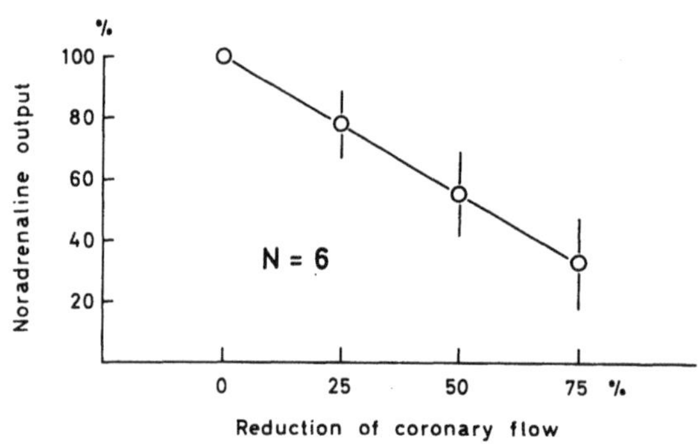

Fig. 6

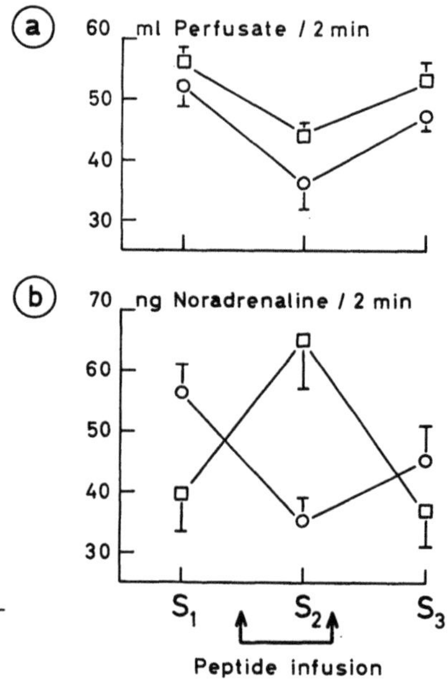

Fig. 7

Fig. 6. Influence of mechanical flow changes on noradrenaline output during sympathetic stimulation. $x \pm s_{\bar{x}}$. N = 6

Fig. 7. Influence of angiotensin and vasopressin on perfusate volume and noradrenaline output during nerve stimulation (S). $x \pm s_{\bar{x}}$, N = 6. □ = angiotensin 100 ng/min. o = Arg^8-vasopressin 5 mU/min

Our results are in agreement with the concept that in general a decrease of the perfusion rate lowers the output of noradrenaline. This will be possibly achieved by increasing the efficiency of the reuptake mechanism as already discussed by Rosell et al. as well as by Paton and Gillis.

From the presented data we concluded that the influence of angiotensin on the output of noradrenaline is the result of at least two opposite effects: first, a diminution of output induced by vasoconstriction and second, an increase of output by an interaction with the function of the adrenergic nerve terminals.

If this conclusion is right, it should be expected that isolated hearts perfused with constant flow would react more sensitive towards angiotensin than those perfused with constant pressure.

Fig. 8 shows the result of such a comparative study. Under the conditions of constant flow perfusion such small amounts as 3.2 ng angiotensin/min or 128 pg/ml were capable to increase significantly the noradrenaline output induced by sympathetic stimulation, whereas under perfusion with constant pressure more than 3 times higher doses were required to obtain similar effects.

The high sensitivity towards angiotensin should be kept in mind, if the physiological significance of our observations will be discussed. In this connexion it is noteworthy that for example the blood of healthy human beings containes 95 pg (Massani et al.) and that of normal dogs 400 pg angiotensin/ml (Finkielman et al.). Since we found 128 pg/ml as the smallest angiotensin concentration causing significant effects, I should like to state that this concentration is in the physiological range. Furthermore, in favour of a physiological importance is the observation that the potentiation of noradrenaline output can be achieved several times during the infusion of angiotensin.

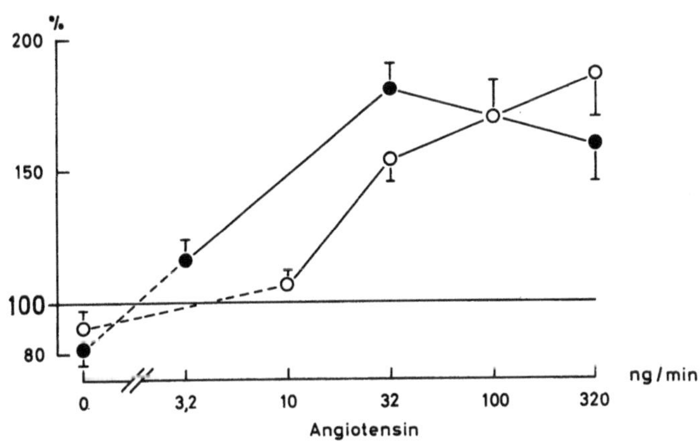

Fig. 8. Effect of angiotensin on noradrenaline output during nerve stimulation. Perfusion with constant pressure (o) or constant flow (●). $x \pm s_{\bar{x}}$, N = 3—9. Output of preceding control stimulation = 100%. (After Starke et al.)

After all, the question arose whether or not the increase of noradrenaline output induced by angiotensin was accompanied by an increased response of the pacemaker or of the myocardium to sympathetic stimulation. Although heart rate and height of contraction were measured in all experiments with constant pressure and constant flow perfusion, we obtained never a significant increase of frequency. Only on hearts perfused at constant flow during infusion of 32 ng/min of angiotensin the positive inotropic effect of sympathetic stimulation was significantly augmented. Obviously, the function of the increased amine output can only be observed under special circumstances, that is after angiotensin concentrations with small direct effects and optimal conditions for noradrenaline release.

The mechanism underlying the angiotensin induced potentiation of noradrenaline output is not yet clear. Inhibition of uptake or reuptake may be responsible as well as potentiation of noradrenaline liberation.

Summary

In the isolated heart perfused with constant pressure, angiotensin in low concentrations—32 to 100 ng/min—caused considerable positive inotropic effects without changing significantly the heart rate. High concentrations decreased heart rate and the increase in the height of contraction was preceded by a transient phase of inhibition. The positive inotropic effect of angiotensin was not accompanied by the appearance of increased amounts of noradrenaline in the perfusate.

Angiotensin caused a dose-dependent increase of the noradrenaline output induced by sympathetic nerve stimulation, while heart rate and contractility were not affected.

Angiotensin diminished coronary flow even after infusion of low concentrations. — In experiments with variation of the perfusion volume the degree of flow reduction and noradrenaline output induced by nerve stimulation decreased proportionally. — Vasopressin had the same effect as the mechanical reduction of flow. Angiotensin, on the contrary, in spite of decreasing the flow volume increased the noradrenaline output.

According to these results the vasoconstrictor effect of angiotensin should counteract its own noradrenaline releasing effect. The peptide was therefore most potent in experiments with constant flow perfusion, whereby its vasoconstrictor effect was eliminated. The threshold concentration of angiotensin was 128 pg/ml, it increased significantly the noradrenaline output by 38%. More than three times higher concentrations were needed to obtain similar results in hearts perfused with constant pressure.

These results favour the assumption that angiotensin is able to modulate the function of peripheral sympathetic nerves.

Acknowledgement

The author is grateful to Miss B. Rawe for skilful technical assistance.

References

Finkielman, S., Worcel, M., Massani, Z. M., Nahmod, V. E., Paladini, A. C., Agrest, A.: Angiotensin blood levels in hypovolemic shock during osmotic diuresis. Amer. J. Physiol. **215**, 308—313 (1968).

Fowler, N. O., Holmes, J. C.: Coronary and myocardial action of angiotensin. Circulat. Res. **14**, 191—201 (1964).

Hertting, G., Schiefthaler, Th.: Beziehung zwischen Durchflußgröße und Noradrenalinfreisetzung bei Nervenreizung der isoliert durchströmten Katzenmilz. Naunyn-Schmiedebergs Arch. exp. Path. Pharmak. **246**, 13—14 (1963).

Hucović, S., Muscholl, E.: Die Noradrenalin-Abgabe aus dem isolierten Kaninchenherzen bei sympathischer Nervenreizung und ihre pharmakologische Beeinflussung. Naunyn-Schmiedebergs Arch. exp. Path. Pharmak. **244**, 81—96 (1962).

Koch-Weser, J.: Nature of the inotropic action of angiotensin on ventricular myocardium. Circulat. Res. **16**, 230—237 (1965).

Krasney, J. A., Paudler, F. T., Hogan, P. M., Lowe, R. F., Youmans, W. B.: Peripheral adrenergic basis for cardioaccelerator action of angiotensin. Amer. J. Physiol. **211**, 1447 to 1450 (1966).

— —, Smith, D. C., Davis, L. D., Youmans, W. B.: Mechanisms of cardioaccelerator action of angiotensin. Amer. J. Physiol. **209**, 539—544 (1965).

Massani, Z. M., Finkielman, S., Worcel, M., Agrest, A., Paladini, A. C.: Angiotensin blood levels in hypertensive and non-hypertensive diseases. Clin. Sci. **30**, 473—483 (1966).

Paton, D. M., Gillis, C. N.: Effect of altered perfusion rates on the retention of noradrenaline by the spleen. Nature (Lond.) **208**, 391—392 (1965).
Rosell, S., Kopin, I. J., Axelrod, J.: Fate of H^3-norepinephrine in skeletal muscle before and following sympathetic stimulation. Amer. J. Physiol. **205**, 317—321 (1963).
Starke, K., Werner, U., Schümann, H. J.: Wirkung von Angiotensin auf Funktion und Noradrenalinabgabe isolierter Kaninchenherzen in Ruhe und bei Sympathicusreizung. Naunyn-Schmiedebergs Arch. Pharmak. **265**, 170—186 (1969).

Prof. Dr. H. J. Schümann
Pharmakologisches Institut
Klinikum Essen
der Ruhr-Universität
D-4300 Essen
Hufelandstraße 55

Discussion

Muscholl: I wonder whether Dr. Thoenen would like to comment on this result because, as Professor Schümann said, it is contrary to the findings of both groups of workers who used the isolated perfused spleen. We are fortunate in having Dr. Thoenen with us.

Thoenen: At the time of our studies on the effect of angiotensin on the norepinephrine output in the isolated perfused spleen, a sufficiently sensitive fluorimetric method for the determination of norepinephrine was not available in our laboratory and we determined the norepinephrine output by bio-assay. On account of the interference with the bio-assay we could use only very low concentrations of angiotensin (1 ng/ml). This concentration of angiotensin was just sufficient to slightly increase the perfusion pressure and to potentiate the effect of sympathetic nerve stimulation on the vascular resistance and to a lesser degree on volume changes. Neither a significant increase in the norepinephrine output, resulting from sympathetic stimulation, nor a change in the uptake of norepinephrine, infused into the arterial inflow, could be detected with this very low concentration of angiotensin. We assumed that the increased response to sympathetic nerve stimulation in the presence of this low concentration of angiotensin was due to an additive effect of angiotensin and norepinephrine on the receptive sites. Our results admit no conclusion as to whether higher concentrations of angiotensin would also produce an increase norepinephrine output in this preparation.

von Euler: I have two comments, one is that the angiotensin does not seem to effect the transmitter re-uptake and release in the isolated nerve granules, so I think this possibility could be excluded as far as one can see now. The second one is that I would like to mention the results by Sjöstrand who noticed, again on the isolated vas deferens, an enhancing effect of the response to nerve stimulation by small doses of angiotensin which might possibly be of the same kind as those mentioned here.

Schümann: I might just add that there exists a paper of Peach Bumpus and Khairallah which came out this summer in which the authors were able to show that very small amounts—in the range of 1 ng/ml—of angiotensin were able to inhibit the noradrenaline re-uptake and so perhaps this inhibition of re-uptake might explain our results. We have tried to repeat these experiments but not with absolutely the same procedure and were not able to confirm their results. We have not yet done experiments with radioactive norepinephrine, so we cannot say exactly what happens. — I personally believe that the action of angiotensin takes place not at the level of the granules because we also found, in accordance with Professor von Euler's observations, that angiotensin is by no means able to influence the uptake or release of noradrenaline from either the sympathetic nerve granules or the granules of the adrenal medulla.

Holtz: I am a little confused by the discrepancy between the higher output, perhaps produced by inhibition of the re-uptake of noradrenaline and the lack of pharmacological action to be expected. If I have understood you correctly, the positive chronotropic action of sympathetic nerve stimulation is not influenced by infusion of angiotensin, although the angiotensin enhances the output by inhibiting the re-uptake. Therefore, I would expect enhancement of the positive chronotropic action of sympathetic nerve stimulation. Why not?

Schümann: I cannot explain the discrepancy at the moment. We tried hard to find an increased heart rate because we thought as you did that an enhancement of noradrenaline output should also be related to an enhanced rate of beat. We got only after stimulation of the sympathetic nerves, an increased noradrenaline output by angiotensin and under these conditions we had already an increased heart rate. Since we had, perhaps, already a nearly maximal stimulation of the heart rate, it might be very difficult to measure the very small differences in frequency.

Holtz: What is the "own" action not caused by liberation of noradrenaline or inhibition of re-uptake—the angiotensin's own action on the frequency?

Schümann: Angiotensin in concentrations of up to 100 ng/min had no significant influence on heart rate, 320 ng/min and higher concentrations exerted a negative chronotropic effect.

Holtz: Opposite effect on the frequency of the heart than noradrenaline.

Carlsson: Well, we do not know the noradrenaline output from the sinus node, do we?

Trendelenburg: It is puzzling that the increase in output was not accompanied by a clear increase in rate of contraction. Would it be possible to calculate how much the rate should have gone up? Could one stimulate the sympathetic nerves at different rates (and thereby vary the amount of transmitter liberated) and then calculate the rate of beat which should have been observed under the influence of angiotensin? Should the rate have gone up by 3 or 10 or 15 beats/min?

Schümann: We have no experiments of that kind.

Muscholl: I would say that I could assist Professor Schümann with the following reasoning—probably the amount of noradrenaline liberated nerve stimulation was already maximal for increasing the heart rate. You can hardly measure the noradrenaline output if you apply a very low rate of stimulation. In order to do these experiments, one has to apply higher rates and one has just a surplus of noradrenaline liberated without having any effect. If the response is maximal you cannot increase it further. Whereas on the ventriculum myocardium, where you can measure, with certain limitations, the effect on contraction amplitude, you may have still not reached the maximum and get some effect. That is our experience with this kind of preparation and with drugs other than angiotensin.

SCHÜMANN: We have stimulated the heart with submaximal stimulation so that we could expect a further increase in contractility by angiotensin and we were able to show that the positive inotropic effect of sympathetic stimulation was significantly augmented by 32 ng/min of angiotensin, as already mentioned.

Bayer-Symposium II, 212—219 (1970)

Neuronal and Hormonal Control of Tyrosine Hydroxylase and Phenylethanolamine N-Methyltransferase Activity

JULIUS AXELROD, ROBERT A. MUELLER, and HANS THOENEN

With 4 Figures

There are a variety of regulatory factors in the control of the biosynthesis of the neurotransmitter noradrenaline and the adrenal medullary hormone adrenaline. Increased sympaticoadrenal activity causes an immediate increase in the formation of noradrenaline from tyrosine in the sympathetic nerve terminals [ALOUSI and WEINER, 1966; SEDVALL and KOPIN, 1967 (1)] and in the adrenal medulla (GORDON et al., 1966). This rapid change in catecholamine synthesis is due to an end-product inhibition of tyrosine hydroxylase (SPECTOR et al., 1967) and/or an increase of cofactor or substrate concentration [SEDVALL and KOPIN, 1967 (2)]. In spite of the more rapid conversion of tyrosine to noradrenaline after increased activity of the sympathetic nervous system, no elevation in the *in vitro* activity of tyrosine hydroxylase activity has been observed (THOA and KOPIN, personal communication). Recent work in our laboratory has shown that the synthesis of tyrosine hydroxylase, the rate limiting step in the catecholamine biosynthesis (LEVITT et al., 1965) and phenylethanolamine N-methyltransferase (PNMT), the final step in adrenaline formation, is regulated by the activity of the sympathetic nervous system and by pituitary and corticoid hormones.

Neuronal Control of Tyrosine Hydroxylase Induction

In a study on the cellular localization of tyrosine hydroxylase it was noted that the administration of 6-hydroxydopamine, a compound that destroys sympathetic nerve terminals (THOENEN and TRANZER, 1968) also caused the complete disappearance of the enzyme in rat hearts [MUELLER et al., 1969 (1)]. At the same time an elevation in tyrosine hydroxylase and PNMT activity was found in this adrenal gland. The fall of tyrosine hydroxylase in the heart indicated that this enzyme is localized entirely in the sympathetic nerves. 6-Hydroxydopamine causes a reduction in blood pressure and presumably results in a reflex stimulation of sympathetic nervous system and an increased activity of splanchnic nerves of the adrenal. Thus the increase in adrenal tyrosine hydroxylase might be the consequence of prolonged sympathetic stimulation of the adrenal medulla. To examine this possibility the effect of reserpine and phenoxybenzamine on adrenal tyrosine hydroxylase activity was examined [THOENEN et al., 1969 (1)]. These drugs lower blood pressure by different mechanisms and produce longlasting reflex increase in nerve impulses to the adrenal gland (IGGO and VOGT, 1960; DONTAS and NICKERSON, 1957). Like 6-hydroxydopamine, reserpine and phenoxybenzamine caused an increase of tyrosine hydroxylase activity in the rat adrenal gland [THOENEN et al.

1969 (1)] in 48 h (Table 1). After the repeated administration of reserpine there was a large (fourfold) increase in enzyme activity after 5 days [Mueller et al., 1969 (2)] (Fig. 1).

Since these drugs lower blood pressure and cause a reflex increase in sympathetic nerve firing, it appeared that increased tyrosine hydroxylase activity might

Table 1. *Effect of drugs that stimulate sympathetic nerve activity on tyrosine hydroxylase activity in the rat adrenal*

Drug	Control	Drug Treated
Reserpine	4.1 ± 0.50	13.3 ± 1.1
Phenoxybenzamine	4.3 ± 0.48	10.5 ± 1.2
6-Hydroxydopamine	3.5 ± 0.36	5.8 ± 0.62

Rats six per group were given 5 mg/kg reserpine (s.c.) 20 mg/kg phenoxybenzamine (IV) or 200 mg/kg 6-hydroxydopamine (IV) 48 and 24 h before they were killed [Thoenen et al., 1969 (1)]. Results are expressed as mμmoles product (^3H_2O) formed enzymatically from ^3H tyrosine per adrenal/per hour mean ± S.E.

be related to a prolonged increase in sympathetic nerve activity. In the following experiments the effect of interrupting nerve impulses to the rat adrenal gland on drug-induced elevation of adrenal tyrosine hydroxylase was examined [Thoenen

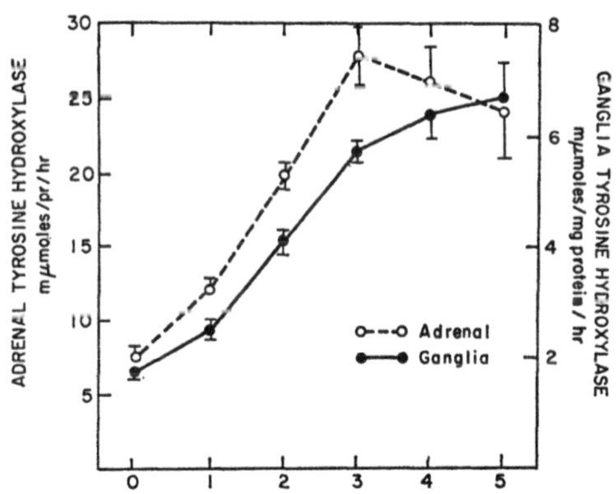

Fig. 1. Increase in tyrosine hydroxylase activity in the rat adrenal and superior cervical ganglia after reserpine. Groups of six rats received 5 mg/kg reserpine for four days s.c. Results are expressed as mean ± S.E. [Mueller, Thoenen and Axelrod, 1969 (2)]

et al., 1969 (1)]. Nerve fibers from the main trunk to the left adrenal were cut under ether anesthesia. Three days later groups of rats were given reserpine, phenoxybenzamine or 6-hydroxydopamine 24 and 48 h before rats were killed. In rats which drug was not given, there was no difference in tyrosine hydroxylase activity on the innervated and denervated side. The drugs caused a two to threefold elevation in tyrosine hydroxylase activity on the innervated adrenal gland but there was no rise in enzyme activity in the denervated adrenal. Effect of denervation on reserpine-induced adrenal tyrosine hydroxylase activity is shown in Fig. 2. These results clearly demonstrated that the elevation in adrenal tyrosine hydroxylase activity is mediated by a prolonged increase in sympathetic nerve activity. It is also possible that there might be hormonal influences in the elevation of adrenal tyrosine hydroxylase activity. To examine such a possibility, 6-hydroxydopamine or reserpine were given to hypophysectomized rats. These drugs caused a rise in adrenal tyrosine hydroxylase in hypophysectomized rats, indicating that the pituitary hormones were not involved in the elevation of tyrosine hydroxylase.

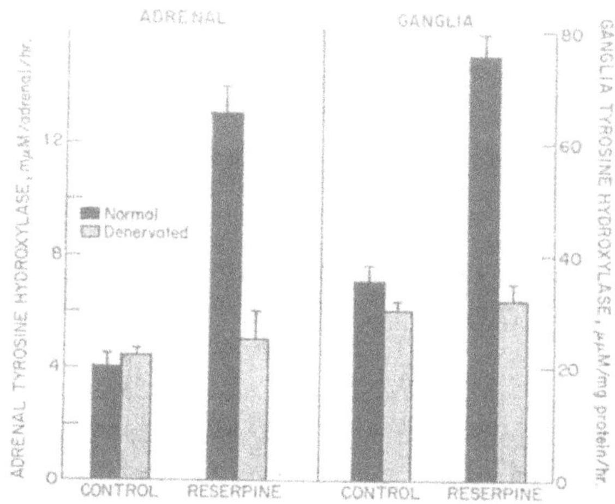

Fig. 2. Trans-synaptic induction of tyrosine hydroxylase in rat adrenal glanf and superior cervical ganglia. Splanchnic fibers leading from the main trunk to the left adrenal or preganglionic fibers leading to the left superior cervical ganglia were transected under ether anesthesia 3 days before reserpine administration. Animals were given 5 mg/kg reserpine s.c. 48 and 24 h before they were killed. Bars represent the mean ± S.E. of 5 or 6 observations

The effect of reserpine on tyrosine hydroxylase in sympathetic ganglia (superior cervical, stellate, lumbar) were also examined [Thoenen et al., 1969 (2)]. Reserpine caused a marked elevation in the superior cervical ganglia (Fig. 1). Similar elevations were found in stellate and lumbar ganglia. The increase in tyrosine hydroxylase activity in the sympathetic ganglia could be the consequence of increased nerve impulses, blood-bourne factors, or a direct effect of the drug. To examine these possibilities, the right superior cervical ganglion was decentralized by transection of the preganglionic trunk 2 days before the reserpine treatment. Decentralization had no effect on untreated rat ganglia but prevented the rise of tyrosine hydroxylase induced by reserpine (Fig. 2), again demonstrating that the rise in the enzyme was caused by nerve activity.

Reserpine also increased tyrosine hydroxylase in the adrenal gland of all mammalian species examined and the rabbit brain stem [Mueller et al., 1969 (2)]. Hypertension caused by sinoaortic denervation (de Quattro et al., 1969), DOCA and salt (Mueller, Thoenen and Axelrod, unpublished observations), and tyrosine and salt (Mueller, Willard and Axelrod, unpublished observations), high doses of isoproterenol, increased psychosocial stimulation (AXELROD, MUELLER and HENRY, unpublished observations), high doses of metapyrone [MUELLER et al., (1)] increase adrenal tyrosine hydroxylase activity. These increases are most likely due to greater firing of the sympathetic nerves to the adrenal gland, produced by these procedures.

The neuronally mediated increase in tyrosine hydroxylase activity could be caused by unmasking hidden enzyme sites or by an increase in enzyme protein. The increase in enzyme activity did not appear to be due to an activator or loss of an inhibitor, since the Km for tyrosine hydroxylase from reserpine treated rats were not different from those of control animals [MUELLER et al., 1969 (2)]. To determine whether the rise in tyrosine hydroxylase activity caused by reserpine could be due to increase in the amount of enzyme, inhibitors of protein and RNA synthesis were examined [MUELLER et al., 1969 (3)]. The administration of cycloheximide, a protein synthesis inhibitor, or actinomycin D, an inhibitor of RNA synthesis, prevented the increase in tyrosine hydroxylase caused by reserpine in the rat adrenal gland and superior cervical ganglia. All these observations indicated that increased nerve impulses results in a trans-synaptic induction of tyrosine hydroxylase. Preliminary studies showed that the induction of tyrosine hydroxylase was not only manifest in the adrenal gland and cell body, but also in the axon and nerve terminals (THOENEN, MUELLER and AXELROD, unpublished observations). Ganglionic blocking agents were also found to prevent the reserpine-induced elevation of tyrosine hydroxylase in the sympathetic ganglia [MUELLER et al. (2)]. Ganglionic blocking agents also caused an increase in adrenal tyrosine hydroxylase which was blocked by cutting the splanchnic nerve.

Hormonal Regulation of Tyrosine Hydroxylase Activity

In a previous study it was noted that tyrosine hydroxylase activity was reduced in the adrenal gland of hypophysectomized rats (WURTMAN and AXELROD, 1966). These results suggested that the pituitary and/or adrenal cortex hormones might be involved in the maintenance of tyrosine hydroxylase activity. Rats were hypophysectomized and the tyrosine hydroxylase activity was measured over a period of 2 weeks [MUELLER et al. (1)]. Adrenal hydroxylase activity was found to decline with a $t_{1/2}$ of about 11 days. The effect of splanchnic denervation on adrenal tyrosine hydroxylase was examined in rats with an intact pituitary. With this procedure adrenal tyrosine hydroxylase declined at a much slower rate $t_{1/2}$ of 21 days. Enzyme activity decreased at a $t_{1/2}$ of 8 days after hypophysectomy and splanchnic nerve transection. The regulation of turnover of noradrenaline in the rat heart requires both glucocorticoids and thyroid hormones (LANDSBERG and AXELROD, 1968). Thus, after hypophysectomy, rats were given either ACTH or thyroxine [MUELLER et al. (1)]. ACTH, but not thyroxine, prevented the decline in adrenal tyrosine hydroxylase activity (Fig. 3). The potent glucocorticoid dexamethasone, or large doses of corticosterone could not be substituted for ACTH although these compounds restored PNMT activity in the hypophysectomized rats (WURTMAN and AXELROD, 1966; WURTMAN, 1966). To examine whether neural integrity was necessary for ACTH to have its effect on tyrosine hydroxylase, the hormone was given to hypophysectomized rats with a splanchnic transection.

In denervated hypophysectomized rats, enzyme activity was lower than hypophysectomized rats with an intact adrenal. ACTH increased tyrosine hydroxylase activity in innervated and denervated adrenals to about the same extent [Mueller et al. (1)]. The administration of large doses of ACTH did not elevate adrenal tyrosine hydroxylase above normal levels. It was concluded from these experiments that ACTH has a direct effect in maintaining the adrenal medullary tyrosine hydroxylase, but, unlike nerve stimulation, it cannot elevate this enzyme above normal levels.

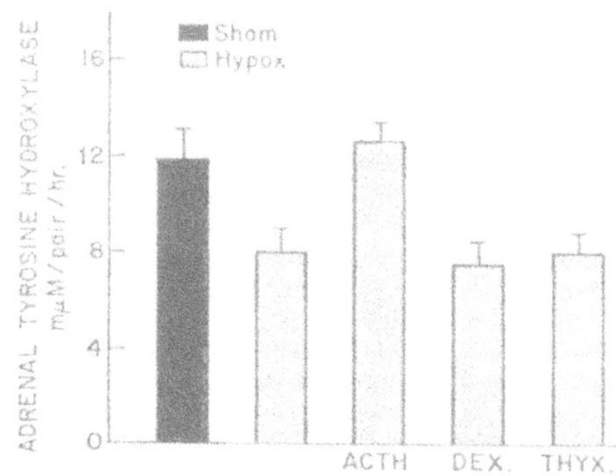

Fig. 3. Effect of hormones on tyrosine hydroxylase activity in rat adrenal after hypophysectomy. Tyroxine 10 mg/day, ACTH, 4 units/day s.c. or dexamethasone 1 mg/kg/day was administered 2 days after hypophysectomy for 7 days. Results are expressed as mean ± S.E.

Neuronally Dependent Elevation of Adrenal PNMT

The final step in the synthesis of adrenaline is the conversion of noradrenaline to adrenaline by PNMT. This enzyme is highly localized in the mammalian adrenal medulla (Axelrod, 1962). PNMT requires S-adenosylmethionine as the methyl donor (Kirschner and Goodall, 1957) and methylates β-hydroxylated phenylethylamine derivatives as well as noradrenaline (Axelrod, 1962).

As previously described, chemical sympathectomy by 6-hydroxydopamine causes a twofold increase in the *in vitro* activity of adrenal tyrosine hydroxylase after 2 days [Mueller et al., 1969 (1)]. In addition to this increase in adrenal tyrosine hydroxylase, a smaller (25%) but statistically significant increase in adrenal PNMT was also found after 6-hydroxydopamine. To examine whether this increase in PNMT activity was due to nerve impulse or hormones, rats were either hypophysectomized or their splanchnic nerves cut before the administration of 6-hydroxydopamine (Thoenen, Mueller and Axelrod, in press). Hypophysectomy caused a marked reduction of adrenal PNMT while 6-hydroxydopamine caused an elevation of PNMT in these rats. When the splanchnic nerves were cut, there was no increase in adrenal PNMT after 6-hydroxydopamine. These observations demonstrated that the increase in PNMT after chemical sympathectomy is due to nerve impulses and not to hormonal factors. The increase in adrenal sympathetic nerve activity could elevate PNMT by a direct effect on the adrenal medulla or by an increase synthesis and/or release of adrenal corticoid hormones to the medulla. This is unlikely since cells of the adrenal cortex have no autonomic innervation.

Hormonal Control of PNMT Activity

It has long been suspected that the adrenal cortex can influence the activity of the adrenal medulla. In species such as rats and man where the adrenal cortex completely surrounds the medulla, the main catecholamine is adrenaline (COUPLAND, 1953). In species such as the dogfish and snake, where the chromaffin tissue is not surrounded by adrenal cortex, the catecholamine is exclusively noradrenaline (COUPLAND, 1953; WURTMAN et al., 1967). These findings suggested that the cortex of the adrenal gland is involved in the methylation of noradrenaline. The development of a sensitive assay for PNMT (AXELROD, 1962) enabled us to examine the relationship between the adrenal cortex and medulla. Since the adrenal cortex is one of the target organs of the pituitary, rats were hypophysectomized and PNMT activity was examined 3 weeks later (WURTMAN and AXELROD, 1966). A marked fall in adrenal PNMT was found after the removal of the pituitary gland (Fig. 4). This experiment indicated that the most likely pituitary hormone affecting adrenal PNMT is ACTH. Thus, in the following experiment ACTH was administered for 6 days to rats which had been hypophysectomized 11 days earlier. This caused the PNMT activity to return almost to normal (Fig. 4). To examine whether ACTH is acting directly, or by increasing the synthesis and secretion of adrenal glucocorticoids, hypophysectomized rats were treated with the synthetic glucocorticoid, dexamethasone. This treatment caused a return of PNMT activity to normal (Fig. 4). These results are in contrast to those obtained with tyrosine hydroxylase (see above), where ACTH but not dexamethasone overcame the effect of hypophysectomy on tyrosine hydroxylase. Treatment of rats with intact pituitary with large doses of ACTH or dexamethasone could not raise PNMT above normal levels (WURTMAN and AXELROD, 1966; LEACH and LIPSCOMB, 1969).

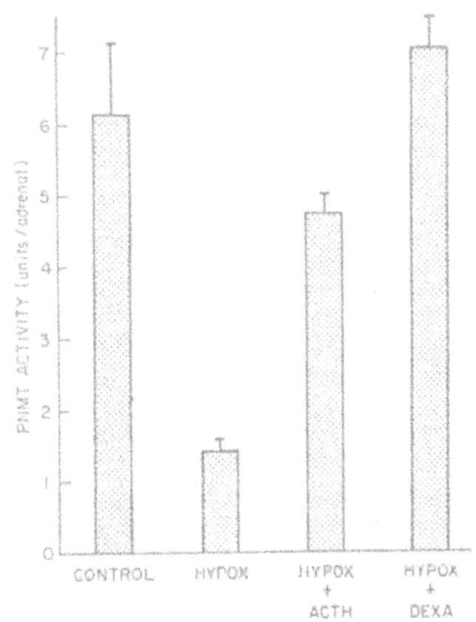

Fig. 4. Effect of hypophysectomy, ACTH and dexamethasone on rat adrenal PNMT activity. Two weeks after hypophysectomy rats were given ACTH 4 units/day or dexamethasone 1 mg/day for 6 days. Results are expressed as mean ± S.E.

It was also found that no other hormones (FSH, LIT, prolactin, TSH, growth hormone estrogen, testosterone or mineralocorticoids) could substitute for ACTH (WURTMAN, 1966). In addition, large doses of glucocorticoid equivalent that delivered to the medulla from the veins of the cortex were necessary to maintain PNMT activity.

ACTH and corticosterone has also been shown to be involved in the rapid development of PNMT in newborn rats' adrenals (MARGOLIES et al., 1966) and in the methylation of noradrenaline in extra-adrenal chromaffin tissues (COUPLAND, 1968).

Adrenaline and the adrenaline-forming enzyme is present in the brain, heart and adrenal gland of the frog (Wurtman et al., 1968). Unlike mammals, hypophysectomy does not reduce PNMT in the frog adrenal gland. Glucocorticoids could elevate PNMT activity directly or by increasing the synthesis or decreasing the breakdown of enzyme protein. Glucocorticoids were found to have little direct effect on enzyme activity *in vitro*.

Hypophysectomized rats were given protein synthesis inhibitors, actinomycin D or puromycin (Wurtman and Axelrod, 1966). Both compounds prevented the increase in PNMT activity caused by dexamethasone, indicating that glucocorticoids elevate PNMT activity by increasing the synthesis of enzyme protein.

Summary

From these observations it is apparent that two different enzymes, tyrosine hydroxylase and PNMT have similar control mechanisms for their maintenance and induction. The normal levels of PNMT in the adrenal gland are maintained by ACTH and glucocorticoids, while those of tyrosine hydroxylase are maintained by ACTH. Both enzymes are elevated above normal levels by increased nerve stimulation, tyrosine hydroxylase being quantitatively more affected than PNMT.

References

Alousi, A., Weiner, N.: The regulation of norepinephrine synthesis in sympathetic nerves: Effect of nerve stimulation, cocaine and catecholamine-releasing agents. Proc. nat. Acad. Sci. (Wash.) **56**, 1491—1496 (1966).

Axelrod, J.: Purification and properties of phenylethanolamine N-methyltransferase. J. biol. Chem. **237**, 1657—1660 (1962).

Coupland, R. E.: On the morphology and adrenaline-noradrenaline content of chromaffin tissue. J. Endocr. **9**, 194—203 (1953).

— Corticosterone and methylation of noradrenaline by extradrenal chromaffin tissue. J. Endocr. **41**, 487—490 (1968).

De Quattro, V., Maronde, R., Nagatsu, T., Alexander, N.: Altered norepinephrine synthesis and storage in the hypertensive buffer denervated rabbit. Circulat. Res. **24**, 545—555 (1969).

Dontas, A. S., Nickerson, M.: Central and peripheral components of the action of ganglionic blocking agents. J. Pharmacol. exp. Ther. **120**, 147—159 (1957).

Gordon, R., Spector, S., Sjoerdsma, A., Udenfriend, S.: Increased synthesis of norepinephrine and epinephrine in the intact rat during exercise and exposure to cold. J. Pharmacol. exp. Ther. **153**, 440—447 (1966).

Iggo, A., Vogt, M.: Preganglionic sympathetic activity in normal and reserpine-treated cats. J. Physiol. (Lond.) **150**, 114—133 (1960).

Kirschner, N., Goodall, McC.: The formation of adrenaline from noradrenaline. Biochim. biophys. Acta (Amst.) **24**, 658—659 (1957).

Landsberg, L., Axelrod, J.: Influence of pituitary, thyroid and adrenal hormones on norepinephrine turnover and metabolism in the rat heart. Circulat. Res. **22**, 559—571 (1968).

Leach, C. S., Lipscomb, H. H.: Adrenal cortical control of adrenal medullary function. Proc. Soc. exp. Biol. (N. Y.) **130**, 448 (1969).

Levitt, M., Spector, S., Sjoerdsma, A., Udenfriend, S.: Elucidation of the rate-limiting step in norepinephrine biosynthesis in the perfused guinea pig heart. J. Pharmacol. exp. Ther. **148**, 1—8 (1965).

Margolies, F., Roffi, J., Jost, A.: Norepinephrine in fetal rats. Science **154**, 275—276 (1966).

Mueller, R. A., Thoenen, H., Axelrod, J.: (1) Compensatory increase in adrenal tyrosine hydroxylase activity after chemical sympathectomy. Science **163**, 468—469 (1969).

— — — (2) Increase in tyrosine hydroxylase activity after reserpine administration. J. Pharmacol. exp. Ther. **169**, 74—79 (1969).

— — — (3) Inhibition of trans-synaptically increased tyrosine hydroxylase activity by cycloheximide and actinomycin D. Molec. Pharmacol. **5**, 463—469 (1969).
— — — (1) Effect of pituitary and ACTH on the maintenance of basal tyrosine hydroxylase activity in the rat adrenal gland. Endocrinology (in press).
— — — (2) Inhibition of neuronally induced tyrosine hydroxylase by nicotinic receptor blockade. Europ. J. Pharmacol. (in press).
SEDVALL, G. C., KOPIN, I. J.: (1) Acceleration of norepinephrine synthesis in the rat submaxillary gland *in vivo* during sympathetic nerve stimulation. Life Sci. **6**, 45—52 (1967).
— — (2) Influence of sympathetic denervation and nerve impulse activity of tyrosine hydroxylase in the rat submaxillary gland. Biochem. Pharmacol. **16**, 36—46 (1967).
SPECTOR, S., GORDON, R., SJOERDSMA, A., UDENFRIEND, S.: End-product inhibition of tyrosine hydroxylase as a possible mechanism for regulating norepinephrine synthesis. Molec. Pharmacol. **3**, 549—555 (1967).
THOENEN, H., MUELLER, R. A., AXELROD, J.: (1) Transsynaptic induction of adrenal tyrosine hydroxylase. J. Pharmacol. exptl. Ther. **169**, 249—254 (1969).
— — — (2) Increased tyrosine hydroxylase activity after drug induced alteration of sympathetic transmission. Nature (Lond.) **221**, 1264 (1969).
— — — Neuronally dependent induction of adrenal phenylethanolamine N-methyltransferase by 6-hydroxydopamine. Biochem. Pharmacol. (in press).
— TRANZER, J. P.: Chemical sympathectomy by selective destruction of adrenergic nerve endings with 6-hydroxydopamine. Pharmak. U. exp. Path. **261**, 271—278 (1968).
WURTMAN, R. J.: Control of epinephrine synthesis in the adrenal medulla by the adrenal cortex: hormonal specificity and dose response characteristics. Endocrinology **79**, 608—614 (1966).
— AXELROD, J.: Control of enzymatic synthesis of adrenaline in the adrenal medulla by adrenal cortical steroid. J. biol. Chem. **241**, 2301—2305 (1966).
— — TRAMEZZANI, J.: Distribution of the adrenaline-forming enzyme in the adrenal gland of a snake *Xenodon merremii*. Nature (Lond.) **215**, 879—880 (1967).
— — VESELL, E., ROSS, G.: Species differences in inducibility of phenylethanolamine N-methyltransferase. Endocrinology **82**, 584—590 (1968).

Dr. J. AXELROD
National Institute of Mental Health
Department of Health,
Education and Welfare
Bethseda 14, MD 20014/USA

Discussion

GEFFEN: With respect to your experiments on the trans-synaptic induction of tyrosine hydroxylase, these are very elegant experiments and I am left in no doubt that there is a trans-synaptic induction. What I am not sure about, and I do not think you have necessarily committed yourself here, is whether the inducer acts trans-synaptically. It seems to me that the terminals of the post-synaptic neuron could feed back to the cell body, perhaps by somatopetal transport of an inducer.

AXELROD: That's the crux of the whole problem and we cannot answer your question. The trans-synaptic inducer could be acetylcholine but preliminary experiments indicate that this is unlikely. Another possible inducer is noradrenaline. This neurotransmitter might repress the synthesis of tyrosine hydroxylase. Upon increased nerve stimulation more noradrenaline is released so that the synthesis of tyrosine hydroxylase is derepressed.

We have also found that the activity of PNMT and dopamine-β-oxidase is also increased after reserpine. It is possible that all of the enzymes for noradrenaline biosynthesis are located on the same operon and noradrenaline is acting as a repressor.

CARLSSON: May I put in here just one thing—how would your reserpine data fit in?

AXELROD: Reserpine can remove the noradrenaline by two mechanisms. It can release it from storage vesicles and it can increase sympathetic nerve activity by a reflex action, thus causing noradrenaline to be reduced from the cell body and terminals.

STJÄRNE: Did not Dr. WEINER show some time ago that there are two mechanisms for synthesis acceleration on increased nerve activity—one which is end product inhibitable, and one which is not?

AXELROD: Dr. WEINER did an important experiment showing that for 30 min after sympathetic nerve stimulation there is an increase in the conversion of tyrosine to noradrenaline. After giving a protein synthesis inhibitor, puromycin, the increase was blocked. From this he concluded that increased nerve stimulation accelerates the synthesis of tyrosine hydroxylase within a short time, 30 min. However, Dr. WEINER did not measure the activity of tyrosine hydroxylase. Drs. KOPIN and THOA confirmed Dr. WEINER's results but did not find an increase in tyrosine hydroxylase activity, 30 min after nerve stimulation.

The shortest time in which we found an increase in tyrosine hydroxylase after reserpine was not earlier than 6 h so that induction of the enzyme takes many hours.

KIRSHNER: In our first experiments on release of dopamine-β-hydroxylase in the rabbit, Dr. VIVEROS also followed the activity of PNMT and tyrosine hydroxylase over a period of up to 6 days. At 24 h there were no changes in the tyrosine hydroxylase but at 48 h he found a doubling of tyrosine hydroxylase. Throughout the 6-day period he found no change in PNMT. About one and a half years ago, DE QUATTRO et al. reported at Federation Meetings that they had found increases in tyrosine hydroxylase in the adrenal medulla and in the heart, following sino-aortic denervation. Under those conditions there were little or no changes in the noradrenaline or adrenaline in the adrenal gland—but he was still getting the increase in tyrosine hydroxylase. So it would seem that noradrenaline as the operon effector of modifier is questionable.

AXELROD: We have always obtained a consistent but small elevation of PNMT, about 25% in the adrenal gland after 6-hydroxydopamine and reserpine. This rise can be blocked by cutting the sympathetic nerve.

DAHLSTRÖM: Firstly, I just want to mention a fact which I think is very familiar to you all but I wish to bring it up, that is, it has been demonstrated that in all kinds of neurons when you stimulate them you can get an increased synthesis of all kinds of proteins. Therefore, after reserpine—since it has been demonstrated that reserpine treatment causes an increase in nerve activity in the preganglionic neuron—you would really expect to find an increase in all proteins in these neurons, enzymes, as well as in chromogranins. The second thing I want to mention is in connection with your observation of an increased tyrosine hydroxylase activity in the medulla oblongata of the rat after reserpine. This observation is in correlation to the observation made by FUXE and myself of supranormal amounts of noradrenaline-induced fluorescence in the cell bodies of noradrenaline neurons in the rat medulla oblongata after reserpine. The noradrenaline appeared to be bound to storage granules or some storage protein, sensitive to a second dose of reserpine. This observation may also indicate an increased protein synthesis in central noradrenaline neurons after reserpine.

The third thing I should like to ask is: Did you find any accumulation of tyrosine hydroxylase when you ligated the nerve? LADURON and BELPAIRE claimed that they did not find any accumulation of this enzyme after nerve ligation, indicating none or, if any, a very low transport of tyrosine hydroxylase in peripheral adrenergic nerves. Did you do any ligation experiments using the sciatic nerve?

AXELROD: We repeated the experiment several times and obtained a small, but significant, elevation of tyrosine hydroxylase in the proximal segments of the sciatic nerve before an elevation was apparent in the distal segments. We did not do any ligation experiments on the sciatic nerve.

GLOWINSKI: I would just like to make some additional comments on the change in tyrosine hydroxylase activity observed in the central nervous system by Dr. AXELROD and Dr. THOENEN after reserpine treatment. Change in the activity of the enzyme of the major limiting step of norepinephrine in the central nervous system can also be observed in physiological situation in which the activity of central noradrenergic neurons has been increased. Last year, we found with Dr. KETY that series of electroshoks applied once a day for 10 days were increasing the turnover of central norepinephrine, this could be observed even 24 h after the last session of electroshoks. In similar experimental conditions, we have recently observed with Dr. MUSACCHIO that tyrosine hydroxylase activity was increased in various structures of the central nervous system, the effect being maximal in the medulla oblongata.

AXELROD: Well, its a claim for priority now!

VON EULER: How long does this increase last?

AXELROD: In the cell body there is a rise of tyrosine hydroxylase which reaches a peak at 3 days and then falls to base line levels at 7 days after reserpine. In the nerve terminals there is a lag of 2 days, followed by a gradual elevation of tyrosine hydroxylase which reaches a peak at 6 days and then falls gradually.

TRENDELENBURG: It is puzzling that chlorisondamine failed to prevent the enzyme induction in the adrenal medulla. Since there is evidence that some species can transmit impulses to the adrenal medulla through muscarinic mechanisms, could you try to repeat these experiments by administering the chlorisondamine together with atropine?

AXELROD: We did that.

VON EULER: Do you think that this very interesting effect could be of some value, or have some effect in connection with physical training?

AXELROD: We have done an interesting experiment with Dr. HENRY at the University of Southern California. He has raised groups of mice in isolation and who were deprived of psychosocial stimulation and groups of mice who were exposed to intense psychosocial stimulation. The weight of the adrenal glands was increased in both groups of mice. Both the PNMT and tyrosine hydroxylase were decreased in the deprived mice and markedly elevated in the stimulated mice.

GEFFEN: I just want to return to the experiments which Dr. DAHLSTRÖM mentioned about ligation axons, because sooner or later you will have to decide whether the newly synthesised tyrosine hydroxylase is rapidly transported down the axon.

AXELROD: I do not think the complete protein is transported because we showed the final assembly takes place in the terminal.

GEFFEN: Whatever the deficiencies of the ligation experiment are, it is one test of this problem. The evidence that tyrosine hydroxylase did not increase above a ligature can be explained on methodological grounds because it was expressed as a concentration. Since the axons become intensely swollen the amounts of tyrosine hydroxylase was probably increased above the ligature. I think Dr. DAHLSTRÖM has come to the same conclusion.

GLOWINSKI: I would like to know if you have some similar data on dopamine-β-carboxylase?

AXELROD: No, but we plan to do this.

Bayer-Symposium II, 223—233 (1970)

Session 4c

Effects of Drugs on Uptake and Release of Catecholamines (III)

Chairman: J. Axelrod

Effect of Drugs on Amine Uptake Mechanisms in the Brain

ARVID CARLSSON

With 7 Figures

It is a favoured view today that thymoleptics owe their antidepressant activity to blockade of a mechanism which we like to call the membrane pump (see CARLSSON, 1966) by means of which biogenic amines released from nerve endings are taken up again. Blockade of the membrane pump should lead to an increase in the amine concentrations at receptor sites in the brain. SIGG et al. (SIGG, 1959; SIGG, SOFFER and GYERMEK, 1963) demonstrated the ability of imipramine and related drugs to potentiate noradrenaline and sympathetic nerve stimulation. The first biochemical evidence came from MARSHALL, STIRLING, TAIT and TODRICK (1960) who demonstrated the blocking action of imipramine on 5-hydroxytryptamine (5-HT) uptake by platelets. Soon afterwards a similar action of imipramine on noradrenaline (NA) uptake by adrenergic nerves was demonstrated (AXELROD, WHITBY and HERTTING, 1961; HERTTING, AXELROD, KOPIN and WHITBY, 1961). Blockade by imipramine of NA uptake by brain tissue was also demonstrated *in vitro* (DENGLER, SPIEGLER and TITUS, 1961) and *in vivo* (GLOWINSKI and AXELROD, 1965). For review, see e.g. SULSER and DINGELL (1966).

The membrane pumps are resistent to reserpine and should be distinguished from the reserpine-sensitive uptake mechanisms of the intraneuronal storage granules (see CARLSSON, 1966).

Techniques for the Investigation of the Membrane Pumps of Central Monoamine Neurons

In the brain the investigation of uptake mechanisms for amines is difficult for several reasons. The amines do not readily penetrate through the blood brain barrier, which thus has to be somehow circumvented. Both neuronal and non-neuronal elements exist in the brain, which are capable of taking up monoamines, and it is difficult to identify the structures responsible for uptake using conventional techniques.

Histochemical fluorescence techniques have contributed significantly to the cellular localization of amine uptake mechanisms in brain. Thus it has been possible to demonstrate active uptake mechanisms of DA, NA and 5-HT carrying neurons, as well as selective blockade of these mechanisms (HAMBERGER and MASUOKA, 1965; CARLSSON, FUXE, HAMBERGER and LINDQVIST, 1966; HAMBERGER, 1967; FUXE and UNGERSTEDT, 1967, 1968). However, there still remained an obvious need of a quantitative biochemical technique for investigating separately the uptake mechanisms of DA, NA and 5-HT neurons. Such a technique has been developed and has been in use for some time in our laboratories. The technique depends on the development of certain monoamine analogues with highly specific properties (CARLSSON et al., 1970). These analogues (Fig. 1) are sufficiently lipid soluble to penetrate the blood-brain barrier, yet they are polar enough to

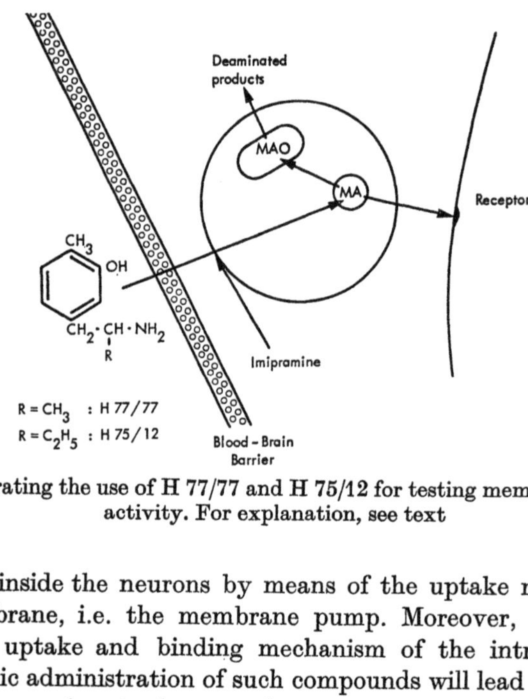

Fig. 1. Model illustrating the use of H 77/77 and H 75/12 for testing membrane-pump blocking activity. For explanation, see text

be concentrated inside the neurons by means of the uptake mechanism located at the cell membrane, i.e. the membrane pump. Moreover, they have a high affinity for the uptake and binding mechanism of the intraneuronal storage granules. Systemic administration of such compounds will lead to displacement of intraneuronal monoamines in the central nervous system as well as the peripheral adrenergic system. When the membrane pump is blocked e.g. by an imipramine-like agent, this displacement of endogenous monoamines is prevented. The effect can be conveniently assessed quantitatively by measuring endogenous catecholamine and 5-HT levels.

Fig. 2 illustrates how the experiment is carried out. In this case H 77/77 is used as the displacing agent, and the blocking action of protriptyline on the displacement of NA in brain and heart is investigated. When given alone in two doses of 12.5 mg/kg i.p. to mice at a two-hour-interval, examination 2 h after the second dose reveals a decrease of the NA level to somewhat less than half in the brain and even less in the heart. Protriptyline (PTP), given alone in two doses of 6.26 and 3.13 mg/kg, respectively, has no significant effect. However, when PTP is given in the same dosage 30 min before each dose of H 77/77 the NA depleting action of this agent is largely prevented. The percentage inhibition can be easily calculated, and thus a dose response curve can be obtained. Fig. 3 shows such a curve, in this case using desipramine as the blocking agent. From such a curve ED 50 can be obtained. For dopamine and NA displacement H 77/77 is used, and for 5-HT the closely related H 75/12.

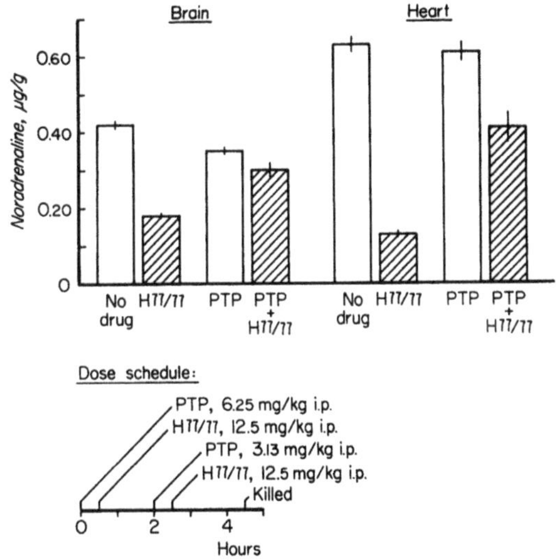

Fig. 2

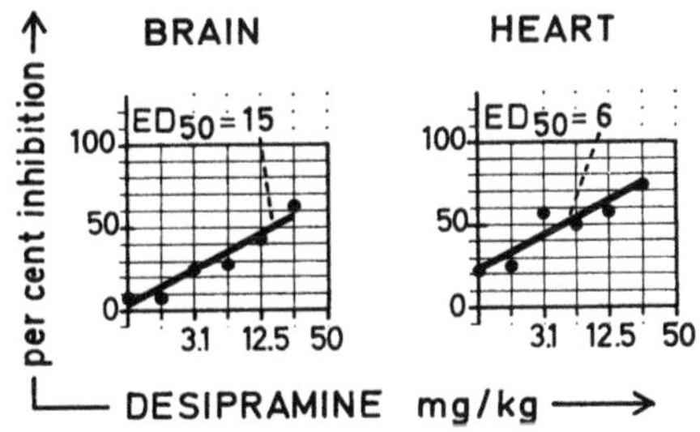

Fig. 3

Fig. 2. Example of test for membrane-blocking activity, using H 77/77. PTP = protriptyline. For explanation, see text. The data are means ± S.E.m.

Fig. 3. Dose-response curves for desipramine, obtained by means of the H 77/77 test for membrane-pump blockade of NA neurons. [From Carlsson et al. 1969 (2)]

Blockade of Membrane Pumps by Antidepressant Drugs, Antihistamines and Analgesics

Using this technique a large number of established and potential thymoleptics have been investigated.

None of the thymoleptics examined so far had any detectable effect on dopamine displacement [Carlsson, Corrodi, Fuxe and Hökfelt, 1969 (1)]. This is in agreement with earlier histochemical and biochemical studies showing that the membrane pump of the dopamine neurons is resistant to thymoleptics in general (Carlsson, Fuxe, Hamberger and Lindqvist, 1966; Hamberger, 1967; Jonason and Rutledge, 1968). The membrane pumps of both NA and 5-HT neurons are sensitive to these drugs. The sensitivities of central and peripheral NA neurons are not identical. Only protriptyline blocks NA displacement about equally in brain and heart; most of the other drugs investigated act more strongly on the peripheral than central neurons. The difference is especially marked in the case of the bicyclic phthalane derivative Lu 3-010, where the ED 50 is about 10 times higher centrally than peripherally. This may be related to the poor lipid solubility of this drug (Carlsson, Fuxe, Hamberger and Malmfors, 1969). Thus data obtained from the peripheral adrenergic system, using biochemical or pharmacological techniques, e.g. NA potentiation, do not permit extrapolation to the brain (Table 1).

Both central and peripheral NA neurons are generally more sensitive to secondary than tertiary amines among the thymoleptic drugs. The reverse holds true for central 5-HT neurons and, according to Yates, Todrick and Tait (1964), 5-HT uptake by platelets. Among the thymoleptic drugs in clinical use chlorimipramine proved most active on 5-HT neurons [Carlsson, Corrodi, Fuxe and Hökfelt, 1969 (2)].

Table 1. *Effect of thymoleptics on 5-HT displacement in brain by H 75/12 and on NA displacement in brain and heart by H 77/77: ED 50 (mg/kg)*

	Brain 5-HT	Brain NA	Heart NA
Imipramine HCl	20	> 25	12
Desipramine HCl	≧ 50	15	6
Chlorimipramine HCl	7	> 25	20
Amitriptyline HCl	12	> 25	14
Nortriptyline HCl	20	> 25	15
Protriptyline HCl	> 25	4	4
Lu 5-003	> 25	8	3
Lu 3-010	> 25	40	3

Table 2. *Effect of antihistamines on 5-HT displacement in brain by H 75/12 and on NA displacement in brain and heart by H 77/77: ED 50 (mg/kg)*

	Brain 5-HT	Brain NA	Heart NA
Chlorpheniramine maleate	8	20	< 12.5
Recipavrin HCl	15	> 25	15
Diphenhydramine HCl	20	—	—
Tripelennamine HCl	> 25	—	—
Phenindamine tartrate	> 25	20	15

Several other agents have been investigated. Among analgesics pethidine and methadone proved moderately active on 5-HT neurons (Carlsson and Lindqvist, 1969). Several antihistamines were active (Table 2). Among these chlorpheniramine and brompheniramine were most active, particularly on 5-HT neurons. The dextrorotatory enantiomer of chlorpheniramine was considerably more potent than the levo form. The halogen-substituted pheniramines compared favourably with chlorimipramine. Several antihistamines were earlier known to block the amine pump of peripheral adrenergic neurons (Isaac and Goth, 1967).

The structural similarity between the tricyclic thymoleptics and some membrane pump blocking antihistamines will be apparent from Fig. 4. Further analysis of structure-activity relationships is in progress.

In the dosage employed H 77/77 caused some peripheral sympathomimetic action and central stimulation. These effects were partly or wholly prevented by the thymoleptic agents when given in doses sufficiently high to prevent NA displacement. They are therefore probably due to release of NA peripherally and centrally. H 75/12, which unlike H 77/77 causes displacement also of 5-HT, induced behavioural changes which clearly differed from the actions of H 77/77. The animals showed a tendency to abduction and extension of the limbs, lordosis and general excitation with stereotype head movements and tremors. The similarity with the syndrome induced by 5-hydroxytryptophan (5-HTP) was evident. These behavioural actions induced by H 75/12 are thus probably due to release of 5-HT.

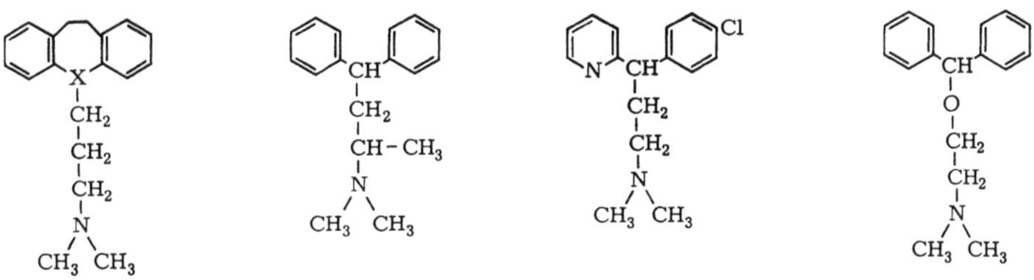

Tricyclic thymoleptic Recipavrin Chlorpheniramine Diphenhydramine

Fig. 4. Chemical structure of some membrane-pump blocking agents

In support of this assumption it was observed that thymoleptics given in doses which prevented 5-HT displacement, also prevented the 5-HTP-like behavioural changes.

Overflow of 5-HT Induced by Membrane Pump Blockade

It was found that those thymoleptics which prevented the 5-HT displacement induced by H 75/12 also potentiated the action of 5-HTP. For example, chlorimipramine was very active in this respect, imipramine less so but still clearly active, whereas desipramine caused but slight 5-HTP potentiation (Carlsson, Jonason, Lindqvist and Fuxe, 1969). Thus chlorimipramine and imipramine block the 5-HTP-like syndrome induced by H 75/12 but potentiate the induction of the same syndrome after 5-HTP administration. This paradox strongly supports the contention that these thymoleptics block the membrane pump of 5-HT neurons. Prevention of the H 75/12 induced syndrome will thus be explained by blockade of the membrane pump dependent intraneuronal accumulation of H 75/12, potentiation of 5-HTP by blockade of re-capture of the excess 5-HT leaking out through the cell membranes of 5-HT neurons. It is difficult to conceive of any alternative explanation.

It is known that thymoleptics also potentiate dopa (Everett, Wiegand and Rinaldi, 1963). In preliminary experiments we have compared the dopa-potentiating activity of several thymoleptics (Carlsson and Lindqvist, unpublished data). The potentiation profile was clearly different from that of 5-HTP potentiation. For example, desipramine was, if anything, more potent than imipramine and chlorimipramine. This suggests that blockade of the membrane pump of NA neurons is involved. However, we found a remarkably slight potentiating activity of protriptyline, the most potent blocker of the membrane pump of central NA neurons known so far. The mechanism of dopa potentiation by thymoleptics evidently needs further investigation.

As is well known, the combined treatment of animals and man with monoamine oxidase inhibitors and tricyclic thymoleptics (or pethidine) may result in dramatic,

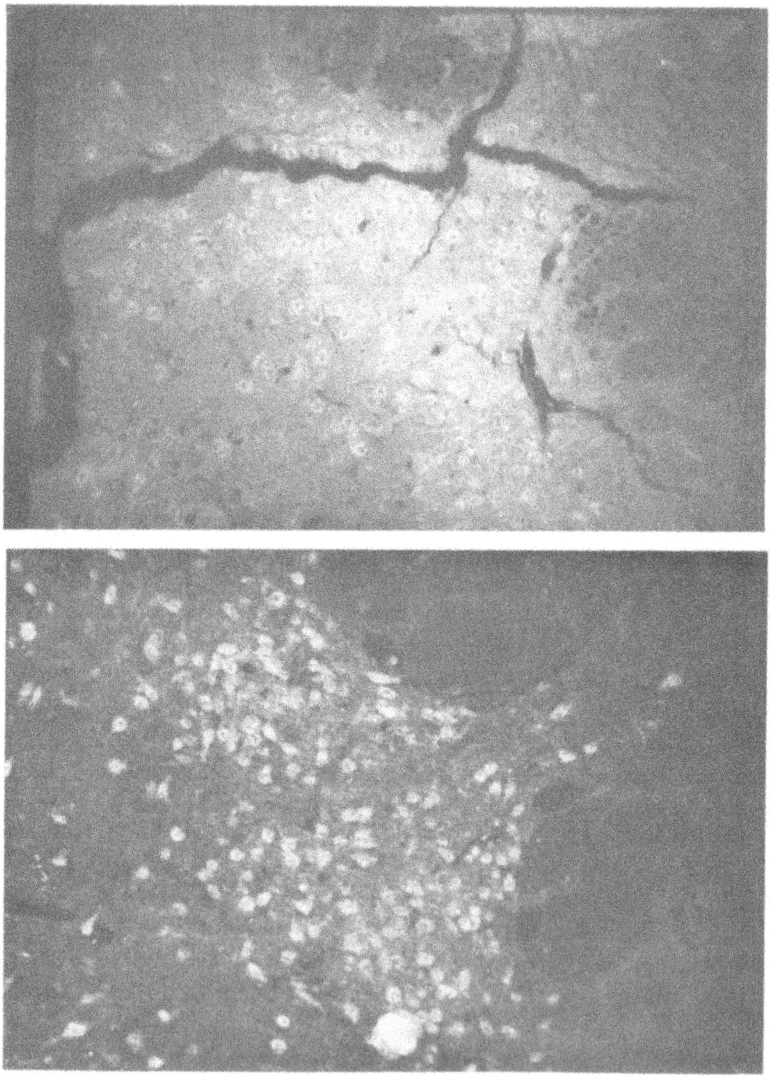

Fig. 5. Top: Nuc raphe dorsalis of a rat treated with reserpine (10 mg/kg, i.p., 24 h before killing) and nialamide (500 mg/kg, i.p., 5 h before killing). There is a marked accumulation of yellow fluorescence in the 5-HT cell bodies of this nucleus after this treatment. × 120. Bottom: The same area and reserpine-nialamide treatment. However, chlorimipramine has also been injected (25 mg/kg, i.p., 30 min before killing). The injection of this drug results in a marked accumulation of yellow fluorescence in the extraneuronal space and a loss of fluorescence from the 5-HT cell bodies. This picture is probably due to a rapid outflow of 5-HT from the cell bodies. × 120

even fatal reactions, involving hyperthermia and various signs of central stimulation. This phenomenon will be explained if we assume that the excess monoamine accumulating in the intraneuronal cytoplasm following monoamine oxidase inhibition will leak out into the extraneuronal space. Normally the membrane pump can cope with such leakage but if the pump is blocked, overt leakage will occur onto receptors, resulting in various functional disturbances. We compared the ability of several thymoleptics to potentiate the action of the monoamine oxidase inhibitor nialamide. We found that those thymoleptics which have a strong action on 5-HT neurons, when given to nialamide-pretreated animals, caused a syndrome characterized by hyperthermia and an excitation syndrome, which included some 5-HTP-like effects, such as stereotype head movements, abduction and extension of limbs and tremors. The activity correlated generally well with the membrane pump blocking activity on 5-HT neurons. Thus chlorimipramine and chlorpheniramine were very potent. Desipramine which has a strong action on the membrane pump of NA neurons but a weak effect on 5-HT neurons induced a mild syndrome characterized mainly by tenseness and other signs of increased wakefulness.

Table 3. *Effect of chlorimipramine following nialamide pretreatment on release of 5-HT from brain slices and on body temperature and behaviour*

Treatment	Rise in body temp. °C/30 min	5-HT, μg/g Incub. med.	5-HT, μg/g Brain tissue	Effect on behaviour
Nialamide 100 mg/kg i.p. + chlorimipramine 12.5 mg/kg i.p.	2.2 ± 0.47 (n = 7) $p < 0.005$	0.24 ± 0.038 (n = 10)	0.56 ± 0.046 (n = 10)	+
Nialamide 100 mg/kg i.p.	0.2 ± 0.13 (n = 7)	0.05 ± 0.012 (n = 10)	0.53 ± 0.049 (n = 10)	0
diff.	$p < 0.005$	$p < 0.001$	—	

According to our interpretation a drug like chlorimipramine when given to nialamide pretreated animals, will cause marked overflow of 5-HT onto receptors. We now attempted to demonstrate this overflow directly, using both histochemical and biochemical techniques (Carlsson, Jonason, Lindqvist and Fuxe, 1969). The result of the histochemical study is illustrated in Fig. 5. The leakage of 5-HT out of the cell bodies into the extraneuronal space is evident. The same phenomenon was seen at 5-HT nerve terminals. For the biochemical demonstration of 5-HT overflow we treated animals (rabbits and mice) with nialamide and chlorimipramine. After various intervals we killed the animals, prepared slices of brain tissue and incubated them in an oxygenated Krebs-Henseleit solution. As shown in Table 3, considerable release of 5-HT into the incubation medium took place, as compared to controls treated with nialamide alone. There was a considerable individual variation of the response among the rabbits. A correlation was found between the effect on 5-HT release and on behaviour and body temperature.

We also compared some thymoleptics with respect to 5-HT releasing activity (Carlsson, Jonason and Lindqvist, 1969) (Fig. 6). Chlorimipramine proved most potent. Imipramine and amitriptyline were moderately active (though with a paradoxical dose-relationship in the case of imipramine), whereas desipramine was hardly active. Thus there seems to be a correlation between 5-HT releasing activity and membrane-pump blocking activity on 5-HT neurons.

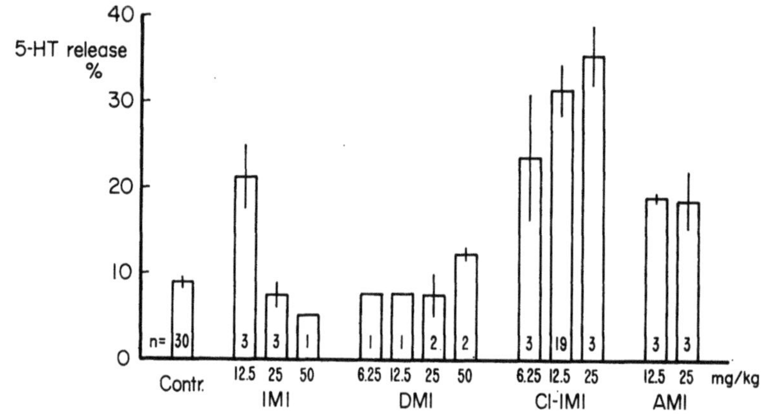

Fig. 6. Effect of thymoleptics on the 5-HT release from rabbit cortical slices. The data are means ± S.E. m. IMI = imipramine. DMI = desipramine. Cl-IMI = chlorimipramine. AMI = amitriptyline. The animals were treated with nialamide alone 100 mg/kg, i.p., 16 h beforehand (controls), or in combination with various doses of antidepressant drugs. The animals were killed 30 min after the injection of the antidepressant drugs, and cortical slices were prepared. Incubation period 40 min. (Data from Carlsson, Jonason and Lindqvist, 1969)

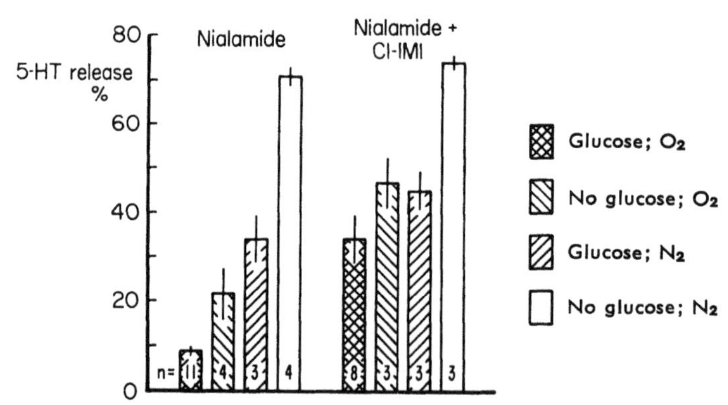

Fig. 7. Effect of glucose deprivation and/or anoxia on the 5-HT release from rabbit brain cortical slices. The data are means ± S.E.m. For experimental conditions see legend of Fig. 6. Dose of chlorimipramine 12 mg/kg i.p. (Data from Carlsson, Jonason and Lindqvist, 1969)

The membrane pumps have been shown to be active, energy-dependent mechanisms (see Hamberger, 1967). Thus we wanted to investigate whether the ability of brain tissue from nialamide-treated animals to retain 5-HT was dependent on energy supply. This proved to be the case (Fig. 7). Deprivation of both oxygen and glucose caused considerable release, whereas deprivation of either component separately had an intermediate effect. The effect of chlorimipramine could be demonstrated only in the presence of energy supply. The data support the view that chlorimipramine acts by blocking an energy-dependent mechanism.

Concluding Remarks

The present data demonstrate a high degree of structural specificity of the different monoamine neurons with respect to blockade of their membrane pumps. It may very well be possible to develop agents which block the membrane pump of one type of monoamine neuron without affecting the other two. But even now the selectivity reached is sufficient to make the membrane-pump blockers very interesting tools for elucidating the function of the different neurons. For example, the fact that 5-HTP is potentiated specifically by those thymoleptics which are potent blockers of the membrane pump of 5-HT neurons e.g. chlorimipramine, lend additional support for the view that the characteristic 5-HTP-induced syndrome is due to overflow of 5-HT from neurons normally storing 5-HT. As is well known this view might otherwise be questioned on the basis that 5-HTP decarboxylase and dopa decarboxylase are identical or at least closely related. Thus, 5-HT may be formed from exogenous 5-HTP also in catecholamine neurons. In this connection the strong potentiation of a monoamine oxidase inhibitor by e.g. chlorimipramine, with the appearance of 5-HTP-like symptoms, is interesting. In this case the evidence is strong that overflow occurs selectively from 5-HT storing neurons. The data thus support the view that 5-HT neurons are involved in the control of body temperature, as proposed by FELDBERG and MYERS (1963, cf. FELDBERG, 1968; CORRODI, FUXE and HÖKFELT, 1967), in the control of muscle tone and movements, as indicated by the abduction and extension of limbs tremors and head movements induced by 5-HT overflow, and finally in the general central excitability. The data also invite the speculation that 5-HT is specifically involved in the control of mood. It was found that the tertiary amines among the thymoleptics, e.g. chlorimipramine, were particularly potent in blocking the membrane pump of 5-HT neurons. These drugs are known to elevate mood selectively. The central NA neurons seem rather to stimulate drive and alertness, as indicated by the selective action of secondary amines like protriptyline and desipramine on these neurons. Clinically these thymoleptics are characterized by a prominent action on drive, resulting in relief of the inhibition generally seen in depressed patients. A role of 5-HT and catecholamines in depression has earlier been proposed on different grounds (see COPPEN, 1967; LEHMANN, 1966; ROOS and SJÖSTRÖM, 1969).

If thymoleptics elevate mood by increasing the concentration of 5-HT at receptor sites—as a consequence of membrane pump blockade—how can we explain that the direct activation of 5-HT receptors brought about by LSD, as indicated by the data of ANDÉN, CORRODI, FUXE and HÖKFELT (1968), induces a picture of overt psychosis? The difference may be quantitative rather than qualitative. The actions of a membrane pump blocker appears to be self limited owing to a feedback mechanism which inhibits neuronal activity (cf. CORRODI and FUXE, 1968; CORRODI, unpublished data; ANDÉN, CARLSSON and HÄGGENDAL, 1969). This self limitation, which cannot operate in the case of direct receptor-stimulating agents, may prevent psychotic reactions.

Summary

The dopamine (DA), noradrenaline (NA) and 5-hydroxytryptamine (5-HT) neurons of the brain have been shown to possess amine uptake mechanisms at the level of the cell membranes, so-called membrane pumps. These mechanisms are resistant to reserpine and should be distinguished from the reserpine-sensitive uptake mechanisms of the intraneuronal storage granules. Tricyclic thymoleptics block the membrane pumps of NA and 5-HT neurons but not that of DA neurons. Tertiary amines, e.g. imipramine, in general exert their chief action on 5-HT neurons, whereas secondary amines, e.g. desipramine act predominantly on NA neurons. Marked differences in sensitivity of central and peripheral NA neurons exist. Certain antihistamines and analgesics also block membrane pumps of 5-HT and NA neurons.

Marked potentiation of 5-hydroxytryptophan and monoamine oxidase inhibitors was demonstrated after blockade of the membrane pump of 5-HT neurons. Potentiation was shown to be due to 5-HT overflow into the extraneuronal space.

The functional implications of the data are discussed. It is suggested that 5-HT neurons are involved in thermoregulation, control of muscular tone and movements, and mood. NA neurons appear to be control drive and wakefulness.

Acknowledgements

This work has been supported by the Swedish Medical Research Council (Grant Nos. B70-14X-155-06A, B70-14X-2157-04B).

References

Andén, N.-E., Carlsson, A., Häggendal, J.: Adrenergic mechanisms. Ann. Rev. Pharmacol. **9**, 119—134 (1969).

Axelrod, J., Whitby, L. G., Hertting, G.: Effect of psychotropic drugs on the uptake of ^{3}H-norepinephrine by tissues. Science **133**, 383—384 (1961).

Carlsson, A.: Physiological and pharmacological release of monoamines in the central nervous system. In: Mechanisms of release of biogenic amines, pp. 331—346. (v. Euler, U. S., Rosell, S., Uvnäs, B., Eds.). Oxford 1966.

— Corrodi, H., Fuxe, K., Hökfelt, T.: (1) Effects of some antidepressant drugs on the depletion of intraneuronal brain catecholamine stores caused by 4,α-dimethyl-meta-tyramine. Europ. J. Pharmacol. **5**, 367—373 (1969).

— — — — (2) Effect of antidepressant drugs on the depletion of intraneuronal brain 5-hydroxytryptamine stores caused by 4-methyl-α-ethyl-meta-tyramine. Europ. J. Pharmacol. **5**, 357—366 (1969).

— Fuxe, K., Hamberger, B., Lindqvist, M.: Biochemical and histochemical studies on the effects of imipramine-like drugs and (+)-amphetamine on central and peripheral catecholamine neurons. Acta physiol. scand. **67**, 481—497 (1966).

— — —, Malmfors, T.: Effect of a new series of bicyclic compounds with potential thymoleptic properties on the reserpine-resistant uptake mechanism of central and peripheral monoamine neurones. Brit. J. Pharmacol. **36**, 18—28 (1969).

— Jonason, J., Lindqvist, M.: On the mechanism of 5-hydroxytryptamine release by thymoleptics. J. Pharm. Pharmac. **21**, 769—773 (1969).

— — —, Fuxe, K.: Demonstration of extraneuronal 5-hydroxytryptamine accumulation in brain following membrane-pump blockade by chlorimipramine. Brain Res. **12**, 456—460 (1969).

— Lindqvist, M.: Central and peripheral monoaminergic membranepump blockade by some addictive analgesics and antihistamines. J. Pharm. Pharmac. **21**, 460—464 (1969).

— —, Wysokowski, J., Corrodi, H., Junggren, U.: Substituted metatyramines as brain monoamine depletors. Acta Pharmaceut. Suec. 1970 (in press).

COPPEN, A.: The biochemistry of affective disorders. Brit. J. Psychiat. **113**, 1237—1264 (1967).
CORRODI, H., FUXE, K.: The effect of imipramine on central monoamine neurons. J. Pharm. Pharmacol. **20**, 230—231 (1968).
— — HÖKFELT, T.: Central serotonin neurons and thermoregulation. Advanc. Pharmacol. **6B**, 49—54 (1967).
DENGLER, H. J., SPIEGEL, H. E., TITUS, E. O.: Uptake of tritium-labeled norepinephrine in brain and other tissues of cat in vitro. Science **133**, 1072—1073 (1961).
EVERETT, G. M., WIEGAND, R. G., RINALDI, F. V.: Pharmacologic studies of some nonhydrazine MAO inhibitors. Ann. N. Y. Acad. Sci. **107**, 1068—1080 (1963).
FELDBERG, W.: The monoamines of the hypothalamus as mediators of temperature responses. Rec. Advanc. Pharmacol. **1968**, 349—397.
— MYERS, R. D.: A new concept of temperature regulation by amines in the hypothalamus. Nature (Lond.) **200**, 1325 (1963).
FUXE, K., UNGERSTEDT, U.: Localization of 5-hydroxytryptamine uptake in rat brain after intraventricular injection. J. Pharm. Pharmacol. **19**, 335—336 (1967).
— — Histochemical studies on the effect of (+)-amphetamine, drugs of the imipramine group and tryptamine on central catecholamine and 5-hydroxytryptamine neurons after intraventricular injection of catecholamines and 5-hydroxytryptamine. Europ. J. Pharmacol. **4**, 135—144 (1968).
GLOWINSKI, J., AXELROD, J.: Effect of drugs on the uptake release and metabolism of norepinephrine in the rat brain. J. Pharmacol. exp. Ther. **149**, 43—49 (1965).
HAMBERGER, B.: Reserpine-resistant uptake of catecholamines in isolated tissues of the rat. A histochemical study. Acta physiol. scand. Suppl. **295**, 1—56 (1967).
— MASUOKA, D.: Localization of catecholamine uptake in rat brain slices. Acta pharmacol. (Kbh.) **22**, 363—368 (1965).
HERTTING, G., AXELROD, J., KOPIN, J. J., WHITBY, L. G.: Lack of uptake of catecholamines after chronic denervation of sympathetic nerves. Nature (Lond.) **189**, 66 (1961).
ISAAC, L., GOTH, A.: The mechanism of the potentiation of norepinephrine by antihistamines. J. Pharmacol. exp. Ther. **156**, 463—468 (1967).
JONASON, J., RUTLEDGE, C. O.: Metabolism of dopamine and noradrenaline in rabbit caudate nucleus *in vitro*. Acta physiol. scand. **73**, 411—417 (1968).
LEHMANN, J.: Mental disturbances followed by stupor in a patient with carcinoidosis. Recovery with tryptophan treatment. Acta psychiat. (Kbh.) **42**, 153—161 (1966).
MARSHALL, E., STIRLING, G. S., TAIT, A. C., TODRICK, A.: The effect of iproniazid and imipramine on the blood platelet serotonin level in man. Brit. J. Pharmacol. **15**, 35—41 (1960).
ROOS, B.-E., SJÖSTRÖM, R.: 5-Hydroxyindoleacetic acid (and homovanillic acid) levels in the cerebrospinal fluid after probenecid application in patients with manic-depressive psychosis. Pharmacol. Clin. **1**, 153—155 (1969).
SIGG, E. B.: Pharmacological studies with tofranil. Canad. Psychiat. Ass. J. **45**, 75—85 (1959).
— — SOFFER, L., GYERMEK, L.: Influence of imipramine and related psychoactive agents on the effect of 5-hydroxytryptamine and catecholamines on the cat nictitating membrane. J. Pharmacol. exp. Ther. **142**, 13—20 (1963).
SULSER, F., DINGELL, J. V.: On the role of adrenergic mechanisms in the mode of action of tricyclic antidepressants. Biochemical and metabolic considerations. In: Antidepressant drugs of non-MAO inhibitor type, pp. 1—19. (EFRON, D. H., KETY, S. S., Eds.). Proceedings of a workshop. Bethseda 1966.
YATES, C. M., TODRICK, A., TAIT, A.: Effect of imipramine and some analogues on the uptake of serotonin by human blood platelets *in vitro*. J. Pharm. Pharmac. **16**, 460—463 (1964).

Prof. Dr. A. CARLSSON
Department of Pharmacology
University of Göteborg/Sweden

Discussion

Blaschko: Did I understand correctly, that these substances are selective depletors? This table showed that there was also some selectiveness in the blocking of this effect. If you took a non-selective depletor, would you then expect a blocking agent which is selective, also preferentially, either to deplete one or the other neuron? I do not know whether my question is quite clear.

Carlsson: I understand what you mean—well, that could occur theoretically but I do not think that could explain all our findings because as I showed in the latter part of this presentation, you could correlate what we have done with these depleting agents with various other phenomena, i.e. 5-HT potentiation and release as observed histochemically, and biochemically, following for instance nialamide and antidepressant drugs and again you have the same profile.

Glowinski: If I have understood correctly, you observed in an histochemical experiment that an overflow of 5-HT occurred from 5-HT cell bodies of the raphe nuclei. I would like to know what are your comments on the possibility of the existence of a release process at the level of cell bodies of aminergic neurons?

Carlsson: Yes, it has been shown several years ago that the membrane pump occurs not only in the nerve terminals but also in the axons and also in the cell bodies. So, I think, this is what you would expect from these studies.

Glowinski: That was not my question: I know that the uptake process occurs in various parts of central aminergic neurons, but do you think that similarly you may have a release of the amine from various parts of the neurons and particularly from cell bodies in physiological conditions. What would be the function of the release at this level?

Carlsson: That could very well be. It is a very interesting possibility that when you depolarize there is not only release at the nerve terminals, but also from the dendrites and so forth that could result in a very interesting dialogue between the neurons.

Philippu: Is there any evidence of the uptake of imipramine or desipramine by the cells?

Carlsson: I do not think there is any evidence of selective uptake by the cell. I think these are just the kind of compounds that we were discussing yesterday. For instance, autoradiography of these characteristic compounds shows an uncharacteristic distribution that perhaps 50% of all drugs have in common. There is probably a general tendency to accumulate in cell membranes and in other biological membranes.

Palm: Do you think that the actions of all these drugs are inhibiting only at the membrane itself? I think Brodie's group and Schildkraut have shown that there is an intraneuronal site of action of these drugs. For instance, they prevent a release of noradrenaline, at least in the heart, produced by reserpine. This would mean that there is a site of intraneuronal action if Dr. Brodie is right, and Schildkraut has shown this in the central nervous system.

CARLSSON: Well this could be a very long discussion, I think, but to make it short. I have not read the paper myself but somebody told me there is a recent paper by BRODIE in which he states that he does no longer believe in this. In any event, BRODIE's coworker, COSTA, has now switched over to the more conventional view about these things.

PHILIPPU: We have also found that desipramine completely blocks the uptake of noradrenaline by isolated hypothalamic vesicles in vitro. I think that Professor VON EULER also observed the inhibition of the amine uptake by desipramine.

CARLSSON: Yes this is the problem of in vitro versus in vivo pharmacology. I think in vitro you can show many things that you cannot show in vivo. I know of no in vivo evidence really that would suggest that you have an action on the intra-neuronal granules using reasonable doses of these drugs—it could be that if you pushed up the doses such things might occur. For instance, block of uptake by storage granules would mean a reserpine-like action and we do not see this. It is true that if you really push doses then many of these drugs cause a certain decrease in brain noradrenaline but the mechanism of that has not been elucidated. It might be secondary to the membrane pump blockade but it may also be something else. It has also been found that if you give radioactive tyrosine followed by protriptoline, there is a decrease in the accumulation of noradrenaline. This is a very interesting observation. I do not know how to explain it but it could very well again be secondary to blockade of the membrane pump.

CORRODI: I think that this is the same phenomenon we discussed yesterday, that if you expose such a highly lipophilic substance to tissue it is taken up quite unspecifically and I think if you take a drug which has no action on depressions or on the membrane pump you will have the same uptake and the same blockade of the granules.

VON EULER: I was going to say something similar and ask Professor CARLSSON whether he thinks desipramine inhibits its own uptake, so to speak.

CARLSSON: Well, these compounds they have a very high lipid solubility so even if they may utilize the membrane pump in order to get inside, I do not believe it would result in any accumulation of desipramine in the cytoplasminside membrane, because these compounds have such a high lipid solubility that they would leak out of the membrane immediately.

VON EULER: In that case, if you think they can pass in through the membrane, would it not be possible for it to be concentrated in the granules?

CARLSSON: Yes, I suppose that a molecule of this configuration would be present in high concentration in many biological membranes in the body.

PHILIPPU: Dr. CORRODI, I believe that these substances are lipophilic. However, you get an inhibition of the noradrenaline uptake in vitro with a concentration of 1×10^{-8} to 1×10^{-9} M. Furthermore, you can abolish the inhibitory effect of desipramine by adding high concentrations of noradrenaline to the incubation medium. It seems, therefore, plausible that a competition exists between noradrenaline and desipramine.

Corrodi: No, I do not dispute that and I only want to show that you should also try membrane pump inactive substance of the same structural type and I think you might find the same blockade or same interaction with noradrenaline. Have you done that or have you only taken active substances?

Philippu: We tried with chlorpromazine but we did not find any significant inhibition of the noradrenaline uptake.

Bayer-Symposium II, 237—248 (1970)

Release of Monoamines in the Central Nervous System

J. GLOWINSKI

With 3 Figures

Introduction

During the last few years, changes in turnover of monoamines in the central nervous system (CNS) have been observed in various pharmacological or physiological situations. These turnover estimations, obtained with the help of amine synthesis inhibitors or by labelling the central aminergic neurons, allowed the detection of large modifications in the rate of global utilization of monoamines. But, changes in turnover do not provide direct and precise information on the rapid fluctuations of quantities of amines released from terminals in central synapses, and thus available at receptor sites. Limitations in the sensitivity of the spectrofluorimetric techniques used to estimate monoamines and practical difficulties to collect amines released from nerve terminals in deep structures of the CNS explain the few attempts which have been made to directly evaluate the release of monoamines (VOGT, 1969). Nevertheless, we have tried during the last 2 years in in vivo or in vitro studies, to demonstrate the release of radioactive catecholamines (CA) or serotonin (5-HT) synthesized from their original precursors: ^{3}H tyrosine and ^{3}H tryptophane respectively. Precursors of high specific activity were used in order to detect very small amounts of ^{3}H amines released ($\neq$ 5 pg). Besides its high sensitivity, this experimental approach offers three main advantages:

1. The ^{3}H amines released spontaneously or after activation of aminergic neurons are liberated, as the respective endogenous amines, specifically from catecholamine or 5-HT containing neurons.

2. The relationships between release and biosynthesis processes occurring during or after the neurons activation can be studied dynamically.

3. Finally, the involvement of the different intraneuronal storage forms of monoamines in the processes of spontaneous or evoked release of transmitters from nerve terminals can be examined and evaluated.

Preliminary results on the various problems mentioned above have been obtained, some of them will be described as examples.

1. Demonstration of Spontaneous and Pharmacologically Induced Release of Monoamines

As reported by many workers on the basis of data obtained with histochemical, biochemical or autoradiographic methods, the most important pathway of dopamine (DA) containing neurons is represented by the dopaminergic nigro-striatal system. The caudate nucleus which contains numerous terminals of DA neurons appeared to be an appropriate preparation for our first investigations on release of ^3H DA endogenously synthesized from ^3H tyrosine. The first studies of my coworkers, Drs. BESSON, FELTZ and CHERAMY were made on curarized and unanaesthesized cats by a modification of the cup technique used by MITCHELL (1966)

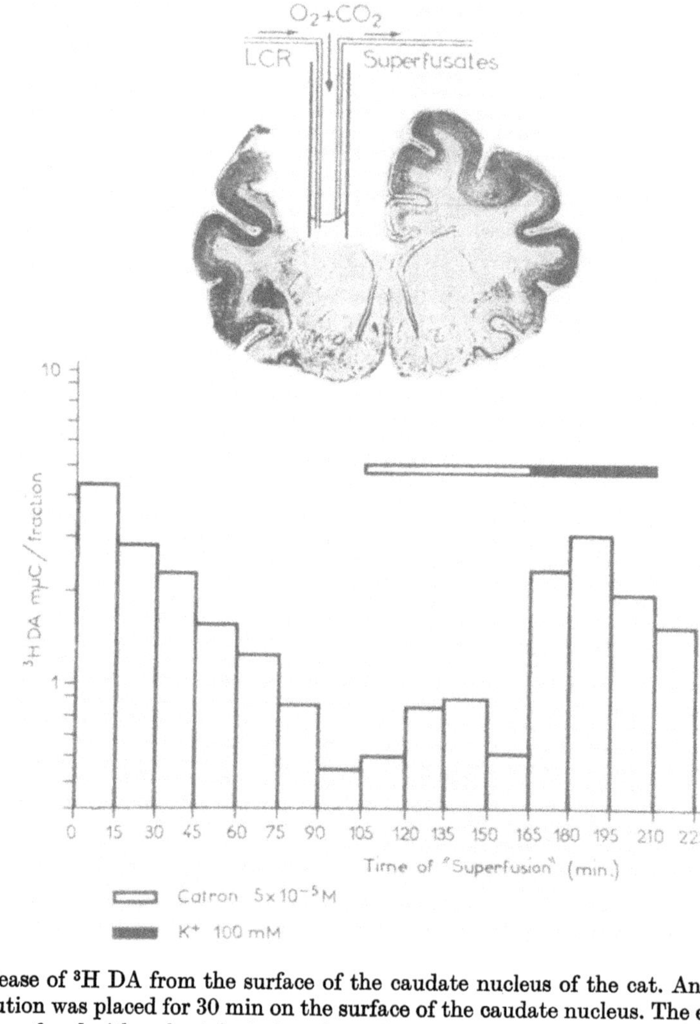

Fig. 1. Release of ^3H DA from the surface of the caudate nucleus of the cat. An ^3H tyrosine (37 μC) solution was placed for 30 min on the surface of the caudate nucleus. The exposed area was then superfused with a physiological medium. ^3H DA was estimated in individual successive 15 min fractions. The MAO inhibitor catron and K were added in the superfusing medium during 60 and 45 min respectively

to demonstrate the cortical release of acetylcholine. A small area of the surface of the caudate nucleus was exposed by removing a part of the cortex. A cup was then placed stereotaxically on the surface of the caudate nucleus and ^3H tyrosine dissolved in an artificial cerebrospinal fluid had been introduced in the cup for 30 min. This solution was then removed and the exposed area of the caudate nucleus had

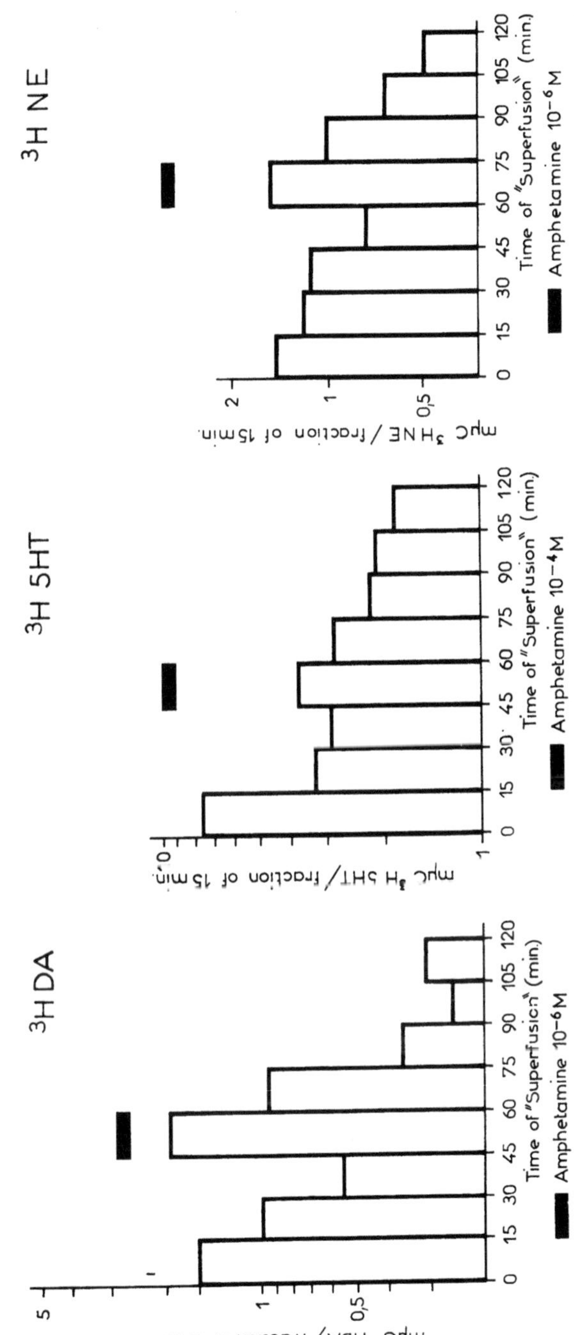

Fig. 2. Effects of D.amphetamine on the release of ^{3}H 5-HT, ^{3}H NE and ^{3}H DA from isolated structures of the rat brain. Isolated striatum were incubated with 16 μC of ^{3}H tyrosine or 60 μC of ^{3}H tryptophan and the isolated medulla oblongata was incubated with 30 μC of ^{3}H tyrosine. After a 30 min incubation period, isolated structures were superfused with a physiological medium, and 15 min fractions were collected. ^{3}H DA, ^{3}H 5-HT and ^{3}H NE released from striatum and medulla oblongata respectively were estimated in individual fractions. Amphetamine was added for 15 min in the respective superfusing medium

been superfused with a physiological oxygenated medium continously for 3 h. Successive fractions of the superfusates were collected. ^{3}H DA spontaneously released could be identified in these various superfusates fractions after acetylation and cochromatography of the acetyl derivative of ^{3}H DA. As shown in Fig. 1, the output of ^{3}H DA increased when a MAO inhibitor or the depolarizing ion K was introduced during a short time period in the superfusing medium. Amounts of ^{3}H DA as low as 5 pg could be thus detected in the superfusing medium.

Further studies were made on the isolated striatum of the rat in order to evaluate the specificity and the characteristics of the process of DA release, but also to test its sensitivity to various drugs and other substances which likely act as transmitters in this structure. As already described [Besson et al., 1969 (1)], the isolated striatum can be incubated in a small chamber for 30 min with ^{3}H tyrosine and then superfused continuously with an artificial medium. As in the in vivo study on the cat caudate nucleus, ^{3}H DA endogenously synthesized in DA terminals of the rat striatum was released spontaneously and could be identified in the successive collected fractions obtained by superfusion of the tissue. Similarly, a MAO inhibitor or the addition of potassium increased the output of ^{3}H DA. An enhanced liberation of ^{3}H DA could also be observed during a "field" electrical stimulation of the striatum. Substances, normally contained in high concentrations in the striatum, such as 5-HT and acetyl-choline were also very active, they increased the efflux of ^{3}H DA when added in low concentration (10^{-7} M) in the superfusing medium. The action of acetyl-choline could only be detected after inhibition of cholinesterase by eserine. This drug was almost not effective when used alone.

The specificity of the observed modifications in ^{3}H DA release from DA nerve terminals was examined in two ways: (1) The action of acetyl-choline on ^{3}H DA was compared to its effects on ^{3}H tyrosine, the precursor of the ^{3}H amine; we also examined the effect of acetyl-choline on ^{14}C urea and ^{3}H inuline previously distributed by diffusion in the extracellular space. Acetyl-choline was not able in any case to modify the exit rate of these three labelled compounds from striatal tissue. (2) In the second experiment, the effects of D-amphetamine on the release of various ^{3}H amines synthesized from their precursors in aminergic neurons were compared: as for ^{3}H DA in the striatum, we observed that ^{3}H NE and ^{3}H 5-HT could be released spontaneously from isolated medulla oblongata and striatum of rats previously incubated for a short time period with ^{3}H tyrosine and ^{3}H tryptophan respectively [Besson et al., 1969 (2, 3)]. As indicated in Fig. 2, amphetamine in low concentration (10^{-6} M) markedly increased the quantities of ^{3}H DA and ^{3}H NE in striatum and medulla oblongata superfusates respectively (135% and 87% when compared with the ^{3}H amines content in the preceding fraction). But this drug had almost no effect on ^{3}H 5-HT release; during the amphetamine superfusion, the quantities of the ^{3}H amine were only increased by 13% when compared with the ^{3}H amine content of the previous fraction, even with high concentration of the drug (10^{-3} M). In conclusion, it appears that changes in ^{3}H DA efflux from DA terminals observed with acetyl-choline or amphetamine represent specific phenomenons. They are likely due to a direct effect of these substances on dopaminergic endings.

2. Simultaneous Effects of Drugs on Synthesis and Release of ^{3}H Amines

By using the in vitro experimental model, described above, we observed that the ^{3}H DA release from DA terminals of the rat striatum was more pronounced just after the end of the incubation with ^{3}H tyrosine than one or 2 h later. This could be visualized by estimating the relative proportion of ^{3}H DA in superfusates and tissues as a function of time. This ratio decreased from 14.3 to 0.7 during the 2 h which followed the incubation period, indicating that the newly synthesized ^{3}H DA was preferentially released spontaneously. This is in agreement with the observations of KOPIN et al. (1968) who found that newly synthesized ^{14}C NA was preferentially released from splenic sympathetic nerve endings during nerve stimulation.

We were thus convinced that it was necessary to study the release of newly synthesized transmitters. This was done by incubating slices of isolated structures

Table 1. *Effects of various drugs on synthesis and release of monoamines in various structures of the CNS*

^{3}H 5HT (hypothalamus)	Imipramine	Drug administered in vivo			Drug added in incubating medium		
		Time before (min)	Tissue %	Medium %		Tissue %	Medium %
		30	61[b]	—	10^{-4} M	22[c]	200[c]
^{3}H CA or ^{3}H NE (medulla oblongata)	Desmethylimipramine	90	114	183[c]	10^{-4} M	226[c]	165[c]
	Amphetamine	30	103	178[c]	—	—	—
^{3}H DA (striatum)	Desmethylimipramine	90	99	101		—	—
	Amphetamine	30	52[c]	184[c]	10^{-4} M	46[c]	—
	Thioproperazine	180	146[a]	149[b]	10^{-4} M	107	111
	Amphetamine +	90	95	322[c]		—	—
	Thioproperazine	60					

[a] $P < .05$; [b] $P < 0.01$; [c] $P < 0.001$

Drugs were administered in vivo (5 mg/kg i.p.) or added in incubating medium (in vitro). Animals were killed at various time intervals after drugs administration. Structures were dissected, cut in slices and incubated for 15 min with ^{3}H precursors. ^{3}H amines were estimated in slices and incubating medium. Results are the mean of groups of 6 to 8 animals ± SEM and are expressed in per cent of respective control values.

of the CNS with ^{3}H precursors for 15 min and by estimating ^{3}H amines quantities in incubating medium and in slices at the end of the incubation period. The simultaneous effects of various drugs on synthesis and release of ^{3}H amines in various structures of the brain of rats were thus analyzed. The animals were pretreated with the drug at various time before their sacrifice or the drug was directly added in the incubating medium 10 to 15 min before the labelled precursor. From the results represented in Table 1, in which effects of various drug treatments on synthesis and release of ^{3}H amines in the hypothalamus, medulla oblongata and striatum are reported, a few remarks can be made:

1. The absolute values of ^{3}H amines found in incubating medium may slightly vary with the incubation conditions: size of the slices or volume of the incubating medium. The quantities of ^{3}H amines detected in the medium are generally relatively small as compared to those found in tissues, they correspond to the sum of two opposite processes: release and reuptake. Therefore, modifications of ^{3}H

amines content in medium observed with drugs may result from their effect on one of these processes or on both.

2. Imipramine markedly enhanced the liberation of ^{3}H 5-HT newly synthesized from ^{3}H tryptophan in the hypothalamus (Jeanne-rose, Hamon and Javoy, unpublished observations). Desmethylimipramine (DMI) had a similar effect on ^{3}H NE in the medulla oblongata (Thierry and Blanc, unpublished observations) and finally amphetamine induced a similar potent effect respectively on ^{3}H CA and ^{3}H DA in the medulla oblongata and the striatum [Besson et al., 1969 (3)]. These antidepressive drugs effects observed after in vivo pretreatment or addition in medium are likely related to their direct inhibiting action on the reuptake of the liberated ^{3}H amines. Imipramine, DMI and amphetamine have been shown to inhibit the specific uptake processes acting at the neuronal membrane level for 5-HT, NE and DA in corresponding aminergic neurons respectively in in vivo or in vitro studies (Glowinski et al., 1966; Fuxe and Ungerstedt, 1968; Carlsson et al., 1968; Snyder et al., 1968). The pronounced effect of these drugs on the output of ^{3}H 5-HT or ^{3}H CA (about 100% increased in all cases) underlined the potency of the specific inactivating reuptake processes in aminergic neurons. Furthermore, it appears that at least 25 to 40% of the newly synthesized ^{3}H amines are rapidly liberated and stay in the vicinity of the receptor sites for longer time period after the administration of these drugs to the animals. For example, the data obtained with amphetamine on noradrenergic and dopaminergic neurons agree closely with the increased release of NE (Carr and Moore, 1969) and of DA (McKenzie and Szerb, 1968) detected in in vivo studies respectively in the hypothalamus and in the caudate nucleus of the cat after administration of this drug.

3. DMI has been shown to affect specifically the uptake process for NE and to be ineffective on DA uptake process (Glowinski et al., 1966; Fuxe and Ungerstedt, 1968; Snyder et al., 1968). Our present results clearly demonstrate that DMI has no effect on DA reuptake, furthermore they show the ineffectiveness of this drug on DA synthesis as well. The clear cut difference obtained with DMI on NE and DA neurons provides a further indication of the specificity and sensitivity of the experimental approach used in this study.

4. A good appreciation of the ^{3}H amines quantities synthesized during the 15 min incubation period can be obtained by adding ^{3}H amines accumulated in tissues and in incubating medium (the ^{3}H amines content in medium differs for the 5-HT, NE and DA neurons, but represents about 10 to 25% of the ^{3}H amines content in tissues in control animals). Drugs which induced similar stimulating effects on the output of ^{3}H amines in incubating medium acted differently on the ^{3}H amines synthesis. The effects of imipramine and amphetamine on 5-HT and DA terminals are comparable: imipramine given in vivo or in vitro is a potent inhibitor of ^{3}H 5-HT synthesis from ^{3}H tryptophan. This confirms and extends the findings of Corrodi and Fuxe (1969) who, with the help of 5-HT synthesis inhibitors, observed that imipramine and some of its derivatives decreased the 5-HT turnover in brain. Similarly, amphetamine given in vivo [Besson et al., 1969 (3)] or added in vitro diminished the rate of ^{3}H DA synthesis from ^{3}H tyrosine in DA terminals of the dopamine nigro-striatal system. More recent experiments in which the rate of conversion of ^{3}H tyrosine to ^{3}H DOPA have been estimated, indicate that the inhibition of synthesis occurred at the first step of the amine synthesis.

The mechanisms by which imipramine and amphetamine inhibited the synthesis of 5-HT and DA are not yet well understood. The fact that these effects can be observed by direct contact of the drug with slices of isolated structures suggests that these compounds act directly on the regulation of these amine synthesis in their respective terminals. It likely excludes an interneuronal feed-back effect which could be induced by the increased amines amounts in synapses observed with these drugs.

The stimulating effects of amphetamine and DMI on ^{3}H CA or ^{3}H NE formation from ^{3}H tyrosine in the noradrenergic neurons of the medulla oblongata are

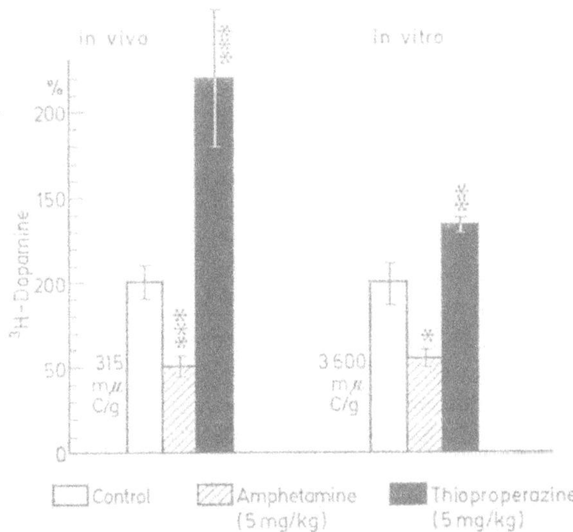

Fig. 3. Effects of D.amphetamine and thioproperazine on the in vitro and in vivo synthesis of ^{3}H DA from ^{3}H tyrosine in the rat striatum. Groups of 8 rats received an intraperitoneal injection of D.amphetamine or thioproperazine 90 and 180 min respectively before the stereotaxic microinjection of ^{3}H tyrosine (2 μC) in the striatum (in vivo) or their sacrifice (in vitro). In the in vivo experiment rats were killed 15 min after ^{3}H tyrosine administration. In the in vitro experiment, striatum were dissected out in slices and incubated with 25 μC of ^{3}H striatum for 15 min. ^{3}H DA was estimated in tissues. Results are expressed in % of respective control values and correspond to the mean ± SEM. Number correspond to absolute levels of ^{3}H DA for control animals. *P < .02; **P < .01; ***P < 0.001

also comparable. The effects were observed by adding ^{3}H amines accumulated in tissues and in medium. The increased ^{3}H NE synthesis detected in this structure after amphetamine treatment is in agreement with and confirms our previous findings: the NE turnover in the brain stem hasbeen shown to be accelerated after amphetamine treatment in normal or stressed rats (Javoy et al., 1968). DMI, like amphetamine, stimulated NE synthesis; this effect was particularly striking when the drug was added directly in the incubating medium of the medulla oblongata slices. The combined actions of both antidepressive drugs on reuptake processes and synthesis in noradrenergic neurons contribute to elevate the NE levels in synapses. It remains to demonstrate if these two drugs effects on NE synthesis are linked to their effect on reuptake process. Furthermore, the opposite action of amphetamine on NE and DA synthesis in noradrenergic and dopaminergic neurons respectively may reveal differences in the regulatory mechanisms of amine synthesis in these two types of neurons.

5. Neuroleptics such as chlorpromazine have been shown to accelerate the turnover of DA or its synthesis in DA terminals of the nigrostriatal system of DA neurons (Nybäck and Sedvall, 1968). We observed in our laboratory that thioproperazine (Mageptil) a potent neuroleptic, had a similar but more pronounced effect (Cheramy et al., 1969). As indicated in Table 1, the increased synthesis of ^{3}H DA from ^{3}H tyrosine was associated with a marked enhancement of the release of the newly synthesized amine (Cheramy et al., 1969). Both effects on synthesis and release were still detectable in amphetamine pretreated animals although thioproperazine blocked the pharmacological syndrom of amphetamine. This provides a good evidence for an indirect action of neuroleptics on DA neurons. The lack of effect of thioproperazine on ^{3}H DA synthesis and release after in vitro addition of the drug is a further argument in favor of the induction of an interneuronal activation of DA neurons following its in vivo administration. Thus, the increased output of ^{3}H DA from slices with thioproperazine cannot be compared to the effects of antidepressive drugs on aminergic neurons. The increased release of the newly synthesized DA, seen after thioproperazine treatment is consecutive to a general and sustained activation of DA metabolism in DA neurons as revealed by the acceleration of ^{3}H DA synthesis in DA terminals and DA cell bodies of the substantia nigra (Javoy et al., 1969).

6. The effects of amphetamine and thioproperazine pretreatment on DA synthesis described on the basis of data obtained with the new experimental approach mentioned above were confirmed in in vivo study by measuring the accumulation of ^{3}H DA in the striatum of rat 15 min after the local injection of ^{3}H tyrosine. As shown in Fig. 3 comparable effects on DA synthesis could be detected (Javoy et al., 1969).

3. Intraneuronal Forms of Storage of Monoamines and Release Processes

We have previously mentioned that newly synthesized DA was preferentially released spontaneously as compared to DA stored for longer time period. This suggests that this amine is not homogenously stored in DA terminals. Some experimental facts support this hypothesis in the case of DA terminals, but in those of NE and 5-HT as well.

In a previous study, we have reported that the half-life for the rate of ^{3}H DA utilization in the isolated rat striatum was about 60 min, 1 h after ^{3}H DA formation from ^{3}H tyrosine [Besson et al., 1969 (1)]. This rate of DA utilization is similar to estimations made in in vivo turnover studies (Iversen and Glowinski, 1966; Costa and Neff, 1965). However, the rate of utilization of the newly synthesized ^{3}H DA as measured by the decline of ^{3}H DA output from isolated striatum immediately after the ^{3}H amine synthesis from ^{3}H tyrosine was much faster (half-life $\neq$ 15 minutes). The discrepancy between these rates of ^{3}H DA utilization can be explained if ^{3}H DA is assumed to be distributed in two pools or compartments corresponding to different storage forms of the ^{3}H amine. The newly synthesized ^{3}H DA may be localized in a small pool in which DA is utilized much more rapidly than the rate of utilization of ^{3}H DA synthesized an hour earlier. SEDVALL et al. (1968) reported that the rate of brain NE synthesis estimated from the content of ^{14}C NE accumulated at the end of a short constant intravenous infusion of ^{14}C tyrosine largely exceeded the rate calculated in turnover studies in which NE was assumed for simplification to be localized in one single pool. These data lead them to conclude that NE was likely stored in two pools presenting different characteristics. We arrived at the same conclusion for 5-HT neurons when we compared the rate of ^{3}H 5-HT synthesis from ^{3}H tryptophan and the turnover rate of 5-HT calculated on the assumption of a single compartment model (MACON, SOKOLOFF and GLOWINSKI, unpublished observations).

In fact, careful studies on the disappearance rate of ^{3}H NE previously taken up and stored in rat brain NE neurons (GLOWINSKI et al., 1965; GLOWINSKI and IVERSEN, 1966) or on ^{3}H NE endogenously synthesized from ^{3}H tyrosine in the

Table 2. *Effects of stress on the utilization of various intraneuronal forms of NE in the brain stem of rats*

Time (min)	^{3}H NE (% decrease)	p.	Time (min)	^{3}H NE (% decrease)	p.
2	27	< .05	20	9	ns
5	20	< .01	180	7	ns

Groups of 8 rats received 23 μC of ^{3}H tyrosine intracisternally and were stressed during 15 min by electrical shocks applied to the feet at various time intervals (time) after the ^{3}H precursor administration. ^{3}H NE was estimated in tissues of stressed rats and in those of control animals killed at similar time intervals. Results represent the percentage of ^{3}H NE decrease in tissues of stressed animals as compared to the respective control values.

brain stem revealed the existence of a multiphasic disappearance of the amine [THIERRY et al., 1970 (1)]. Similarly, we observed that ^{3}H 5-HT endogenously synthesized from ^{3}H TRY disappeared in a multiexponential fashion in the brain stem of rats (THIERRY et al., 1969). In both cases an early short and rapid phase of disappearance of the ^{3}H amines was followed by a slower phase lasting for many hours. These indications as well as pharmacological (HÄGGENDAL and LINDQVIST, 1964; WEISSMAN et al., 1966; GLOWINSKI et al., 1966) or physiological observations (WISE and STEIN, 1969; KORDON and GLOWINSKI, 1969) lead us to believe that the three monoamines: NE, DA and 5-HT are heterogenously stored in their respective terminals. We have thus assumed for simplification, as a working hypothesis, the coexistence of two main "compartments" or "pools" of amines. The early short and rapid phase of disappearance of ^{3}H amines endogenously synthesized may correspond in part to the rapid utilization of the ^{3}H amines in a small "functional pool" from which the transmitter is rapidly and preferentially released. The second and slower phase of disappearance of the ^{3}H amines represents likely a main "storage pool" in equilibrium with the "functional pool".

This hypothesis was tested again in further studies on noradrenergic neurons of the rat brain stem. As previously reported (THIERRY et al., 1968) we have examined in earlier studies the effect of a stress induced by electrical shocks applied to the feet on the rate of utilization of ^{3}H NE, which had been taken up and stored in noradrenergic neurons 2 h before the stress session. A marked increase in the rate of NE utilization could only be detected after the application of six stress sessions of 10 min each spaced over a total period of 3 h. It was therefore interesting to test the sensitivity of the newly synthesized NE, localized in part in the "functional pool" to a single mild stress of short duration. The following experiment was done: rats of various groups received an intracisternal injection of tracer doses of ^{3}H tyrosine and were stressed for 15 min by electrical shocks applied to the feet at various time intervals after the ^{3}H precursor administration. ^{3}H NE levels were estimated in tissues of stressed animals and were compared with those of control rats killed at similar time intervals after ^{3}H tyrosine administration. As indicated in Table 2, it appears clearly that the newly synthesized amine was preferentially utilized during the short and mild stress session as compared to the ^{3}H amine stored for longer time period. No changes in endogenous NE were detected [Thierry et al., 1970 (2)]. These results are in agreement with the hypothesis of a greater availability of the newly synthesized transmitter for evoked release. The ^{3}H amine stored for longer time period may be used in more dramatic stress situation. It remains to demonstrate and to explain the morphological basis of this various forms of intraneuronal storage for NE as well as other amines.

4. Conclusion

Two main approaches can be used in order to understand the involvement of particular systems of aminergic neurons in various functions of the CNS or in the mechanisms of action of psychotropic drugs. (1) The first possibility is to modify pharmacologically the levels of monoamines in a local area of the brain and to correlate these effects to specific changes in physiological functions or in behavioral effects of psychotropic drugs. (2) The second approach requires the estimations of changes in levels of amine released from terminals of well defined systems of aminergic neurons. Some of the experiments and results which have been decribed are in the line of this research direction. Moreover, they emphasized the importance of newly synthesized transmitter in release processes occurring in physiological or pharmacological conditions. They also pointed out the complexity of the interelationship between the processes of release and reuptake and the regulatory mechanisms of amine synthesis.

Summary

Release of labelled monoamines from central aminergic terminals has been studied in in vivo or in vitro studies. In all cases, ^{3}H catecholamines (CA) or ^{3}H serotonin (5-HT) have been synthesized endogenously from their respective precursors, ^{3}H tyrosine or ^{3}H tryptophan (TRY).

The spontaneous and pharmacologically induced release of ^{3}H DA from terminals of the caudate nucleus of the cat has been described. Similar observations were made for ^{3}H DA, ^{3}H NE and ^{3}H 5-HT, endogenously synthesized, on isolated structures of the rat brain.

The effects of various psychotropic drugs on synthesis, release and reuptake processes in central aminergic terminals have been examined in slices of isolated structures of brain of animals pretreated with the drug by estimating simultaneously ^{3}H amines formed from ^{3}H precursors and accumulated in tissues and incubating medium. Finally, experimental evidence in favor of the existence of intraneuronal storage forms of monoamines and for a preferential release of the newly synthesized amines have been given.

References

Besson, M. J., Cheramy, A., Feltz, P., Glowinski, J.: (1) Demonstration of release of newly synthesized dopamine from dopamine-containing terminals in the striatum of the rat. Proc. nat. Acad. Sci. (Wash.) **62**, 741 (1969).

— — — — (2) Mise en évidence de la libération de la dopamine (DA) et de la sérotonine (5HT) au niveau du noyau caudé chez le rat. J. Physiol. (Paris) **61**, 90 (1969).

— — Glowinski, J.: (3) Effects of amphetamine and desmethylimipramine on amines synthesis and release in central catecholamine containing neurons. Europ. J. Pharmacol. **7**, 111 (1969).

Carlsson, A., Fuxe, K., Ungerstedt, U.: The effect of imipramine on central 5-hydroxytryptamine neurons. J. Pharm. Pharmacol. **20**, 150 (1968).

Carr, L. A., Moore, K. E.: Norepinephrine: Release from brain by d-amphetamine in vivo. Science **164**, 322 (1969).

Cheramy, A., Besson, M. J., Glowinski, J.: Increase release of dopamine from striatal dopaminergic terminals in the rat after treatment with a neuroleptic: thioproperazine. Europ. J. Pharmacol. (in press).

Corrodi, H., Fuxe, K.: Decreased turnover in central serotonin nerve terminals induced by antidepressant drigs of the imipramine type. Europ. J. Pharmacol. **7**, 56 (1969).

Costa, E., Neff, N. H.: Isotopic and non isotopic measurements of the rate of catecholamines biosynthesis. Proceedings of the second symposium on the Parkinson's disease information and research center, p. 14. (Costa, E., Cote, L. J., Yahr, M. D., Eds.). New York: Raven Press 1965.

Fuxe, K., Ungerstedt, U.: Histochemical studies on the effects of (+) amphetamine, drugs of the imipramine group and tryptamine on central catecholamines and 5-hydroxytryptamine neurones after intraventricular injections of catecholamine and 5-hydroxytryptamine. Europ. J. Pharmacol. **4**, 135 (1968).

Glowinski, J., Axelrod, J., Iversen, L.: Regional studies of catecholamines in the rat brain. IV. Effects of drugs on the disposition and metabolism of ^{3}H norepinephrine and ^{3}H dopamine. J. Pharmacol. exp. Ther. **153**, 30 (1960).

— Iversen, L.: Regional studies of catecholamines in the rat brain. I. The disposition of ^{3}H norepinephrine, ^{3}H dopamine ^{3}H DOPA in various regions of the brain. J. Neurochem. **13**, 665 (1966).

— — Axelrod, J.: Storage and synthesis of norepinephrine in the reserpine treated rat brain. J. Pharmacol. exp. Ther. **151**, 385 (1966).

— Kopin, I., Axelrod, J.: Metabolism of ^{3}H norepinephrine in the rat brain. J. Neurochem. **12**, 25 (1965).

Häggendal, J., Lindqvist, M.: Disclosure of labile monoamine fractions in brain and their correlation to behaviour. Acta physiol. scand. **60**, 351 (1964).

Iversen, L., Glowinski, J.: Regional studies of catecholamines in the rat brain. II. Rate of turnover of catecholamines in various regions. J. Neurochem. **13**, 671 (1966).

Javoy, F., Hamon, M., Glowinski, J.: Disposition of newly synthesized biogenic amines in cell bodies and terminals of central aminergic neurons. I. Effect of amphetamine and thioproperazine on the metabolism of catecholamines in the caudate nucleus, the substantia nigra and the ventromedial nucleus of the hypothalamus. Europ. J. Pharmacol. (in press).

— Thierry, A. M., Kety, S. S., Glowinski, J.: The effect of amphetamine on the turnover of brain norepinephrine in normal and stressed rats. Comm. Behav. Biol. Part A **1**, 43 (1968).

Kopin, I. J., Breeze, G. R., Krauss, K. R., Weise, U. K.: Selective release of newly synthesized norepinephrine from the cat spleen during sympathetic nerve stimulation. J. Pharmacol. exp. Ther. **161**, 271 (1968).

KORDON, C., GLOWINSKI, J.: Selective inhibition of superovulation by blockade of dopamine synthesis during the "critical period" in the immature rat. Endocrinology 85, 924 (1969).
MCKENZIE, G. M., SZERB, J. C.: The effect of dihydroxyphenylalanine pheniprazine and dextro-amphetamine on the in vivo release of dopamine from the caudate nucleus. J. Pharmacol. exp. Ther. 162, 302 (1968).
MITCHELL, J. F.: Acetyl-choline release from the brain. In: Mechanisms of release of biogenic amines, Vol. 5, pp. 425—438 (VON EULER, U. S., ROSELL, S., UVNÄS, B., Eds.). Venner Gren Center Int. Symp. S. Pergamon Press 1966.
NYBÄCK, H., SEDVALL, G.: Effect of chlorpromazine on accumulation and disappearance of catecholamines from tyrosine ^{14}C in brain. J. Pharmacol. exp. Ther. 162, 294 (1968).
SEDVALL, G. C., WEISE, U. K., KOPIN, I. J.: The rate of norepinephrine synthesis measured in vivo during short intervals. Influence of adrenergic nerve impulse nerve activity. J. Pharmacol. exp. Ther. 159, 274 (1968).
SNYDER, S. H., GREEN, A. I., HENDLEY, E. D.: Kinetics of ^{3}H norepinephrine accumulation into slices from different regions of the rat brain. J. Pharmacol. exp. Ther. 164, 90 (1968).
THIERRY, A. M., BLANC, G., GLOWINSKI, J.: Preferential utilization of newly synthesized norepinephrine in the brain stem of stressed rats. Europ. J. Pharmacol. (in press).
— — — (1) Effects of stress on the metabolism of norepinephrine, dopamine and serotonin in the central nervous system of the rat. III. Modifications of various intraneuronal storage form of monoamine under stress condition (1970) (in preparation).
— FEKETE, M., GLOWINSKI, J.: Effects of stress on the metabolism of noradrenaline, dopamine and serotonin in the central nervous system of the rat. II. Modifications of serotonin metabolism. Europ. J. Pharmacol. 4, 384 (1968).
— JAVOY, F., GLOWINSKI, J., KETY, S. S.: Effects of stress on the metabolism of norepinephrine, dopamine and serotonin in the central nervous system of the rat. I. Modifications of norepinephrine turnover. J. Pharmacol. exp. Ther. 163, 163 (1968).
VOGT, M.: Release from brain tissue of compounds with possible transmitter functions: Interaction of drugs with those substance. Brit. J. Pharmacol. 37, 325 (1969).
WEISSMAN, A., KOE, B. K., TENEN, S. S.: Antiamphetamine effects following inhibition of tyrosine hydroxylase. J. Pharmacol. exp. Ther. 151, 339 (1966).
WISE, C. D., STEIN, L.: Facilitation of brain self stimulation by central administration of norepinephrine. Science 163, 299 (1969).

Dr. J. GLOWINSKI
College de France
Laboratoire de Neurophysiologie
Generale Unite de Neuropharmacologie
Biochimique
11 Place Marcelin-Berthelot
F-75 Paris V^e^

Discussion

AXELROD: Dr. GLOWINSKI, in the very last experiment you said that there was an increase synthesis in vitro, that is after you stimulate. Have you any idea as to the cause of this, whether it is due to increase in the enzyme or increased capacity?

GLOWINSKI: An increased synthesis of norepinephrine from tyrosine can be seen on slices of rat medulla oblongata in vitro. Slices which have been obtained, for example, from the rat and which have been stressed by electrical shocks applied to the feet with rest periods over a total time of 3 h. I do not think that in this case you will be able to see an increased tyrosine hydroxylase activity.

However, as I mentioned yesterday, such an increase activity can be induced if the animals are submitted chronically to a stressful situation once a day for a few days.

AXELROD: What would your thinking be as to the mechanisms of this accelerated synthesis.

GLOWINSKI: One possible explanation for the mechanism by which synthesis is increased in a stressful situation, and is still detectable in vitro, may be related to a modification of the initial step of synthesis, initiated by a diminution of intraneuronal norepinephrine. This decrease in norepinephrine may occur in a site in which it normally regulates the conversion of tyrosine to dopa according to the hypothesis of synthesis regulation by end product inhibition. This mechanism of synthesis regulation which has been demonstrated in central noradrenergic neurons appears also to be present in central serotoninergic neurons. Thus, J. MACON has shown in our laboratory that the rate of conversion of ^{3}H-tryptophan to ^{3}H-serotonin was diminished in brain of rats which had been previously treated with an MAO inhibitor, indicating that high endogenous levels of serotonin are inhibiting the rate of conversion of ^{3}H-tryptophan to ^{3}H-serotonin.

VON EULER: I see you have expressed the idea here of two pools one of which you call functional pool, which may have something to do with the available pool and some subsequently described similar pools, and you also gave some figures for the relative size of these pools which were most interesting. I wonder whether you have any available figures for the noradrenaline pools too.

GLOWINSKI: I have no figure for the size of the so-called "functional pool" in norepinephrine containing neurons. In the case of the serotoninergic neurons, this has been calculated on the basis of the curve of ^{3}H-serotonin disappearance in brain tissues. The ^{3}H-amine being previously synthetized from ^{3}H-tryptophane. Nevertheless, I believe that the size of this pool in noradrenergic neurons is relatively small. This is also suggested by the amount of endogenous norepinephrine in animals which had been previously treated with reserpine 24 h earlier and which had recovered from the reserpine syndrome.

VON EULER: At any rate you assume that it is going to be a great deal smaller than the total granular pool, as we have also assumed.

GLOWINSKI: It is likely that the size of the "functional pool" of norepinephrine is smaller than the amount of norepinephrine which is found in particules after homogenization. It is likely that this form of amine, stored in central noradrenergic neurons, is more labile and not tightly bound. According to data obtained with Drs. AXELROD and SNYDER a few years ago, on the subcellular distribution of brain, ^{3}H-norepinephrine as a function of time after its intraventricular administration, it appears that the newly stored ^{3}H-amine, which can be preferentially utilized, for example, in stress situation, is found after homogenization to a greater extent in the soluble fraction than the ^{3}H-amine stored in tissues for a longer period of time. This does not mean that the newly synthesized amine is not localized in vesicles. The binding of this amine is, perhaps, not so important than the binding of the "older" amine in vesicles.

PHILIPPU: Did you try to determine the dopamine metabolites after perfusing with dopamine-releasing agents, as for example, acetylcholine or amphetamine?

GLOWINSKI: No, we have not estimated the ^{3}H-dopamine metabolites in these release studies.

CARLSSON: In our laboratory, PERSSON and WALDECK also have some data indicating a very rapid preferential metabolism of newly synthesised monoamines from precursors, radioactive tyrosine or dopa, but there is one thing I would like to ask you about. Are we necessarily dealing here with a specific compartment, could not the whole phenomenon be rather a geometric one in the sense that when radioactive tyrosine enters a nerve terminal, the specific activity of the products formed from it would be higher, close to the surface, than in the centre. Preferential release of superficial high-activity transmitter would then occur without necessarily postulating the presence of a well-defined compartment with high turnover.

GLOWINSKI: It is difficult to present good experimental evidence against the possibility of a preferential localization of the newly synthesized labelled amine near the membrane, as a consequence of geometric distribution. However, such distribution may also occur for endogenous tyrosine taken up in noradrenergic neurons and, therefore, the newly synthesized norepinephrine will also be near the membrane and thus preferentially exposed to release. The autoradiographic data which have been obtained after ^{3}H-norepinephrine administration do not give any indication, as far as I know, of a preferential localization of the radioactive material near the membrane.

I am not convinced that the effects on release which we obtained with newly synthesized ^{3}H-norepinephrine from ^{3}H-tyrosine under stress condition are mainly due to geometric distribution of the labelled precursor and consequently labelled amine in nerve ending. The existence of various forms of storage and of the "functional" characteristics of the newly synthesized amine may be given by other experimental approaches. Recently, we have been able to demonstrate, with Dr. KORDON, that inhibition of dopamine synthesis in the median eminence of rat was leading to the blockade of the ovulatory process. The effect was induced by injecting α-methyl-p-tyrosine.

The inhibition of ovulation was seen immediately after the injection of the inhibitor, by which time the endogenous levels of amine in the tissues were almost decreased. This indicates that the lack of release of newly synthesized dopamine was affecting immediately at the hypothalamic level, the complex process of ovulation. It demonstrates the preferential ability of newly formed amine to be physiologically active at the receptor site.

It should be added that this rapid effect of this norepinephrine synthesis inhibitor was seen only when the drug was given during the critical period of ovulation and was not observed when the drug was given a few hours earlier. Thus, it appears in this later situation, that the dopamine stored for a longer period of time could be released instead of the newly synthesized amine and thus active at the receptor site. This experiment is in agreement with the hypothesis of the existence of more than one form of storage and with the idea of a preferential release of the newly synthesized catecholamine.

GEFFEN: There are many clear cut results here but there is one basic thing I am confused about. This is, the spontaneous release of radioactivity into the superfusate which declines with time and which you can reverse with various drugs. Is this efflux of radioactive amines, occurring across membranes, which are impermeable to circulating amines? The second problem is, are they being released by nerve impulses or is this some sort of spontaneous release? Is it possible that you have overloaded the storage capacity of these neurons and this is diffusion out of the neurons? The reason I ask this is that when you added monoamine oxidase inhibitors, you could increase the release, whereas in the peripheral adrenergic nerves it is very difficult to show any increased release of noradrenaline in the presence of monoamine oxidase inhibitors, whose action one assumes is intracellular.

GLOWINSKI: It is difficult to answer the question about the exact mechanism of release of the newly synthesized radioactive amine in our in vitro preparation and release by nerve impulse or spontanenous release. However, this is unlikely to be the consequence of an overloading of amine in the neuron. The specific activity of ^{3}H-tyrosine is very high, 30 Ci/mM, and the quantities of ^{3}H-amine in the tissues negligible compared to the non-labelled amine. Furthermore, the output of ^{3}H-amine in medium does not appear to be due to physical diffusion process: the relative amount of ^{3}H-amine in medium compared to the tissue, is different for serotoninergic neurons, dopaminergic neurons or noradrenergic neurons and is affected differently by various drugs.

Bayer-Symposium II, 252—255 (1970)

The Effect of Different Neuroleptic Drugs on the Monoamine Neurons of the Rat Brain

H. CORRODI

Recently it has been shown that psychoactive substances induce changes in the neuronal activity in the monoamine neurons of the rat brain leading to changes in the turnover of the respective monoamine. Such a turnover change can be revealed by blocking the biosynthesis of the corresponding amine. For references see ANDÉN et al. (1968).

As inhibitor of the hydroxylation of tyrosine, the rate limiting step in the biosynthesis of dopamine and noradrenaline H 44/68 (methylester hydrochloride of dl-α-methyltyrosine) and as inhibitor of tryptophan hydroxylase H 22/54 (α-propyldopacetamide) were used [CORRODI and HANSON, 1966; ANDÉN et al., 1966; CARLSSON et al., 1963; CORRODI et al., 1967 (1)].

The results obtained with the neuroleptic drugs haloperidol and chlorpromazine have already been published [CORRODI et al., 1967 (2); ANDÉN et al., 1967]. They demonstrated that these neuroleptics accelerate the disappearance of noradrenaline in the rat brain after an inhibition of its biosynthesis, whereas an acceleration of the disappearance of dopamine was observed only after high doses or repeated administration of the neuroleptic drug. Since the rate of synthesis of these amines is dependant on intact neuronal impulse flow (ANDÉN et al., 1966; CORRODI and MALMFORS, 1966) these results would mean that after chlorpromazine and haloperidol the noradrenaline neurons are strongly activated whereas the dopamine neurons are activated only to a limited extent. This activation has been ascribed to a compensatory feedback mechanism used by the organism to try to neutralize the postsynaptic receptor blockade caused by these neuroleptic drugs (CARLSSON and LINDQVIST, 1963; ANDÉN et al., 1964; ROOS, 1965).

In the present work a couple of new clinically used neuroleptics are compared with haloperidol and chlorpromazine. namely spiroperidol, perphenazine and clothiapine (Entumin).

Methods and Results

Male rats (Sprague-Dawley, body weight 180 to 220 g) received the test substance intraperitoneally and 15 min later 250 mg/kg of H 44/68 was injected i.p. Two resp. 4 h later the animals were killed and noradrenaline and dopamine in the brains determined fluorimetrically (BERTLER et al., 1958; CARLSSON and LINDQVIST, 1962). As controls in every experiment a) untreated animals, b) rats receiving the test substance only and c) rats receiving H 44/68 only were used. The body temperature of the rats was monitored frequently during the experiment and by adjustment of the environmental temperature kept at 37 to 38°. This is important, since the combination of a tyrosine hydroxylase inhibitor and a neuroleptic agent has a hypothermic effect which would make the interpretion of the results more difficult. The statistical significance between the differences found was checked by the t-test. For details see tables.

Table 1. *Noradrenaline and dopamine in the rat brain* $2^1/_4$, *resp. 2;* $4^1/_4$, *resp. 4 h after spiroperidol (1 mg/kg i.p.) and H 44/68 (250 mg/kg i.p.). In per cent of normal values (noradrenaline: 450 ± 9 ng/g; dopamine: 650 ± 24 ng/g; n = 10). Means ± s.e.m. (4 experiments)*

4 h	noradrenaline	dopamine
controls	100.0 ± 2.2	100.0 ± 3.7
spiroperidol	98.7 ± 4.2	98.7 ± 6.1
H 44/68	46.8 ± 3.1 a)	28.7 ± 3.8 c)
spiroperidol ± H 44/68	38.0 ± 1.9 b)	15.9 ± 0.9 d)
Difference a)—b) P ≅ 0.05		
c)—d) P < 0.02		
2 h		
spiroperidol	107.5 ± 3.2	80.9 ± 2.3
H 44/68	71.9 ± 5.0 a)	52.1 ± 3.7 c)
spiroperidol + H 44/68	66.9 ± 4.5 b)	32.2 ± 1.9 d)
Difference a)—b) N.S.		
c)—d) P < 0.005		

Table 2. *Noradrenaline and dopamine* $4^1/_4$, *resp. 4 h after clothiapine (Entumin; 1 mg/kg i.p.) and H 44/68 (250 mg/kg i.p.). In per cent of normal values. Means ± s.e.m. (4 experiments)*

	noradrenaline	dopamine
controls	100.0 ± 2.2	100.0 ± 3.7
clothiapine	99.7 ± 4.4	74.3 ± 3.1
H 44/68	49.2 ± 5.5 a)	29.6 ± 0.3 c)
clothiapine + H 44/68	45.5 ± 1.9 b)	17.9 ± 2.6 d)
Difference a)—b) N.S.		
c)—d) P < 0.005		

Table 3. *Noradrenaline and dopamine* $4^1/_4$, *resp. 4 h after perphenazine (1 mg/kg i.p.) and H 44/68 (250 mg/kg i.p.). In per cent of normal values. Means ± s.e.m. (4 experiments)*

	noradrenaline	dopamine
controls	100.0 ± 2.2	100.0 ± 3.7
perphenazine	101.4 ± 5.7	99.3 ± 9.7
H 44/68	51.6 ± 1.9 a)	27.9 ± 1.6 c)
perphenazine + H 44/68	37.7 ± 2.5 b)	20.7 ± 1.1 d)
Difference a)—b) P < 0.005		
c)—d) P < 0.01		

None of these neuroleptics investigated here did change the turnover of 5-hydroxytryptamine in the rat brain, not even after high doses. For these experiments the neuroleptic was given 15 min before the tryptophan hydroxylase inhibitor H 22/54 (500 mg/kg i.p.) and the animals were killed 3 h later. An effect of these neuroleptic drugs on the activity of the cerebral 5-hydroxytryptamine neurons seems not to be probable.

Discussion

The results described show that the neuroleptic drugs spiroperidol, perphenazine and clothiapine tested in this model differ considerably in their influence on the disappearance from the rat brain of the catecholamines after inhibition of their biosynthesis.

Already a couple of years ago it has been shown that chlorpromazine and haloperidol most probably block the postsynaptic dopamine- and noradrenaline receptors in the brain (Carlsson and Lindqvist, 1963; Andén et al., 1964; Roos, 1965; DaPrada and Pletscher, 1966). Whereas the catecholamine contents in the brain remain relatively constant, markedly increased concentrations of the catecholamine metabolites normetanephrine, 3-methoxytyramine and homovanillic acid were found. By a concomitant administration of an inhibitor of the rate limiting step of the biosynthesis of the catecholamines evidence was obtained that chlorpromazine and haloperidol really increase the neuronal activity of the presynaptic catecholamine neurons [Corrodi et al., 1967 (2); Andén et al., 1967].

Table 4. *The noradrenaline and dopamine receptor blocking effect of some neuroleptic drugs. Arbitrary evaluation according to the results obtained with the biochemical test mentioned in this publication.* + + = *strong receptor blockade;* + = *slight receptor blockade;* — = *no receptor blockade at the dose given in brackets (in mg/kg i.p.)*

Substance (mg/kg i.p.)	Noradrenaline receptor	Dopamine receptor
chlorpromazine (25)	+ +	+
haloperidol (5)	+ +	+
perphenazine (1)	+	+
spiroperidol (1)	+	+ +
clothiapine (1)	—	+ +

A plausible explanation of these two observations can be given by postulation of an interneuronal or transsynaptic feedback mechanism which tries to counteract the postsynaptic receptor blockade by the induction of increased firing of the presynaptic neuron. This increased firing leads to an increased release of transmitter and, therefore, to an increased amount of transmitter metabolites and to an increased turnover of the catecholamines which can be unmarked by inhibiting their biosynthesis.

The neuroleptic drugs chlorpromazine, haloperidol, perphenazine, sprioperidol and clothiapine seem to differ in their blocking effect on the brain noradrenaline and dopamine receptors.

In Table 4 an attempt is made to demonstrate these differences. Since it is very difficult to compare the effects of the substances described quantitatively, this table is kept as simple as possible. A more thorough investigation on the effect of neuroleptic drugs on the rat brain monoamine neurons will be published elsewhere (ANDÉN et al., 1970).

It will of course be extremely interesting to learn what effect these different neuroleptics have in psychiatric patients when compared to each other. If differences could be found this might contribute to an enhanced understanding of the physiological importance of the cerebral catecholamine neurons.

Summary

The neuroleptics perphenazine, clothiapine and spiroperidol have been compared to chlorpromazine and haloperidol in a test reveiling changes in rat brain catecholamine turnover. It was shown, that clothiapine and spiroperidol mainly increase the turnover of dopamine whereas haloperidol and chlorpromazine mainly increase the noradrenaline turnover. An interneuronal feedback mechanism is postulated to explain how the wellknown blockade of the postsynaptic catecholamine receptors caused by these neuroleptics might lead to these increases in the neuronal activity in the presynaptic catecholamine neurons.

References

ANDÉN, N. E., BUTCHER, S. G., CORRODI, H., FUXE, K., UNGERSTEDT, U.: Receptor activity and turnover rate of dopamine and noradrenaline after neuroleptics. Europ. J. Pharmacol. (1970) (in press).

— CORRODI, H., DAHLSTRÖM, A., FUXE, K., HÖKFELT, T.: Effects of tyrosine hydroxylase inhibition on the amine levels of central monoamine neurons. Life Sci. 5, 561—568 (1966).

— — FUXE, K.: Turnover studies using synthesis inhibition. In: Metabolism of amines in the brain, 1. Auflage, pp. 38—47. London: Macmillan 1969.

— — — HÖKFELT, T.: Increased impulse flow in bulbospinal noradrenaline neurons produced by catecholamine receptor blocking agents. Europ. J. Pharmacol. 2, 59—64 (1967).

— ROOS, B. E., WERDINIUS, B.: Effects of chlorpromazine, haloperidol and reserpine on the levels of phenolic acids in rabbit corpus striatum. Life Sci. 3, 149 (1964).

CARLSSON, A., CORRODI, H., WALDECK, B.: α-substituierte Dopacetamide als Hemmer der Catechol-O-methyl-transferase und der enzymatischen Hydroxylierung aromatischer Aminosäuren. In den Catecholamin-Metabolismus eingreifende Substanzen. Helv. chim. Acta 46, 2271 (1963).

— LINDQVIST, M.: Effect of chlorpromazine or haloperidol on formation of 3-methoxytyramine and normetanephrine in mouse brain. Acta pharmacol. (Kbh.) 20, 140 (1963).

CORRODI, H., FUXE, K., HÖKFELT, T.: (1) Replenishment by 5-hydroxytryptophan of the amine stores in the central 5-hydroxytryptamine neurons after depletion induced by reserpine or by an inhibitor of monoamine synthesis. J. Pharm. Pharmacol. 19, 433—438 (1967).

— — — (2) The effect of neuroleptics on the activity of central catecholamine neurons. Life Sci. 6, 767—774 (1967).

— HANSON, L.: Central effects of an inhibitor of tyrosine hydroxylation. Psychopharmacologia (Berl.) 10, 116—125 (1966).

— MALMFORS, T.: The effect of nerve activity on the depletion of the adrenergic transmitter by inhibitors of noradrenaline synthesis. Acta physiol. scand. 67, 352—357 (1966).

DA PRADA, M., PLETSCHER, A.: On the mechanism of chlorpromazine-induced changes of cerebral homovanillic acid levels. J. Pharm. Pharmacol. 18, 628 (1966).

ROOS, B. E.: Effects of certain tranquillisers on the level of homovanillic acid in the corpus striatum. J. Pharm. Pharmacol. 17, 820 (1965).

Dr. H. CORRODI
AB Hässle
Fack Göteborg 5
40220 Göteborg/Sweden

Discussion

Axelrod: The drugs which specifically effect the dopamine adrenergic neurons—have you studied the physiological disposition of these drugs in the brain—do they go selectively to the corpus striatum?

Corrodi: That has been discussed by Janssen in Liège in 1969. He said that all these neuroleptics form a monomolecular layer in the brain, but that there might be different rates of disappearance from different areas.

Glowinski: Recently, it has been mentioned by Dr. Janssen in Liège that powerful neuroleptics of the butyrophenone type, particularly active clinically and potent stimulators of dopamine synthesis in the caudate nucleus, have a specific localization in the brain after their intraperitoneal injection. This has been shown by using radioactive neuroleptics and by autoradiographic techniques. Drugs such as chlorpromazine which appear to affect NA turnover, as well as DA turnover, in the central nervous system and is less specific in its pharmacological action presented at the opposite an homogenous distribution throughout the central nervous system.

de Robertis: I also want to say that Professor Janssen has observed that 4 h after injection of a labelled butyrophenone, the only region which still contains the radioactive neuroleptic is the caudate nucleus. We have recently studied the uptake of different drugs in the nerve ending membranes isolated from the brain and found that the α-blocking agents ^{14}C-Sy 28 and ^{14}C-dibenamine, as well as the β-adrenergic drug ^{14}C-propanolol are preferentially bound to the nerve ending membranes isolated from basal ganglia. The neuroleptic ^{3}H-chloropromazine also binds to these membranes, while ^{14}C-mescaline is slightly concentrated in the synaptic vesicles. We have additional evidence which suggests that neuroleptics are bound postsynaptically and probably block receptor sites.

Corrodi: May I comment on this, I do not think that you show exactly what you want to show—you should have included substances not blocking the receptors of the same degree of lipophilicity. Mescaline used by you has a lower lipophilicity than the other drugs mentioned.

de Robertis: These studies were done at very low concentrations of the order of 10^{-7}, 10^{-6}, i.e. in the range of high affinity binding. At higher concentrations non specific binding may take place. We can also displace the specific binding by proper competition experiment in vitro.

Corrodi: If you have a dilute water solution of a drug and you put a drop of oil in it then you get a tremendous concentration of lipophilic substances into the oil drop.

Stjärne: With reference to what Dr. Corrodi mentioned concerning a possible negative feed back regulation, triggered from the receptors and operating on the presynaptic transmitter release. Also on behalf of the interest of my colleague, Dr. Hedqvist, who has shown that prostaglandins may be involved in this kind of regulatory mechanism in sympathetic transmission, I would like to mention that Dr. Floyd Bloom seems to have similar evidence from the central nervous system. Application of exogenous prostaglandin inhibits certain kinds of transmitter release in the brain. Since it seems that release of endogenous prostaglandins is a common phenomenon in many secretory events, these substances could have regulatory functions on transmitter secretion in the central nervous system as well.

Efron: I am a little perplexed. After Dr. Carlsson spoke I had the thought that when I go back to my office I should initiate some studies using two types of drugs, one affecting the adrenergic receptors and the other affecting the cholinergic receptors. Now, from what you are saying, I feel that we should add drugs affecting the dopaminergic receptors. What is your thinking on this point?

Corrodi: According to my records there is only one substance which slightly blocks dopamine uptake, namely, adamantanamine.

Efron: It means that we should use two substances: one affecting norepinephrine and one affecting dopamine.

Corrodi: Yes, in studying depression.

Bayer-Symposium II, 258—267 (1970)

Release of Catecholamines from the Hypothalamus by Drugs and Electrical Stimulation*

ATHINEOS PHILIPPU

With 10 Figures

Recently it was reported that calcium and acetylcholine accelerate in vitro the release of noradrenaline from the intact hypothalamic cells (PRZUNTEK and PHILIPPU) as well as from the subcellular noradrenaline storing vesicles of the hypothalamus (PHILIPPU and PRZUNTEK). It was suggested that both agents may be responsible also for the release of noradrenaline from the nerve endings of the hypothalamus under physiological conditions. In order to prove this hypothesis the cat hypothalamus was perfused in vivo by the method of FELDBERG and MYERS and the effects of calcium and acetylcholine as well as of the electrical stimulation of hypothalamic nuclei on the release of catecholamines were studied. Furthermore the effect of DMPP and atropine on the release of noradrenaline was investigated.

The experiments were performed together with Mrs. GESINE HEYD and Dr. ALBRECHT BURGER.

Uptake and Spontaneous Release of Noradrenaline

The third ventricle was perfused by the method of FELDBERG and MYERS which enables a selective perfusion of the hypothalamus. The hypothalamus was labelled with 5 μc noradrenaline-C^{14} (50 mc/nmole) through a cannula inserted into the third ventricle. After 4 h the aqueduct was also cannulated and the third ventricle was perfused with artificial cerebrospinal fluid or with Ringer solution at a rate of 0.15 ml/min. The perfusates were collected in tubes containing perchloric acid.

In order to prove whether the radioactive noradrenaline is taken up by the specific amine storing vesicles of the hypothalamic cells, the hypothalamus was removed 4 h after labelling and gently homogenized in 0.3 M sucrose solution. The homogenate was layered on the surface of a continuous sucrose gradient which ranged from 0.35 to 1.5 M. The gradient was then centrifuged for 3 h at 64,000 g. About 40 fractions were obtained by piercing the bottom of the centrifuge tube and collecting samples of ten drops in scintillation vials (Fig. 1).

The radioactive substances are found both in the soluble supernatant fraction and in the particulated fraction III, which contains synaptosomal particles (POTTER and AXELROD), while the myelin fraction I, the microsomal fraction II and the mitochondrial fraction IV contain negligible amounts of radioactive compounds. Therefore 4 h after labelling the hypothalamus, i.e. at the beginning of

* This work was supported by the Deutsche Forschungsgemeinschaft.

its perfusion, about 60% of the total radioactivity recovered are bound in particulate fractions of the hypothalamic cells.

During perfusion the release of total radioactivity from the hypothalamus declines gradually (Fig. 2). The released radioactivity is expressed as per cent of the radioactivity of the first effluent. The initial slope of the curve declines with

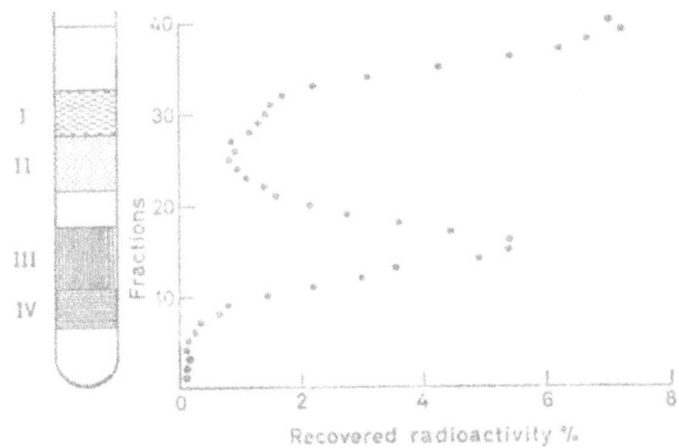

Fig. 1. Subcellular localization of noradrenaline-C^{14} in the cat hypothalamus 4 h after labelling intraventricularly. The experiment was repeated four times

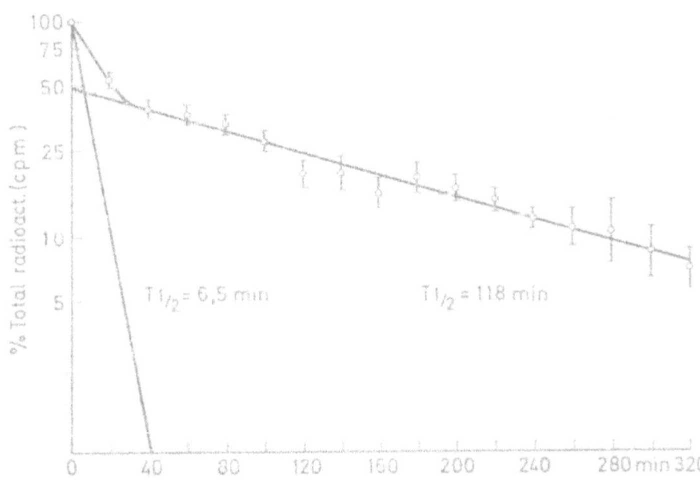

Fig. 2. Efflux of radioactivity from the cat hypothalamus. Between 20 and 320 min the regression coefficient is 0.005887 ± 0.000425/min (S.D.M.). Mean values of 22 experiments ± S.E.M. [PHILIPPU, A., HEYD, G., BURGER, A.: Europ. J. Pharmacol. 9, 52—58 (1970)]

a half-life of about 6.5 min. The slope of the second part of the curve follows a single exponential decrease with a half-life of 118 min. The exponential decline of the second part of the curve suggests that between 40 and 320 min the catecholamines are released only from intracellular compartments. All experiments were performed during this period.

This method enables the continuous determination of noradrenaline and its metabolites released from the hypothalamic cells into the perfusing solution.

Effect of Acetylcholine, DMPP and Calcium on the Release of Catecholamines

The third ventricle was perfused with Ringer solution containing tropolone (50×10^{-6} g/ml) and nialamid (1.5×10^{-6} g/ml). Addition of acetylcholine (2.2×10^{-3} M) to the perfusing fluid increases the release of catecholamines from the hypothalamus (Fig. 3). The releasing effect of acetylcholine can be demon-

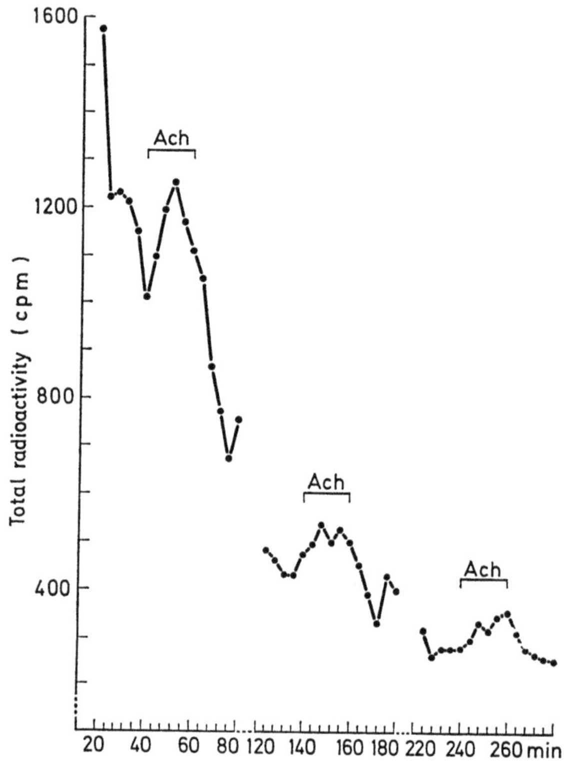

Fig. 3. Effect of acetylcholine on the release of catecholamines from the cat hypothalamus. Ach = Acetylcholine (2.2×10^{-3} M). Total radioactivity of 0.2 ml of the effluents, total volume of each sample 3 ml. [PHILIPPU, A., HEYD, G., BURGER, A.: Europ. J. Pharmacol. 9, 52—58 (1970)]

strated repeatedly at the same cat. Also a smaller concentration of this agent (2.2×10^{-4} M) enhanced the release of catecholamines, although not always.

In order to study the influence of calcium ions on the catecholamine releasing effect of acetylcholine experiments were performed in the presence and in the absence of this ion. The ionized calcium of the hypothalamic region was removed by perfusing the third ventricle with Ringer solution containing EDTA (1×10^{-3} M) for 60 min before the collection period. During perfusion with Ringer solution containing calcium (2.5×10^{-3} to 1×10^{-2} M) acetylcholine (2.2×10^{-3} M) increases the release of catecholamines from the hypothalamus. However acetylcholine is ineffective if the third ventricle is perfused with calcium free solution (Fig. 4). DMPP (3×10^{-4} M) is also able to accelerate the release of amines (Fig. 5).

The effect of calcium ions on the release of catecholamines from the hypothalamus is shown in Fig. 6. During perfusion with calcium free solution the release of total radioactivity declines gradually whereas the addition of calcium

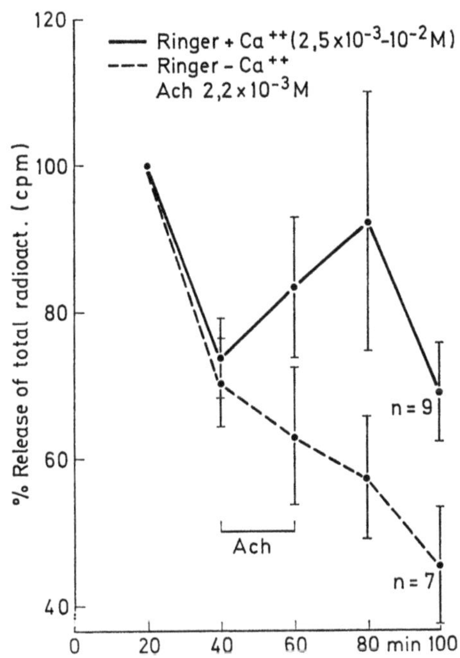

Fig. 4. Effect of acetylcholine on the release of catecholamines from the cat hypothalamus in the presence and in the absence of calcium. Ach = Acetylcholine. Mean values ± S.E.M. [PHILIPPU, A., HEYD, G., BURGER, A.: Europ. J. Pharmacol. 9, 52—58 (1970)]

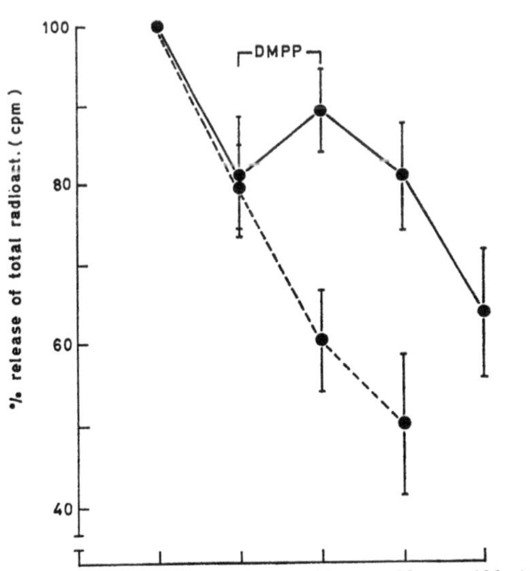

Fig. 5. Effect of DMPP on the release of catecholamines from the cat hypothalamus. — — — Controls, ——— DMPP (1.3 × 10^{-4} M). Mean values of 5 experiments ± S.E.M.

(1×10^{-2} M) provokes a pronounced release of radioactive compounds. Also a smaller concentration of calcium (5×10^{-3} M) was able to enhance significantly the release of catecholamines from the hypothalamus.

These results are in favour with the assumption that also in the central nervous system acetylcholine may evoke the discharge of noradrenaline from the noradrenergic nerve endings by causing calcium ions to penetrate their membrane in high amounts. Since calcium ions are able to increase the release of noradrenaline from the hypothalamic vesicles (Philippu and Przuntek) it may be postulated that also in vivo during depolarization by acetylcholine calcium ions enter the cell and evoke the release of noradrenaline from the subcellular particles.

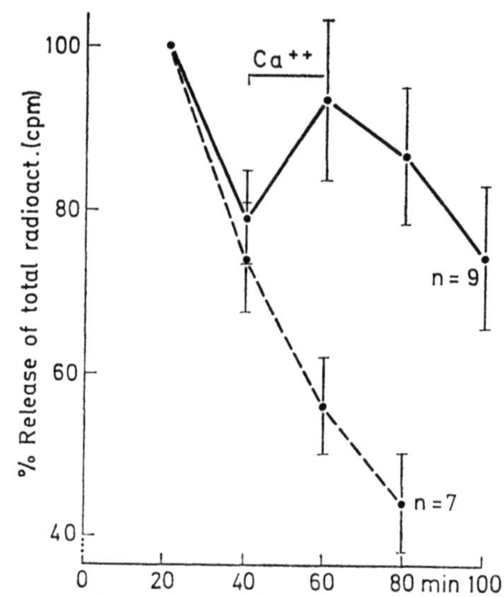

Fig. 6. Effect of calcium on the release of catecholamines from the cat hypothalamus. Ca = Calcium (1×10^{-2} M). Mean values ± S.E.M. [Philippu, A., Heyd, G., Burger, A.: Europ. J. Pharmacol **9**, 52—58 (1970)]

Effect of Atropine of the Release of Catecholamines by Acetylcholine

Bradley et al. have shown that acetylcholine exerts in the medulla excitatory and inhibitory responses and further that excitatory responses have both nicotinic and muscarinic properties while the inhibitory responses are only muscarinic. Since atropine blocks muscarinic responses it seemed likely that it could also potentiate the catecholamine releasing effect of acetylcholine. The cat hypothalamus was therefore perfused with 9×10^{-7} M atropine for 60 min and then 2.2×10^{-5} M acethylcholine were added to the perfusing fluid. This concentration of acetylcholine is normally ineffective (Fig. 7). However in the presence of atropine also this small concentration of acetylcholine increases slightly the release of catecholamines from the hypothalamus. A similar potentiation of the effect of acetylcholine by atropine was shown by Lindmar et al. in the rabbit heart. A higher concentration of atropine (9×10^{-6} M) accelerates the release of catecholamines (Fig. 8) presumably by blocking inhibitory responses. Moreover it is possible that atropine also excites the excitatory muscarinic responses (Bradley et al.).

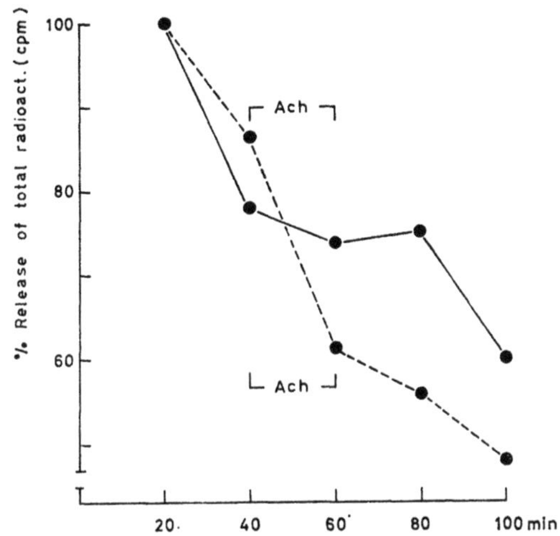

Fig. 7. Effect of atropine on the release of catecholamines by acetylcholine. — — — In the absence, ——— in the presence of atropine (9×10^{-7} M). Ach = Acetylcholine (2.2×10^{-5} M) Mean values of 3 experiments

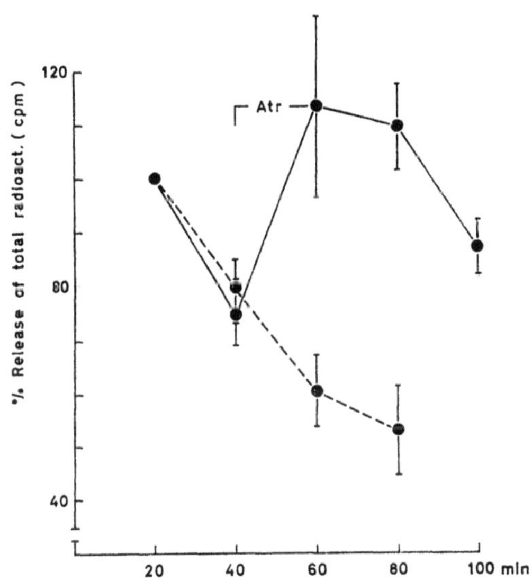

Fig. 8. Effect of atropine on the release of catecholamines from the cat hypothalamus. — — — Controls, ——— atropine (9×10^{-6} M). Mean values of 6 experiments ± S.E.M.

Release of Catecholamines by Electrical Stimulation

The influence of the electrical stimulation on the release of catecholamines was studied by stimulating different nuclei of the hypothalamus, as nucleus anterior medialis, nucleus posterior and nucleus ventralis medialis. These nuclei were chosen because they possess a high density of noradrenergic nerve terminals (Fuxe).

During a collection period of 20 min the nucleus anterior medialis was stimulated ten times for 60 sec. The electrical stimulation produces an enhanced release of catecholamines from the hypothalamus (Fig. 9). Also the stimulation of the nuclei posterior and ventralis medialis provokes an increased release (Fig. 10). Although statistically not significant, it seems that the electrical stimulation of

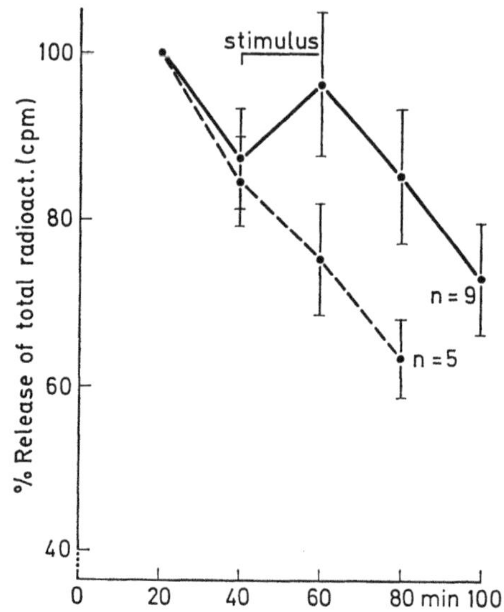

Fig. 9. Effect of electrical stimulation of the nucleus anterior medialis on the release of catecholamines from the cat hypothalamus. Electrical stimulation: 10 V, 6 msec., 40 c/sec. Mean values ± S.E.M. [Philippu, A., Heyd, G., Burger, A.: Europ. J. Pharmacol. **9**, 52—58 (1970)]

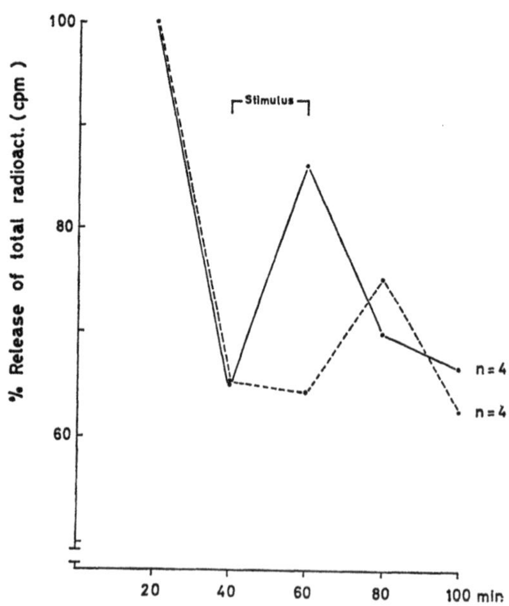

Fig. 10. Effect of electrical stimulation of the nucleus posterior (———) and ventralis medialis (— — —) on the release of catecholamines from the cat hypothalamus. Electrical stimulation: 10 V, 6 msec., 40 c/sec. [Philippu, A., Heyd, G., Burger, A.: Europ. J. Pharmacol. **9**, 52—58 (1970)]

the nucleus posterior causes the most pronounced release of catecholamines. Probably a correlation exists between the releasing effect of the electrical stimulation and the concentration of noradrenaline in the different nuclei of the hypothalamus, since FUXE has demonstrated that the nucleus posterior of the rat contains the highest density of noradrenergic terminals.

The effect of acetylcholine and calcium ions as well as of the electrical stimulation on the relative concentration of the released noradrenaline and its metabolites was also studied. For this purpose the radioactive compounds were separated by paper chromatography (VOGT; LANGER) before and during perfusing with calcium and acetylcholine or electrical stimulation of the hypothalamic nuclei.

Table 1. *Release of noradrenaline and its metabolites by calcium, acetylcholine and electrical stimulation*

	n	NA %	DOMA %	NM %	MOPEG %	VMA %
Controls	3	17.2 ± 2.1	19.6 ± 2.3	17.3 ± 0.4	11.2 ± 2.5	34.8 ± 0.9
Ca^{++}	3	24.7 ± 4.8	19.5 ± 1.2	16.8 ± 1.6	11.0 ± 3.0	27.8 ± 5.6
Controls	4	24.3 ± 2.7	19.8 ± 0.5	15.0 ± 2.0	9.4 ± 1.0	31.5 ± 3.6
Acetylcholine	4	24.5 ± 5.8	16.7 ± 2.7	18.1 ± 3.9	8.6 ± 1.1	32.6 ± 7.0
Controls	6	18.1 ± 0.9	26.6 ± 1.1	14.1 ± 2.5	12.4 ± 1.6	28.9 ± 1.6
Electrical stimulation	8	15.8 ± 1.8	25.8 ± 3.7	12.2 ± 1.9	13.6 ± 2.0	32.6 ± 3.1

The third ventricle was perfused with Ringer solution containing nialamid (1.5×10^{-6} g/ml) and tropolone (50×10^{-6} g/ml). Ca^{++}: 5×10^{-3} M, acetylcholine: 2.2×10^{-3} M. Electrical stimulation: 10 V, 6 msec., 40 c/sec. Mean values ± S.E.M. [PHILIPPU, A., HEYD, G., BURGER, A.: Europ. J. Pharmacol. **9**, 52—58 (1970)].

Table 2. *Noradrenaline-C^{14} and its metabolites in the hypothalamus*

Radioactive compounds	Relative concentration %
NA	48.6 ± 4.8
DOMA	21.7 ± 3.6
NM	13.0 ± 1.2
MOPEG	7.1 ± 1.5
VMA	9.6 ± 1.5

The values are expressed as per cent of the total radioactivity recovered. Mean values of 9 experiments ± S.E.M. [PHILIPPU, A., HEYD, G., BURGER, A.: Europ. J. Pharmacol. **9**, 52—58 (1970)]

About 20% of the radioactivity which is released spontaneously from the hypothalamus is due to noradrenaline, 80% to deaminated and methylated metabolites (Table 1). Neither the perfusion with calcium and acetylcholine nor the electrical stimulation of the hypothalamic nuclei influence significantly the relative concentration of the radioactive compounds. Therefore acetylcholine, calcium and electrical stimulation evoke an enhanced release of the total radioactivity without changing the relative concentration of noradrenaline and its metabolites.

The relative concentration of noradrenaline and its metabolites in the hypothalamus was studied at the end of the perfusion experiments. The relative concentration of noradrenaline is 48.6% (Table 2) i.e. much higher than its concentration in the perfusates. Therefore noradrenaline is further metabolized after its release from the hypothalamic cells, although the perfusing fluid contains both nialamid and tropolone.

Acetylcholine may be a chemical transmitter in the central nervous system. On the other hand the existence of nerve terminals containing catecholamines which are stored in subcellular particles and the possibility to increase the release of catecholamines by electrical stimulation are compatible with the assumption that also noradrenaline may be a mediator in the hypothalamus. Recently BLOOM et al. have reported that all areas of the hypothalamus possess acetylcholine sensitive neurones which respond to acetylcholine either by increasing or by decreasing their spontaneous rate of discharge. These observations together with the finding that acetylcholine enhances the release of noradrenaline could support the hypothesis that in the hypothalamus the transmitter of cholinergic neurones influence the release of catecholamines from the noradrenergic nerve terminals, provided that calcium ions are present.

Summary

The hypothalamus of the cat was labelled with noradrenaline-C^{14}. The fractionation of the hypothalamus showed that the radioactive noradrenaline was taken up by the synaptosomes. During perfusion of the third ventricle noradrenaline is released spontaneously from the hypothalamus into the perfusing fluid. Addition of acetylcholine, calcium, DMPP or atropine to the perfusing solution caused an enhanced release of catecholamines from the hypothalamus. The releasing effect of acetylcholine was abolished in the absence of calcium ions and enhanced in the presence of atropine. The electrical stimulation of the hypothalamic nucleus anterior medialis, nucleus posterior and nucleus ventralis medialis evoked a pronounced release of catecholamines from the hypothalamus. Calcium and acetylcholine as well as electrical stimulation of the nuclei enhanced the release of total radioactivity without influencing the relative concentration of noradrenaline and its metabolites.

Acknowledgement

The author wishes to thank Pfizer GmbH, Karlsruhe, for nialamid and Farbenfabriken Bayer, Leverkusen, for supplying cats. He also thanks Miss B. PIEL and Miss T. SEEBER for skilful technical assistance.

References

BLOOM, M. F. E., OLIVER, A. P., SALMOIRAGHI, G. C.: The responsiveness of individual hypothalamic neurons to microelectrophoretically administered endogenous amines. Int. J. Neuropharmacol. **2**, 181—193 (1963).

BRADLEY, P. B., DHAWAN, B. N., WOLSTENCROFT, J. H.: Pharmacological properties of cholinoceptive neurones in the medulla and pons of the cat. J. Physiol. (Lond.) **183**, 658—674 (1966).

FELDBERG, W., MYERS, R. D.: Appearance of 5-hydroxytryptamine and an unidentified pharmacological active lipid acid in effluent from perfused cerebral ventricles. J. Physiol. (Lond.) **184**, 837—855 (1966).

Fuxe, K.: Evidence for the existence of monoamine neurons in the central nervous system. IV. Distribution of monoamine nerve terminals in the central nervous system. Acta physiol. scand. **64**, Suppl. **247** (1965).
Langer, S. Z.: Release of ^{3}H-noradrenaline from the isolated nictitating membrane of the cat by intramural stimulation. J. Physiol. (Lond.) **197**, 43—44 P (1968).
Lindmar, R., Löffelholz, K., Muscholl, E.: A muscarinic mechanism inhibiting the release of noradrenaline from the peripheral adrenergic nerve fibres by nicotinic agents. Brit. J. Pharmacol. **32**, 280—294 (1968).
Merlis, J. K.: The effect of changes in the calcium content of the cerebrospinal fluid on spinal reflex activity in the dog. Amer. J. Physiol. **131**, 67—72 (1940).
Philippu, A., Przuntek, H.: Noradrenalin-Speicherung im Hypothalamus und Wirkung von Pharmaka auf die isolierten Hypothalamus-Vesikel. Naunyn-Schmiedebergs Arch. Pharmak. exp. Path. **258**, 238—250 (1967).
Potter, L. T., Axelrod, J.: Subcellular localization of catecholamines in tissues of the rat. J. Pharmacol. exp. Ther. **142**, 291—298 (1963).
Przuntek, H., Philippu, A.: Noradrenalin-Freisetzung aus dem Hypothalamus in vitro. Experientia (Basel) (in press).
Vogt, M.: The secretion of the denervated adrenal medulla of the cat. Brit. J. Pharmacol. **7**, 325—330 (1952).

Prof. Dr. A. Philippu
Institut für Pharmakologie und Toxikologie
der Universität
D-8700 Würzburg
Koellikerstraße 2

Discussion

Glowinski: I would like to ask Professor Philippu about the amount of radioactive norepinephrine which was injected in his experiments. Furthermore, what is the approximate concentration of radioactive norepinephrine in the hypothalamus at the time of stimulation?

Philippu: 5 μCi noradrenaline-^{14}C. I did not understand the second question...

Glowinski: As you have been using ^{14}C-norepinephrine which has a relatively low specific activity, perhaps you are not working in tracers conditions and therefore you have an increase concentration of norepinephrine in the tissue.

Philippu: The net uptake of noradrenaline is about 700 ng/g hypothalamus.

Axelrod: You mention giving nialamide and COMT inhibitors—do they effect the change in the metabolites?

Philippu: Yes, if we perfuse with Ringer solution containing tropolone and nialamide we find a higher relative concentration of noradrenaline, while nialamide alone has no effect on the relative concentration of noradrenaline or its metabolites. Vogt has also shown that the noradrenaline content of the cat brain is not increased by MAO-blockers. Furthermore, MAO-blockers do not influence the metabolism of noradrenaline in the cat brain, although MAO is present.

Axelrod: It is very interesting, the species are different because in the rat one never finds BMA's and metabolites, where in the cat you showed considerable concentration.

MUSCHOLL: Your observation, that the proportion of noradrenaline/metabolites in the perfusate is lower than the proportion in the hypothalamus after the experiment, does not necessarily mean that there is a metabolism of noradrenaline after it has been released. It can also mean that noradrenaline is much better bound in the tissue than the metabolites—Would you not agree with this?

PHILIPPU: Yes.

DE ROBERTIS: If I understood correctly, blocking of the muscarine receptor facilitates the output of acetylcholine and the action of catecholamine. Have you tried blocking the nicotinic receptors and followed the action of this blocking?

PHILIPPU: Yes, we have done some preliminary experiments with ganglionic blocking drugs but the number of the experiments is too small to permit any comment.

STJÄRNE: I think it was interesting to see that on stimulation you have an equal mobilization of both noradrenaline and metabolites in your system. It was shown 6 years ago in Dr. AXELROD's laboratory, and it has been repeatedly confirmed, that in various peripheral tissues nerve stimulation does not cause an increased efflux of metabolites, but a selective, or a nearly selective increase in noradrenaline efflux. I just wonder if you think that there could be differences in secretory mechanisms between central and peripheral neurons, or if you think you have evoked release by some different kind of mechanism from that which is normally operating?

PHILIPPU: I don't think so. I would like to define our experiments as semi-acute, since we perfuse the hypothalamus with acetylcholine or calcium ions for 20 min at a rate of 0.15 ml/min. Therefore, there is time enough for the metabolism of noradrenaline, after its release from the storage sites of the hypothalamus.

TRENDELENBURG: Dr. LANGER exposed isolated nictitating membranes to dl-norepinephrine-^{3}H and (after a time interval) applied field stimulation. Of the released radioactive material most was metabolized and only a small fraction represented unchanged norepinephrine. Thus there seem to be pronounced organ and species differences.

THOENEN: I think that the most plausible reason for the large proportion of metabolites appearing in the bathing fluid after stimulation of the isolated nictitating membrane is the fact that the norepinephrine, liberated by sympathetic stimulation, has to diffuse through the nictitating membrane and metabolized on the way out. In contrast, in a perfused organ such as the cat spleen or the gracilis muscle of the dog, the transmitter liberated from the nerve endings is rapidly removed by the perfusion fluid and, therefore, sympathetic nerve stimulation results mainly in an increase of norepinephrine in the perfusate and not of its metabolites.

I expect that in the isolated vas deferens where the liberated transmitter must also diffuse through the tissue into the bathing fluid, the results would be similar to those in the nictitating membrane.

TRENDELENBURG: While I fully agree with Dr. THOENEN's views, I should like to emphasize the following observation by Dr. LANGER. Although the homogenized vas deferens of the rat has at least as much COMT activity as the homogenized nictitating membrane of the cat, pronounced O-methylation of both endogenous or exogenous noradrenaline is observed only with isolated nictitating membranes and not with isolated vasa deferentia. Thus, there seem to be species and/or organ differences even when we exclude the experimental differences mentioned by Dr. THOENEN.

Bayer-Symposium II, 270—282 (1970)

Session 5

General Discussion

Chairman: H. J. Schümann

SCHÜMANN: It is our aim to discuss now a few points which have not been discussed so far, or not intensively enough. Since many participants asked me to continue the discussion on the problem of exocytosis, I first of all would like to give Professor BLASCHKO the word because he wanted to say something about lysolecithin and exocytosis.

BLASCHKO: If one tries to peer into the future, I want to say that as far as the adrenal medulla is concerned, we must not neglect that in addition to dopamine-β-hydroxylase there must be other insoluble protein constituents in the membranes and they may well differ chemically very much from the soluble proteins. May I remind you of the importance of the sulfhydryl groups for amine uptake and release. One of the interesting things about the chromogranin is that there are very few cystine residues in it. A long time ago, Dr. D'IORIO carried out a piece of work on the effect of sulfhydryl reagents on amine release. I think that Professor CARLSSON did some similar work. We remember in this context also the much more recent work of TAUGNER and HASSELBACH in which they have shown that there are some very strategic sulfhydryl groups which they have been able to protect with ATP from reacting with N-ethyl maleimide. I think that here we have the possibility of a biochemical attack on the problem of the membrane protein(s). This work encourages us in our belief that the chemistry of some of these insoluble proteins is very different from the chemistry of the soluble proteins.

The second fact which has occupied us for some time is the presence of lysolecithin. Lysolecithin was discovered in the adrenal medulla some time ago by HAJDER, WEISS and TITUS. My colleagues have shown that this very high lysolecithin content of the adrenal medullary cell is due to its presence in the chromaffin granules. This seems to me to be a challenge: here we have a hydrophilic phospholipid which differs from all the other membrane constituents (if it is a membrane constituent—it does not seem to come out on stimulation). What is its function? When we first found it we thought that the presence of this compound might turn out to be an artifact of the preparation, a postmortal change, and I would not like to rule it out completely, although Dr. WINKLER and his colleagues have shown that the lysolecithin content is equally high in laboratory animals, when the tissues are immediately examined. Is the lysolecithin evidence of a high lecithin turnover or is it a compound that is, in some way, involved in membrane fusion or a similar event? I mention this point because, of course, in the isolation of all these organelles that we study, we encounter artifacts. However, it is also a fact that although the biochemical and cytological evidence for their existence becomes increasingly good, these organelles sometime begin to disappear just as to my mind mitochondria are beginning to disappear when we start talking about their outer and inner membranes. I mean that the outer membrane does not necessarily end on the mitochondrion itself. So there are certain artifacts of isolation which we must keep in mind. But on the whole I would say that to me the evidence in favour of exocytosis looks pretty good as far as the adrenal medulla is concerned.

And if I may say something about the cytology then I think we ought to pay a tribute to Professor DE ROBERTIS who had the courage to put forward this notion for the adrenal medulla at a time when the technique was far less advanced than the techniques available today. Pictures of the kind he put on the blackboard have increasingly been seen by the electronmicroscopists. Why should they be artifactual; after all, the story fits very well.

The work of COUPLAND who has shown very nice pictures, supplements those of Professor DE ROBERTIS. The excellent paper by Madame DINER in Paris on the adrenal medulla of the golden hamster contains a great number of instances of membrane fusion. I think the challenge now is to see to what extent these observations can be useful in the study of adrenergic neuron—to what extent is this picture which one obtains for the adrenal medulla helpful in interpreting what happens at the adrenergic nerve endings and to what extent are things different?

HOLTZ: May I ask one question, that is, would you call exocytosis the secretion of specific proteins which we call enzymes from exocrine glands, e.g. from the excretory glands of the stomach mucosa—pepsin—or diastase and the secretion of renin, all of which are stored in granular-like particles. The particles disappear if you stimulate the gland. Is it true for renin-secretion that there is a parallelism to exocytosis of which we speak about all the time?

DE ROBERTIS: I think the point which has been brought up by Professor HOLTZ is a very important one because the mechanism of exocytosis is very general for most gland cells. One of the best places in which it has been studied is in the pancreas. Secretion of the pancreatic enzymes is by a mechanism very similar to the one of the adrenal. Also in this case the material is synthesized in the endoplasmic reticulum and then transferred to the Golgi which makes the membrane of the zymogen granule. This membrane then fuses with the outer membrane and so forth. In this way there is transfer of material from one region of the cell to the other.

The Golgi membrane is able to fuse with the outer membrane because it is more similar biochemically. In relation to the adrenal medulla I would like to say that when the adrenal cell is stimulated not only are the granules liberated by exocytosis but there is an increase in synthesis of new material which can be seen in the Golgi. In fact after 10 min of stimulation of the splanchnic nerve there are more droplets in the Golgi region which contain catecholamines. The mechanism of exocytosis is now very well proven for the adrenal medulla; what happens in the adrenergic nerve ending is a different story.

Folkow: Just a comment concerning the adrenergic nerve fibres as compared with the adrenal medullary cells, where no doubt a close relationship exists from the *embryological* point of view. However, there is an enormous difference between them with respect to their *dimensions* in the adult organism. For example, the *transport distance* from the perikarion to the site of release is almost a millionfold longer in neurons if one considers the axonal length in the longest adrenergic nerves in man.

Now, if we draw a parallel to the cholinergic neurons, where more information is available and where the same principles for transmitter release are probably valid, we can choose a "tonic" motoneuron, some of which control as many as 200 skeletal muscle fibres. It has been deduced that at each endplate about 200 quantal packets are released per *impulse* and there are 200 such endplates. Neurons of this type often fire at rates of 10 to 20 per sec and this would mean that 400,000 to 800,000 packets can be released per *second*.

Suppose for a moment that each packet really is identical with one entire ACh vesicle, that each vesicle has to be produced in the cell soma and that a vesicle is used only once; then imagine that up towards one million of such vesicles would have to be transported down along the axon *each second*. To me, at least, such an example, based on realistic figures for motor unit organisation and rate of discharge, make it next to impossible to imagine that one quantal packet would be identical with one discharged vesicle (granule).

As an attempt to overcome these difficulties of transport (and vesicular formation), it might then be assumed that the vesicles (granules) discharge their *entire* transmitter content at the endings, but that they are rapidly reloaded *locally* and again used. However, we then face another difficult problem if it is assumed that also the synthesizing enzymes and the binding proteins are discharged in analogy to the situation in the adrenal medulla. This seems to be prerequisite if it is a "true" exocytosis. Again there would be a demand for a tremendous amount of highly specific material produced each second in the cell soma which then has to be as rapidly transported down along the narrow axon. An alternative would be if the axon ramifications had means to synthesize rapidly these huge amounts of released specific proteins but to my knowledge there is no evidence whatsoever for such an enormous *peripheral* synthesis of specific protein.

In my opinion there appears to be only *one* reasonable way out of these great and unavoidable problems of transport and synthesis in neurons, and a very simple one. Suppose that neurons—besides their differentiation in terms of an enormous lengthening compared with gland cells, thus establishing *direct* contact with the effector cells—are differentiated also in *another* respect, closely related to their direct contact to the effectors: There may have occurred a "compartmentalization" of the transmitter vesicles (granules), intended to "save" transmitter by releasing only fractions of the vesicular (granular) content, which appears reasonable since the intimate contact with the effectors hardly necessitates release of large transmitter amounts. If one does not want to consider such an alternative one has to face the above mentioned enormous demands for axonal transport because this is a logistic problem that cannot be by-passed. I must admit that I seriously doubt that such enormous transports exist in neurons; most studies indicate that axonal transport is a fairly *slow* process.

Schümann: May I just ask first of all to discuss the conditions of the adrenal medulla because it would be easier, perhaps, from there to come to a comparison. Therefore, I would first like to ask Dr. Kirshner to show a slide about catecholamine DBO which he already showed yesterday during his presentation and which he wanted to discuss in more detail.

Kirshner: I should like to show this first slide (Fig. 1) to clarify any impressions that I may have left that I have some doubts about the stoichiometry of release of dopamine-β-hydroxylase and catecholamines from the adrenal medulla. Just looking at the soluble fractions of the storage vesicles, it would appear that the ratio of dopamine-β-hydroxylase to catecholamines changes after release of catecholamines produced by insulin hypoglycemia. However, when one calculates these ratios in the unstimulated gland, the ratio of dopamine-β-hydroxylase to catecholamines is approximately 0.16; in the stimulated gland the ratio is 0.2 and these are not different when subjected to the Student Test. Now in the 100,000 g supernatant fraction we appear to have an excess of dopamine-β-hydroxylase because in the normal gland the ratio of dopamine-β-hydroxylase to catecholamines is approximately 0.3 which is about twice as much as we find in the lysate of the vesicles. After stimulation this ratio is approximately 0.4. We do not know the source of this extra dopamine-β-hydroxylase but it accounts for about 5 to 8% of the total dopamine-β-hydroxylase. This may come from some dopamine-β-hydroxylase which may have been solubilised from the membrane during the preparation, or perhaps from some vesicles which are being newly synthesised but have not acquired their total catecholamine content.

When one adds the total soluble dopamine-β-hydroxylase and the total soluble catecholamines, this ratio in the control animal is about 0.16, the same as in the vesicle.

In the stimulated glands the ratio of the total soluble amounts is now about 0.25 but this is largely due to the excess amount which is in the 100,000 g supernatant fraction since it contributes a much larger proportional amount to the total soluble fraction than it does in the unstimulated gland.

Philippu: Do you think that the exocytosis takes place also in vitro, as for example during incubation of isolated vesicles?

Kirshner: No, I do not, we have some information on that which we have not yet completed. We have stimulated release from the isolated vesicles by Poisner's method by adding ATP and magnesium in salt solutions and then putting these granules on sucrose density gradients. These vesicles do have a lighter density than vesicles incubated in the absence of ATP or magnesium, so it appears to be not an exocytotic release under those conditions. We have also measured the release of dopamine-β-hydroxylase and catecholamines, you get the same proportion of release as one finds in the intact vesicles but again it does not appear as though it is a quantal release from the vesicles.

Carlsson: It has been shown repeatedly that stimulation of the adrenal medulla, via its secretory nerves may cause an increased catecholamine synthesis to such an extent as to compensate almost completely for the loss through release.

Now the question comes: if we accept that this is so, how do you envisage this phenomenon—if you have at the end of the experiment an almost normal adrenaline content would it not be reasonable to think that this adrenaline is present in a normal manner in the gland—in other words, present in normal vesicles that contain chromogranin, DBO and so forth? Now if this is so, then I think we have to postulate that when you start to stimulate the gland, then the synthesis of catecholamines, chromogranin, ATP, dopamine-β-hydroxylase, all of these constituents start to be synthesized immediately more or less and in such a way as to be formed in stoichiometric amounts. Now, how do you envisage this if this is true, do you think that all these constituents are put into the empty sacks or do you think that entirely new granules are formed? If this is the case, then there should be an increase in the number of empty sacks with a normal number of normal granules. This might be something worthwhile examining in the electron microscope—I would just like to have your reactions to this kind of question.

Kirshner: I think it is questionable whether there is rapid resynthesis during stimulation. There are controversial reports in the literature. Holland and Schümann, and Bygdemann and von Euler have reported rapid resynthesis of the adrenal contents during splanchnic stimulation and after insulin injection. Butterworth and Mann and Eade and Wood found no evidence for rapid resynthesis following i.v. injection of acetylcholine or after splanchnic stimulation. The latter two groups found that the amounts of catecholamines which appeared in the adrenal venous blood were equal to the amounts lost from the glands.

I do not think the vesicles are re-used again, but there is no firm evidence of this. If there was a rapid resynthesis of the entire granules, then we should see an increase in the amount of particulate DBO, provided that the particulate DBO of the old vesicles is not rapidly destroyed. I cannot imagine all of these things being so coincidental that the rate of destruction will exactly balance the rate of synthesis and that the rate of synthesis of the storage vesicle would be equal to the rate at which noradrenaline was released. So I think that there is an incompatibility between the reports on rapid resynthesis and secretion by exocytosis. A theoretical physicist—I do not recall his name—once remarked that if the data don't fit the theory then throw away the data. I think that there is enough support for exocytosis from a variety of sources now that if there are some data which are in contradition to exocytosis then I think we have to look at this contradictory data in a very careful manner.

Carlsson: In an attempt to reconcile these two incompatibilities, could it not be that in the case of the experiments where you found this output of adrenaline and at the same time preservation of levels, that in those experiments the people were not pushing the gland so very hard so that it could go on in a rather physiological way, whereas in these experiments that you have shown, where you knock out 75% of the catecholamines, that then things start to come into play that do not necessarily occur under more gentle conditions?

Kirshner: In some of the experiments we did, the response of the rabbits to insulin was quite variable. In some of those animals we only got a 25 to 50% decrease in the catecholamine content which we assumed was a mild stimulation and under those conditions we also observed the decrease in dopamine-β-hydroxylase which was of the expected order.

BLASCHKO: I should like to say that what Professor CARLSSON has discussed here suggests new experiments and it seems to me important that these should be carried out. Quite a number of people are now working in this field, and such experiments could be done without much difficulty.

STJÄRNE: I just wonder if you could not say that this relatively gentle experiment has been made already by Professor BLASCHKO and his co-workers, when they used the adrenal gland of the calf *in situ*, perfused with its own blood and stimulated via the splanchnic nerves. The only criticism I can think of is that in their publication they only mentioned the result of stimulation at the frequency of 30 per sec, which I think is a bit unfortunate. Maybe other frequencies were tried and maybe they were not successful. It would be very interesting to know from Professor BLASCHKO.

BLASCHKO: The perfusion and stimulation experiments were, in fact, done in Cambridge by COMLINE and SILVER. Dr. COMLINE and Mrs. SILVER were interested in changes in the adrenaline: noradrenaline ratio in the adrenal venous blood of calves and cows when they stimulated the splenic nerves. Dr. SMITH and I went across to Cambridge on two separate occasions and collected some plasma from experiments in which they were measuring the catecholamine content.

SCHÜMANN: In connection with the discussion-remark of Dr. STJÄRNE, I would like to mention that the preparation of the retrograde perfusion of an isolated adrenal medulla is not quite physiological. We used this method, too, for a long time and therefore we have, of course, always to ask ourselves, how good that preparation is and where its limitations are. This is no objection against the method but we have just to be aware of what we are doing. Are there any more questions about the adrenal medulla and the exocytosis, if not, then I will ask Dr. GEFFEN to give his comment.

GEFFEN: I agree that the quantitative evidence suggests that synaptic vesicles are used over and over again. There is no evidence at present that if these proteins are liberated from sympathetic nerves, they are liberated in the same proportions as exist in the intact vesicles. Whatever evidence there is, and there is only data from two laboratories, indicates that the amount of noradrenaline compared to chromogranins is very much greater in the perfusate in spite of the fact that there is re-uptake of noradrenaline. Therefore the ratio in the synaptic gap may even be greater than that presenting in the perfusate. My own impression from the transport of radioactive proteins and from the amounts that can be liberated into the perfusate is that the rapid component of protein transport down sympathetic nerves could replenish the very small amounts of protein which are liberated. But total exocytosis of synaptic vesicles appears to me at present to be unlikely.

FOLKOW: This is an entirely different thing from the exocytosis which is seen in the adrenal gland and I fully agree with this view; it indicates that the release situation is quantitatively a different one at nerve endings as compared with adrenal medullary cells.

Stjärne: I would just like to mention that we have tried to make the old trick of labelling the adenine nucleotides of the amine granules, both in adrenal medulla and sympathetic nerves. This has been tried previously and people have failed. I think this is because the period of treatment was far too short. So we did something really excessive, we shot a millicurie of inorganic phosphate into cats daily for one week. In some cases we started the experiment by exercising the amine storage particles by giving the cat insulin. If after a week's treatment with radiophosphate you denervate one adrenal gland and then give the cat insulin you get a decrease in the labelled ATP of the specific amine granules of the innervated gland but not of the denervated adrenal, and not in any other fraction. If you perfuse the isolated gland you get each time you induce secretion by carbachol, an increase in amine efflux and simultaneously an increase in efflux of phosphorus labelled material. In the spleen taken from the same animal and perfused in the same manner electrical nerve stimulation, causing the spleen to contract, also gives an increased efflux of phosphorus labelled material. But if you prevent the contraction by giving phenoxybenzamine and thus get a 10-fold increase in the noradrenaline secretion in response to nerve stimulation, you do not see any increase in phosphorus labelled material leaving the organ. On the basis of this we have concluded that, although there may be qualitative similarities, there certainly are also extremely important quantitative differences between the secretory mechanisms of the adrenal medulla and of sympathetic nerves. In view of the data concerning concomitant protein and amine release from nerves that we now have, I would regard it as worthwhile to consider that neurotransmitter secretion could be based on partial exocytosis, allowing repeated re-use of the nerve granules.

Helle: I should like to say something in relation to exocytosis in the nerve. I think there may be a misunderstanding in our interpretation of the term "exocytosis". Does exocytosis in the nerve imply a release of the soluble constituents of the granule only, or does it also include a release of the whole granule material?

As far as we know the chromogranin in the nerve granule appears to be very insoluble. We do not know what proportion of chromogranin is soluble and whether it is only the soluble fraction of chromogranin which has been set free as a result of nervous stimulation.

The evidence for an exocytotic release mechanism in the nerve would be of great importance for the view on how the catecholamines leave the nerve: Are the granules approaching the membrane and there emptying their content directly into the synaptic gap, or does the transmitter pass through the cytoplasm before it diffuses through the presynaptic membrane? I think the release of the small amount of chromogranin detected after nervous stimulation is favouring the idea that the granules empty on the surface of the presynaptic membrane. The amount of released protein is not so important in this discussion.

Schümann: May I say, if I understood correctly, that you mean by the expression "exocytosis", more a "microexocytosis"?

Helle: Yes.

von Euler: I think that one should not forget that nerve granules from the nerve trunk are of the same kind as those obtained from a homogenate of an organ like the heart. They have a property of releasing very rapidly their catecholamines in contra-distinction to the catecholamines of the chromaffin cell granules. This is, of course, only an indirect support of the idea that the catecholamines can leave the granules, so to speak, within the nerve terminal, a function which can be modified and modulated by the concentration of amines around them, by ATP and by several other factors. I believe this speaks in favour of a release from the granules—one does not necessarily have to assume exocytosis.

SCHÜMANN: May I just add one question to Dr. HELLE before we go on with the discussion and that is: What do we know about the possibility of a storage complex with ATP, chromogranin or other proteins. Is there any reason to believe that there exists a complex with catecholamines, ATP and chromogranin or not?

HELLE: In work that has been published by KIRSHNER—and I have also repeated his results—dilute solutions of purified chromogranin have only been shown to hold small amounts of catecholamines. However, I think that these results are not quite relevant to the conditions in the intact granule. We know that the concentration of stored constituents are very high there. Before we can exclude a binding complex involving chromogranin, we shall have to repeat these experiments with highly concentrated protein and without removal of the phospholipids; without dissociating the protein from its macromolecular complex with other constituents of the granule.

SCHÜMANN: I agree.

CARLSSON: I should like to elaborate a little bit further on Dr. HELLE's remark —the earlier one she made. I think that there is rather strong evidence that the amines have to get into the granules, or vesicles, in order to become available for the nerve impulse. For instance, if you block the uptake mechanism of the granules by means of reserpine then it is possible to load the extragranular cytoplasm by means of amines, either α-methylated amines or the physiological amines, following monoamine oxidase inhibition. If then the nerves are stimulated this extragranular material cannot be released—it has been shown both histochemically and biochemically. Furthermore, if you follow the time course—the recovery time course after reserpine, it can be seen that as soon as the granular uptake mechanism recovers, then the response to nerve stimulation also returns and this is long before you can really detect an increase in the bulk of the store. So there is again a close correlation between the uptake mechanism of the granules and the transmission machinery. This indicates that the amine has to get into the granule in order to become available for the nerve impulse. However, on this basis alone it is, of course, impossible to decide whether the catecholamines are released alone or together with ATP and/or protein.

DE ROBERTIS: I think that the comments of Professor CARLSSON are most important because the problem of quantal release and synaptic vesicles has so far been studied mainly in cholinergic mechanisms. He mentioned that after reserpine it is necessary to restore the granule to have a functional recovery. I would like to draw your attention again to the fact that the vesicle which contains norepinephrine is the small granulated vesicle of 500 Å present in peripheral adrenergic endings.

Some of these vesicles are close to the membrane or attached to it and represent those involved in the process of releasing the transmitter. This is in relation to what Professor CARLSSON said before about the geometry of the transmitter release. I think that the vesicles which are near to the membrane are the ones which are secreting their products faster. The norepinephrine would be released from them earlier than from the other storage sites and this will make the functional pool. I do not think that it is necessary to postulate a free pool which can be released upon stimulation. Those vesicles which have lost part or all of the norepinephrine, are also the ones that recharge faster. This would make a functional compartment with a very fast re-uptake and re-synthesis.

VON EULER: I am afraid I cannot quite agree with the idea of a binding to granules as absolutely necessary for the release of catecholamines after reserpine. In some experiments on vas deferens we have seen repeatedly that the acute effect of reserpine is to inhibit the response to nerve stimulation and if we use enough concentration this can be done quite quickly. After 10 min of reserpine action we do not really have to assume that the granular stores are depleted—they cannot be in that time—but what we find is that when we add noradrenaline to the bath in rather low concentrations, the function of the nerve is restored. So this seems to indicate, to my mind at least, that the added noradrenaline is taken up by some—I would say again, a readily available-pool, a functional pool, different from that of the granules which probably are still filled, although they cannot make their contents available to the secondary pool. I come back to my scheme of some kind of secondary retail pool which is structurally not known but still may be present, either in the membrane or in organelles and which is being ordinarily refilled from the big stores. At any rate, what is present there can be released probably without affecting the granules.

CARLSSON: I would just like to comment on Professor VON EULER's interesting data here. I think we must distinguish these experiments where you have added reserpine in a concentration that is, I think, about ten times higher than the one necessary to block the ATP dependent uptake mechanism of the granules. If you do experiments *in vivo* where you inject reserpine, I do not think anybody has demonstrated this effect so far, but rather it has been shown that the response to nerve stimulation goes on for several hours after the intravenous injection of reserpine, as long as there are reasonable amounts left in the granules, so I think we are talking about different effects of reserpine.

STJÄRNE: I think it was very clearly pointed out by Dr. HELLE that the nucleus of the concept of secretion by exocytosis is secretion directly from storage granules onto the cell surface. This mechanism for secretion would involve intermittent local functional "destruction" of the granule and cell membranes, though not necessarily the formation of anatomical openings. Now, is there any experimental evidence which would serve to support the assumption that the adrenal medullary granules would discharge all or none in this situation, while the nerve granules would not? Yes, I think there is, in the very simple experiments concerning osmotic lysis. In the adrenal medullary granules osmotic lysis causes immediate and almost complete release of both amines and ATP. In the nerve granules, in the identical situation, you have only fractional amine release. Whether this *in vitro* finding has any relevance for amine secretion at the cellular level, I do not know.

I just want to point out that when the granule membrane is destroyed, under the highly artificial *in vitro* conditions you do get fractional release from one kind of granule but not from the other.

GLOWINSKI: I would just like to make some comments on the discussion between Professor CARLSSON and Professor DE ROBERTIS. I agree with Professor CARLSSON about the question of the recovery of some metabolic characteristics of aminergic neurons 24 h after an acute treatment with reserpine. The recovery of the ability to accumulate exogenous norepinephrine in norepinephrine containing neurons, 24 h or more after reserpine treatment, does not necessarily mean that the exogenous amine will be stored for a long period of time. It appears that this amine which cannot be stored for a long period of time is immediately available for release and can be considered to be localized in the "functional pool". Again, this does not mean that this amine is not localized in vesicles, perhaps on the surface of the vesicle, various forms of storage may coexist in vesicles.

DE ROBERTIS: I think that all the evidence suggests that norepinephrine is inside the vesicle; particularly the histochemical technique with the chromate reaction shows that the reactive points are within the vesicle and not outside. The idea that it may be outside cannot explain also the reaction of monoamine-oxidase. I think that Professor CARLSSON made another interesting point: by using a MAO inhibitor and reserpine the stimulus cannot go through an adrenergic ending, in spite of the fact that there is much catecholamine in a free pool. It is evident that the amine should be in a vesicle to be physiologically released.

THOENEN: Only a short comment on this question which will support the view that only the transmitter(s) stored in the granulated vesicles can be liberated by nerve impulses. Among the phenethylamines which are taken up into the adrenergic nerve endings, only those which are stored in the granulated vesicles act as false transmitters.

HÄGGENDAL: Dr. DAHLSTRÖM and I have for some time tried to correlate different recovery functions after reserpine with an axonal down-transport of newly formed amine storage granules: As Dr. DAHLSTRÖM showed in her lecture, the amount of noradrenaline accumulated for 6 h above a ligation of the rat sciatic nerve at different time periods after a single dose of reserpine started to increase rather early—about 18 h after reserpine. Thereafter, there was a supranormal accumulation which continued for about one week, whereafter the accumulation returned to normal or even somewhat lower.

The recovery of the tissue levels of endogenous noradrenaline after reserpine has been studied in more detail. During the first day only very low levels of noradrenaline were seen after reserpine. But with sensitive techniques the recovery of noradrenaline levels was found to start rather early. 36 h after reserpine the increase was found to be statistically significant; and 48 h after reserpine the increase was statistically highly significant. The time of onset of this recovery seems to be simultaneous with the onset of the observed recovery of the nerve function, e.g. at electrical stimulation.

Furthermore, we are studying the recovery of the capacity of the tissues to take up and store labelled noradrenaline. In fact, the time of onset of this function after reserpine appears to be about the same as for the onset of the recovery of the endogenous levels. In some preliminary experiments we have also performed an unilateral ligation of the rat sciatic nerve at different time intervals after reserpine (24 and 36 h) and kept these animals ligated for 12 h before the labelled noradrenaline was injected. In the tissues on the unligated side there was the same recovery of the capacity to take up labelled noradrenaline as just mentioned. On the side ligated for 12 h, the recovery was not as fast as on the unligated one. In fact, the recovery seemed to be delayed by about 12 h.

We think that for the different recovery functions discussed, a most important role is played by the newly formed granules, down-transported through the axons. Furthermore, these amine storage granules appear to have a rather long life-span of about 3 weeks, where they disintegrated.

However, this does not necessarily mean that all the functions of the granules may correspond to a life-span of about 3 weeks. It may only be that the capacity of the granules to contain endogenous noradrenaline is about 3 weeks, while other functions, e.g. the capacity to take up labelled noradrenaline or release transmitter during normal nerve activity, may be changed when the granules in the nerve terminals become older.

The shape of the curve for the recovery of the endogenous noradrenaline levels in the tissues seems to correspond rather well with the amount of noradrenaline that can be down-transported through the axons at different time intervals after reserpine. Thus, at first there is a fast increase of the endogenous noradrenaline levels, followed after about a week by a slower increase. By about the fourth week there is a drop in the noradrenaline levels. The explanation may be, that during the first week after reserpine there is a supra-normal down-transport of granules. After about 4 weeks these granules are so old that they die. The down-transport of new granules, going on all the time, is now only normal and cannot compensate for this increased loss.

Folkow: Just a short comment to avoid any mistakes concerning my position as regards these problems. I am, of course, fully aware of, and in agreement with, the views that the vesicle (granule) *is* the bearer of the transmitter and that it must come in contact with the nerve membrane to release the transmitter.

My only argument is: Is it really so that in neurons the *entire* content of the vesicles is released? I have outlined the enormous difficulties earlier in the example from a motoneuron. If one considers the mammalian skeletal muscle, when perfused, and explores one of the few studies that can be quantitatively analysed, namely that by MacIntosh and Emmelin and uses their figures for acetylcholine release, one ends up with an ACh content in the assumed packet corresponding to only about 600 molecules. This appears to be only a *small* fraction of what seems to be a reasonable figure for the ACh content of a vesicle.

Not for a moment do I doubt that the vesicle (granule) contains, stores and synthesized the transmitter but I think that most deductions available make it likely that only a *fraction* of the vesicle content is released at nerve endings. I find no difficulty in imagining that the situation can in this respect be a different one in gland cells; they are in many respects very different from nerve cells, and it hardly seems to call for a revolution in cellular function to make transmitter granules "compartmentalized".

von Euler: Just a brief comment, I wonder how you would account for the action of tyramine which blocks its own uptake, so to speak, if it did not work somewhere rather close to the surface of the axon membrane?

Stjärne: I would just like to ask Professor Muscholl to answer one very important point concerning the calcium dependence.

Muscholl: I think one should not confuse the release brought about by tyramine and the release brought about by nerve stimulation because tyramine apparently can release the transmitter if it is free in the axon and not bound. Further, tyramine causes release in the absence of calcium whereas we all know that noradrenaline release by nerve stimulation cannot be brought about in the absence of calcium. Thus, experiments with tyramine, as interesting as they are, cannot add to the understanding of release mechanism brought about by nerve stimulation.

von Euler: I would answer that it is not necessary that the nerve stimulation acts like tyramine but as for the site of available transmitter. I think it is of importance.

Kirshner: I would like to make a comment on reserpine. Although we know many things that reserpine does, we still do not know the molecular basis for its action on any of these things, e.g. we know that reserpine will prevent the incorporation of radioactive noradrenaline into the storage complex. but reserpine itself does not appear to affect the complex once the catecholamines are stored. We and others have tried to get complete inhibition of uptake by reserpine but we never get complete inhibition, it is about 90 to 95%, so apparently there can be another way in which the catecholamines can get into the complex even though the major route is blocked by reserpine.

Stjärne: I am very proud of being given the chance of having the final word. Let me just point out that on the basis of rather simple experiments with isolated splenic nerve granules, in which the synthesis of noradrenaline was studied, we have postulated that newly formed noradrenaline is not bound in the same compartment which stores the main body of the performed, endogenous noradrenaline. We have interpreted the evidence as indicating that the newly formed noradrenaline is bound in, or on the granule membrane, which should, perhaps, please Dr. Glowinski.

Schümann: I am afraid we have now to come to an end. We have discussed very intensively the problem of exocytosis and I believe that not only a number of facts came out but also ideas to continue the work on this problem.

I regret that we had no more time to discuss other important topics here in the general discussion, for instance, the problem of prostaglandin action which I find myself very interesting.

In closing the meeting, I should first like to thank Dr. BRANDAU and Dr. HOFFMEISTER and the staff of the Bayer Company for all they have done for us in organising this attractive Symposium and my thanks are also due to you all who have contributed to the success of this meeting, which I feel has then been achieved if one can leave it with a few, or even one, new idea for further experiments. I wish you a pleasant return journey.

Author Index

Names of principal speakers are set in **boldface,** those of speakers in discussions in *italics*
Figures in *italics* are references to the literature

Acosta-Ferreira, W., see Estable, C. 4, 12, *16*
Aghajanian, G. K., see Bloom, F. E. 23, *33*
Agrest, A., see Finkielmann, S. 207, *208*
— see Massani, Z. M. 207, *208*
Akert, K., see Pfenninger, K. 155, *158*
Alexander, C. S. 195, *196*
Alexander, N., see De Quattro, V. [37] 81, *87;* 214, *218;* 220
Alousi, A., Weiner, N. 212, *218*
Amsterdam, A., Ohad, I., Schramm, M. [43] 82, *88*
Andén, N. E., Butcher, S. G., Corrodi, H., Fuxe, K., Ungerstedt, U. 255, *255*
— Carlsson, A., Häggendal, J. 20, *33;* 100, *108;* 231, *232*
— Corrodi, H., Dahlström, A., Fuxe, K., Hökfelt, T. 252, *255*
— — Fuxe, K. 252, *255*
— — — Hökfelt, T. 231, 252, 254, *255*
— Roos, B. E., Werdinius, B. 252, 254, *255*
Andres, K. H. 14, *15*
Angelakos, E. T., Bloomquist, E. 169, *184*
Angeletti, P. U., see Levi-Montalcini, R. 130, 137, *141*
Armett, C. J., Ritchie, J.M. 153, *156*
Arqueros, L., see Viveros, O. H. 46, 51, 54, *56;* 58, *72;* [34, 35, 36] 80, 81, *87;* [41, 44, 45] 82, 83, *88;* 115, 121, 122, *127*
Austin, L., Livett, B. G., Chubb, I. W. [33] 80, *87*
— see Livett, B. G. 40, *41;* 58, *71*
Axelrod, J. 37, 38, 72, 77, 212, 216, 217, *218;* 219, 220, 221, 222, 248, 249, 256, 267, 268
— Hertting, G., Potter, L. 148, *156*
— Mueller, R. A., Henry 215
— Snyder 37, 249
— Whitby, L. G., Hertting, G. 223, *232*
— see De Champlain, J. 138, *141*
— see Glowinski, J. 223, *233;* 242, 245, *247*
— see Hertting, G. 223, *233*
— see Iversen, L. L. 137, *141*
— see Landsberg, L. 215, *218*
— see Mueller, R. A. [38, 39] 82, *87;* 133, 140, *141;* 212, 213, 214, 215, 216, *218*
— see Potter, L. T. 52, *56;* [56] 85, *88;* 258, *267*
— see Rosell, S. 100, 102, *109;* 121, *126;* 206, *209*
— see Thoenen, H. 28, *35;* [40] 82, *88;* 140, *141;* 212, 213, 214, 215, 216, *219*
— see Wurtmann, R. J. 215, 217, 218, *219*

Bak, I. J. 188
—, see Grobecker, H. 189, 190, 191, 192, 194, *197*
Balzer, H., Makinose, H., Fiehn, W., Hasselbach, W. 193, *196*
— — Hasselbach, W. 193, *196*
Banks, P. 54, *55* [4] 78, *86*
— Helle, K. B. 46, 53, *55* [5] 79, *86;* 121, *124*
— — Major, D. 47, 51, 52, *55;* 67, *71;* 121, *124*
— Mangnall, D., Mayor, D. 22, *33*
Barnett, A., Benforado, J. M. 180, 181, *184*
Belleau, B. 152, *156*
— Martel, R., Lacasse, G., Menard, M., Weinberg, N. L., Perron, Y. G. 148, *156*
Belpaire, F., see Laduron, P. 20, 22, *35;* 43, 46, 50, 51, *56;* [57] 85, *88;* 221
Benforado, J. M., see Barnett, A. 180, 181, *184*
Bennett, H. S., see De Robertis, E. 14, 15, *16*
Bennet, M. R., Burnstock, G. 93, *96;* 98
Bensch, K. G., see Malawista, S. E. 32, *35*
Bentley, G. A. 183, *184*
Beran, see Nedergaard 186
Berneis, K. H., Pletscher, A., Da Prada, M. 61, *71*
Bertler et al. 252
Besson, M. J., Cheramy, A., Feltz, P., Glowinski, J. 238, 240, 244, *247*
— — Glowinski, J. 240, 242, *247*
— see Cheramy, A. 244, *247*
Bhagat, B. 180, 181, 182, *184*
— Bovell, G., Robinson, I. M. 182, *184*
— Robinson, I. M., West, W. L. 182, *184*
Bialy, H. S., see Seeman, P. M. 190, 192, *197*

Bigler 140
Birmingham, A. T., Wilson, A. B. 149, *156;* 182, 183, *184*
Blakeley, A. G. H., Brown, G. L., Geffen, L. B. 70, *71*
Blanc, G., see Thierry, A. M. 242, 245, 246, *248*
Blaschko, H. 1, 2, 44, 55, 75, 142, 234, 270, 275
— Comline, R. S., Schneider, F., Silver, M., Smith, A. D. 45, 54, *55;* [18] 79, *87;* 121, *124*
— Helle, K. B. 45; [10] 79, *86*
— Smith, A. D., Winkler, H. [12] 79, *86*
Bloom, F. 256
Bloom, F. E., Aghajanian, G. K. 23, *33*
Bloom, M. F. E., Oliver, A. P., Salmoiraghi, G. C. *266*
Bloomquist, E., see Angelakos, E. T. 169, *184*
Blümcke, S., Niedorf, H. R. 14, *16*
Blumenthal, M. R., see McGiff, J. C. 179, *185*
Booker, B., see Levi-Montalcini, R. 130, *141*
Boullin, J., Costa, E., Brodie, B. B. 100, 101, 103, *108*
Bovell, G., see Bhagat, B. 182, *184*
Boyle, F. C., Gillespie, J. S. 20, *33*
Bradley, P. B., Dhawan, B. N., Wolstencroft, J. H. 262, *266*
Breemen, V. L., Van Anderson, E., Reger, J. F. 14, *16*
Breese, G. R., see Kopin, I. J. 124, *126*
Breeze, G. R., see Kopin, I. J. 241, *247*
Brandau, K. 3, 282
Brender, D., see Wilcken, D. E. L. 150, *158;* 195, *197*
Brodie, B.B. 200, 234, 235
— see Boullin, J. 100, 101, 103, *108*
Bronk, D. W., Pitts, R. F., Larrabee, M. G. 94, *96*
Brown, G. L. 100, *108*
— see Blakeley, A. G. H. 70, *71*
Brundin, J., see Hedqvist, P. 154, *157*
Burch, G. E., see Sun, S.-C. 195, *197*
Burgen, A. S. V., Iversen, L. L. 147, 148, *156*
— see Metcalf, J. C. 192, *197*
Burger, A., Philippu, A., Schümann, H. J. 146, *156;* 188, *196*
— see Philippu, A. 258, 259, 260, 261, 262, 264, 265
Burn, J. H. 155, *156;* 178, 179, *184*
— Rand, M. J. 114, *124;* 147, 152, 155, *156;* 169, 179, *184*
Burns, R. B. P., see McGiff, J. C. 179, *185*
Burnstock, G., Holman, M. E. 91, 93, 94, *96;* 116, 118, *125;* 153, *156*
— — Kuriyama, H. 120, *125*
— Robinson, P. M. 114, *125;* 152, *156*
— see Bennet, M. R. 93, *96*; 98
Butcher, S. G., see Andén, N. E. 255, *255*
Butcher, R. W., see Sutherland, E. W. 144, 145, *158*
Butterworth, Mann, Eade, Ivood 274
Bygdeman, S., Euler, U. S. v. 148, 274
— — Hökfelt, B. 122, *125*
— Johnsen, Ø. 148, *156*
— see Stjärne, L. [49] 85, *88;* 92, *97; 109; 126*

Cannon, W. B., Rosenblueth, A. 130, *141*
Carlsson, A. 20, 24, *34;* 43, 73, 74, 75, 89, 95, *96;* 100, *108;* 117, *125;* 199, 200, 210, 220, 223, *232;* 234, 235, 250, 256, 270, 273, 274, 275, 277, 278, 279
— Corrodi, H., Fuxe, K., Hökfelt, T. 225, 226, *232*
— — Waldeck, B. 252, *255*
— Folkow, B., Häggendal, J. 102, *108*
— Fuxe, K., Hamberger, B., Lindqvist, M. 224, 225, *232*
— — — Malmfors, T. 225, *232*
— — Ungerstedt, U. 242, *247*
— Hillarp, N.-Å., Waldeck, B. 24, *34;* [26] 80, *87;* 103, *108;* 144, 145, 148, *156;* 188. 189, *196*
— Jonason, J., Lindqvist, M. 229, 230, *232*
— — — Fuxe, K. 227, 229, *232*
— Lindqvist, M. 226, 227, *232;* 252, 254, *255*
— — Wysokowski, J., Corrodi, H., Junggren, U. 224, *232*
— see Andén, N.-E. 20, *33;* 100, *108;* 231, *232*
Carr, L. A., Moore, K. E. 242, *247*
Cervoni, P., see Kirpekar, S. M. 115, *126*
Cheramy, A., Besson, M. J., Glowinski, J. 244, *247*
— see Besson, M. J. 238, 240, 242, 244, *247*
Chiang, T. S., Leaders, F. E. 178, *184*
Chornock, F. W., see Korr, I. M. 12, *16*
Chubb, I. W., see Austin, L. [33] 80, *87*
Cohen, S. 130, *141*
Colcolough, H. L., see Sun, S.-C. 195, *197*
Comline, Silver 275
Comline, R. S., see Blaschko, H. 45, 54, *55;* [18] 79, *87;* 121, *124*
Connett, R. J., see Viveros, O. H. 46, 51, 54, *56;* [36] 81, *87;* 82, *88*
Cooper, T. 136, *141;* 169, *184*

— see Jellinek, M. M. 137, *141*
Coppen, A. 231, *233*
Corrodi, H. 143, 200, 231, 235, 236, 252, 256, 257
— Fuxe, K. 231, *233;* 242, *247*
— — Hökfelt, T. 231, *233;* 252, 254, *255*
— Hanson, L. 252, *255*
— Jonsson, G. 20, *34*
— Malmfors, T. 252, *255*
— see Andén, N. E. 231, 252, 254, 255, *255*
— see Carlsson, A. 224, 225, 226, *232;* 252, *255*
Costa, E. 235
— Neff, N. H. 244, *247*
— see Boullin, J. 100, 101, 103, *108*
Coupland, R. E. [30] 80, *87;* [52] *88;* 217, *218;* 271

Dahlström, A. 2, 4, 14, *16;* 18, 19, 20, 21, 22, 25, 26, 29, 31, 33, *34;* 36, 37, 38, 39, 70, *71;* 76, 77, 103, *108;* 110, 128, 129, 143, 200, 220, 222, 279
— Fuxe, K. 20, 25, *34*
— — Hillarp, N.-Å. 25, *34*
— Häggendal, J. 4, 14, *16;* 20, 22, 24, 25, 26, 27, 29, *34;* 39, *41;* 95, 97, *97*
— — Hökfelt, T. 91, 93, *97*
— Hökfelt, T. 21
— Jonason, J. 22, *34*
— see Andén, N. E. 252, *255*
— see Häggendal, J. 28, 29, *34*
— see Hökfelt, T. 31
Da Prada, M., Pletscher, A. 254, *255*
— see Berneis, K. H. 61, *71*
Davies, B. N., Horton, E. W., Withrington, P. G. 153, *157*
— Withrington, P. G. 117, *125*
Davis, L. D., see Krasney, J. A. 202, *208*
Day, M. D. 183, *184*
— Owen, D. A. A. 150, *157*
De Champlain, J., Krakoff, L., Axelrod, J. 138, *141*
De Potter, W. P. 129
— De Schaepdryver, A.F., Moerman, E. J., Smith, A. D. 54, *55;* 67, 68, *71;* 73, 75, 121, 122, *125*
De Quattro, V., Maronde, R., Nagatsu, T., Alexander, N. [37] 81, *87;* 214, *218;* 220
De Robertis, E. 4, 17, 18, 19, 42, 43, 44, 57, 72, 76, 88, 142, 199, 200, 256, 268, 271, 277, 279
— Bennett, H. S. 14, 15, *16*
— Franchi, C. M. 4, *16*
— Pellegrino de Iraldi, A. 12, *16*
— Sotelo, J. R. 4, *16*
— see Pellegrino de Iraldi, A. 4, 5, 7, 11, 12, *16;* 23, 24, *35;* *41*
— see Thornburg, W. 4, *17*
De Robertis, E. D. P., Vaz Ferreira, A. 53, *55;* [20] 79, 80, *87*
De Schaepdryver, A. F., see De Potter, W. P. 54, *55;* 67, 68, *71;* 73, 75, 121, 122, *125*
De Scarries, Droz 43
Del Castillo, J., Katz, B. [46] 85, *88*
Dengler, H. J., Spiegel, H. E., Titus, E. O. 223, *233*
Dhawan, B. N., see Bradley, P. B. 262, *266*
Diner 271
Dingell, J. V., see Sulser, F. 223, *233*
D'Iorio, A. 146, *157;* 270
Dixon, W. E. 169, *184*
Dontas, A. S., Nickerson, M. 212, *218*
Douglas, W. W. [54] 85, *88;* 113, 115, 121, *125*
— Kanno, T., Sampson, S. R. 114, *125*
— Poisner, A. M., Rubin, R. P. [1, 2, 3] 78, *86;* 121, *125*
— Rubin, R. P. 114, *125*
— see Kanno, T. 114, *125*
— see Malamed, S. [24] 80, *87;* 122, *126*
— see Poisner, A. M. [23] 80, *87;* 122, *126*
— see Trifaro, J. M. [22] 79, *87*
Draskóczy, P. R. 161
— Trendelenburg, U. 161, *165*
Droz, B. 23, *34*
— see De Scarries 43
— see Koenig, H. L. 15, *16*
Duch, D. S., Viveros, O. H., Kirshner, N. [29] 80, *87*
Düring, M. v. 14, *16*

Eade, N. R. 45, *55*
Eccles, J. C. 91, *97;* 100, *108*
Efron 159, 257
Ehinger, B., Falck, B. 180, *184*
— — Persson, H., Rosengren, E., Sporrong, B. 114, *125*
Elmqvist, D. 100, *108*
Emmelin et al. 96
Eränkö, O., Härkönen, M. 20, *34*
Estable, C., Acosta-Ferreira, W., Sotelo, J. R. 4, 12, *16*
Eugster, C. H., see Pfenninger, K. 155, *158*
Euler, U. S. v. 1, 2, 3, 4, 72, 73, 75, 110, 111, 123, 128, 129, 144, 152, *157;* 159, 160, 186, 187, 198, 199, 209, 210, 221, 222, 235, 249, 276, 278, 281
— Hedqvist, P. 154, *157*
— Lishajko, F. 123, *125;* 144, 145, 146, 147, 148, 150, 152, *157;* 188, 189, 192, 193, *196*
— — Stjärne, L. 24, *34*
— Persson, N.-Å. 150, *157*
— Stjärne, L., Lishajko, F. 103, *108;* 189, 192, *196*
— see Bygdeman, S. 122, *125;* 148

Euler, U. S. v. see Stjärne, L. 161, *165*
Estelmann, T., see Lang, K. 32, *35*
Everett, G. M., Wiegand, R. G., Rinaldi, F. V. 227, *233*

Falck, B., Owman, C. 65, *71*
— see Ehinger, B. 114, *125;* 180, *184*
Fekete, M., see Thierry, A. M. 245, *248*
Feldberg, W. 231, *233*
— Myers, R. D. 231, *233;* 258, *266*
Feltz, P., see Besson, M. J. 238, 240, 244, *247*
Ferry, C. B. 153, *157*
Fiehn, W. see Balzer, H. 193, *196*
Finkielmann, S., Worcel, M., Massani, Z. M., Nahmod, V. E., Paladini, A. C., Agrest, A. 207, *208*
— see Massani, Z. M. 207, *208*
Flatmark, T. 55
Folkow, B. 91, 93, 97, 98, *97;* 108, 111, 272, 275, 280
— Häggendal, J. 101, 102, 103, 104, *108*
— — Lisander, B. [48] 85, *88;* 91, 92, 95, *97;* 100, 101, 104, 107, *108;* 113, 115, 117, 118, *125;* 155, *157*
— see Carlsson, A. 102, *108*
Fowler, N. O., Holmes, J. C. 202, *208*
Franchi, C. M., see De Robertis, E. 4, *16*
Friedmann, S., see Kaufmann, S. 22, *35*
Furchgott, R. F., Sanchez-Garcia, P. 164, *165*
Furchgott, R. W. 169, *184*
Fuxe, K. 221, 263, 165, *267*
— Ungerstedt, U. 224, *233;* 242, *247*
— see Andén, N. E. 231, 252, 254, 255, *255*
— see Carlsson, A. 224, 225, 226, 227, *232;* 242, *247*
— see Corrodi, H. 231, *233;* 242, *247;* 252, 254, *255*
— see Dahlström, A. 20, 25, *34*

Geffen, L. B. 17, 18, 36, 39, 41, 42, 43, 58, 72, 73, 74, 75, 76, 77, 90, 98, 100, 101, *108;* 127, 166, 219, 222, 251, 275
— Hunter, C., Rush, R. A. 40, *41*
— Livett, B. G. 68, *71*
— — Rush, R. A. 54, *55;* 58, *71*
— Ostberg, A. 40, *41;* 59, *71*
— Rush, R. A. 20, *34;* 39, *41*
— see Blakeley, A. G. H. 70, *71*
— see Livett, B. G. 24, *35;* 40, *41;* 52, *56;* 58, *71;* [58] 85, *88*
Gey, K. F., see Thoenen, H. 130, *141*
Giachetti, A., Shore, P. A. 161, *165*
Giacobini 146
Gillespie, J. S. see Boyle, F. C. 20, *33*
Gillis, C. N., see Paton, D. M. 206, *209*
Glagov, S., see Wolinsky, H. 137, *142*
Glowinski, J. 76, 110, 143, 221, 222, 234, 237, 248, 249, 250, 256, 267, 279, 281
— Axelrod, J. 223, *233*
— — Iversen, L. *247*
— Iversen, L. 245, *247*
— — Axelrod, J. 242, 245, *247*
— Kopin, I., Axelrod, J. 245, *247*
— see Besson, M. J. 238, 240, 242, 244, *247*
— see Cheramy, A. 244, *247*
— see Iversen, L. L. 137, *141;* 244, *247*
— see Javoy, F. 243, 244, *247*
— see Kordon, C. 245, *248*
— see Macon 245
— see Thierry, A. M. 242, 245, 246, *248*
Goldfisher, S., see Novikoff, A. B. 11, *16*
Goodall, McC., see Kirschner, N. 216, *218*
Gordon, R., Spector, S., Sjoerdsma, A., Udenfriend, S. 212, *218*
— see Spector, S. 212, *219*
Goth, A., see Isaac, L. 226, *233*
Graefe, K.-H. 161
Graupner, K. I., see Richardson, H. L. 195, *197*
Green, A. I., see Snyder, S. H. 242, *248*
Grobecker, H. 188
— Holtz, P., Palm, D., Bak, I. J., Hassler, R. 189, 190, 191, 192, 194, *197*
— Palm, D., Schümann, H. J. 193, *197*
Grov, A. 55
Gueudet, R., see Pellegrino de Iraldi, A. 12, *16*
Gustavsson, M. 32, *34*
Gyermek, L., see Sigg, E. B. 223, *233*

Haefely, W., Hürliman, A., Thoenen, H. 117, *125;* 182, *184*
— see Häusler, G. 137, 138, *141;* 175, 178, *184*
— see Thoenen, H. 100, *109;* 130, *141*
Häggendal, J. 33, 38, 75, 76, 91, 92, 97, 100, 103, 108, *108;* 109, 110, 111, 128, 279
— Dahlström, A. 28, 29, *34*
— Johansson, Jonason, Ljung 100
— Lindqvist, M. 245, *247*
— Malmfors, T. 20, *35;* 117, *125*
— see Andén, N.-E. 20, *33;* 100, *108;* 231, *232*

— see Carlsson, A. 102, *108*
— see Dahlström, A. 4, 14, *16;* 20, 22, 24, 25, 26, 27, 29, *34;* 39, *41;* 91, 93, 95, *97*
— see Folkow, B. [48] 85, *88;* 91, 92, 95, *97;* 100, 101, 102, 103, 104, 107, *108;* 113, 115, 117, 118, *125;* 155, *157*
Härkönen, M., see Eränkö, O. 20, *34*
Häusler, G. 130
— Haefely, W., Hürlimann, A. 137, *141*
— — Thoenen, H. 138, *141*
— Thoenen, H., Haefely, W. 138, *141*
— — — Hürlimann, A. 175, 178, *184*
Hager, H., see Schlote, W. 12, *17*
— see Wechsler, W. 12, *17*
Hamberger, B. 110, 224, 225, 230, *233*
— Masuoka, D. 224, *233*
— see Carlsson, A. 224, 225, *232*
Hamon, M., see Javoy, F. 244, *247*
— see Jeanne-Rose 242
Hanson, L., see Corrodi, H. 252, *255*
Hasselbach, W., see Balzer, H. 193, *196*
— see Taugner, G. 146, *158;* 188, *197*
Hassler, R., see Grobecker, H. 189, 190, 191, 192, 194, *197*
Hedqvist, P. 100, 107, *108;* 120, *125;* 151, 154, 155, *157;* 256
— Brundin, J. 154, *157*
— see Euler, U. S. v. 154, *157*
— see Stjärne, L. [49] 85, *88;* 92, *97; 109;* 116, 117, 118, 119, 122, *126; 127;* 193, *197*
Helle, K. B. 2, 44, 45, 46, 47, 53, *55;* 56, 57, [14] 79, *86;* 90, 127, 128, 129, 276, 277, 278
— Serck-Hanssen, G. 49, 52, *55*
— see Banks, P. 46, 47, 51, 52, 53, *55;* 67, *71;* [5] 79, *86;* 121, *124*
— see Blaschko, H. 45, [10] 79, *86*
Hendley, E. D., see Snyder, S. H. 242, *248*
Hennemann, H. 161
Henry 222
— see Axelrod, G. 215
Hertting, G., Axelrod, J., Kopin, J. J., Whitby, L. G. 223, *233*
— Schiefthaler, T. 136, 137, *141;* 206, *208*
— see Axelrod, J. 148, *156;* 223, *232*
Heyd, G., see Philippu, A. 258, 259, 260, 261, 262, 264, 265
Hillarp, N.-Å. 45, *55;* [53] *88;* 92, *97;* 146, 152, *157;* 180, *184*
— Falck et al. 20
— Högberg, B., Nilson, B. 144, *157*
— Lagerstedt, S., Nilson, B. [9] 79, *86*
— see Carlsson, A. 24, *34;* [26] 80, *87;* 103, *108;* 144, 145, 148, *156;* 188, 189, *196*
— see Dahlström, A. 25, *34*
Hinterberger, A., see Wilcken, P. E. L. 195, *197*
Hiscoe, H., see Weiss, P. 20, *36*
Högberg, B., see Hillarp, N.-Å. 144, *157*
Hökfelt, B., see Bygdeman, S. 122, *125*
Hökfelt, T. 20, 21, 23, 24, *35;* 37, 92, 97, *97;* 113, 118, *125*
— Dahlström, A. 31
— see Andén, N. E. 231, 252, 254, *255*
— see Carlsson, A. 225, 226, *232*
— see Corrodi, H. 231, *233;* 252, 254, *255*
— see Dahlström, A. 21, 91, 93, *97*
Hoffmann, E. J., see Hoffmann, F. 169, *184*
Hoffmann, F., Hoffmann, E. J., Middleton, S., Talesnik, J. 169, *184*
Hoffmeister 282
Hogan, P. M., see Krasney, J. A. 202, *208*
Holloway, C., see Kirshner, N. 45, 46, *55;* [11] 79, *86*
Holman, M. E. 94, *97;* 98, 99
— see Burnstock, G. 91, 93, 94, *96;* 116, 118, 120, *125;* 153, *156*
Holmes, J. C., see Fowler, N. O. 202, *208*
Holland, Schümann, H. J. 274
Holtz, P. 1, 2, 142, 210, 271
— Palm, D. 188, *197*
— see Grobecker, H. 189, 190, 191, 192, 194, *197*
Hopwood, D. 51, 53, *55;* 67, *71*
Horton, E. W. 121, *125*
— see Davies, B. N. 153, *157*
Hukovic, S. 149, *157*
— Muscholl, E. 114, *125;* 168, 177, *184;* 202, *208*
Hunter, C., see Geffen, L. B. 40, *41*
Hürliman, A., see Haefely, W. 117, *125;* 182, *184*
— see Häusler, G. 137, *141;* 175, 178, *184*
— see Thoenen, H. 100, *109;* 130, *141*
Hydén, H. 28, *34*

Imaizumi, R., see Izumi, F. 195, *197*
Iggo, A., Vogt, M. 28, *35;* 36, 212, *218*
Isaac, L., Goth, A. 226, *233*
Iversen, L. L. 20, 26, *35;* 95, *97;* 100, 106, *108;* 147, *158*
— Glowinski, J. 244, *247*
— — Axelrod, J. 137, *141*
— see Burgen, A. S. V. 147, 148, *156*

— see Glowinski, J. 242, 245, *247*
— see Lightman 166
— see Uretsky 143
Izumi, F., Oka, M., Yoshida, H., Imaizumi, R. 195, *197*

Järlfors, Smith 24
Jaim Etcheverry, G., Zieher, L. M. 12, *16*
— see Pellegrino de Iraldi, A. 14, *16*
Janssen 256
Javoy, F., Hamon, M., Glowinski, J. 244, *247*
— Thierry, A. M., Kety, S. S., Glowinski, J. 243, *247*
— see Jeanne-Rose 242
— see Thierry, A. M. 245, *248*
Jeanne-Rose, Hamon, M., Javoy, F. 242
Jellinek, M. M. P., Kaye, M. P., Nigh, C. A., Cooper, T. 137, *141*
Johansson, B., Ljung, B. 94, 99, *97*
— see Häggendal, J. 100
Johnsen, Ø., see Bygdeman, S. 148, *156*
Johnson, J., see Ochs, S. 23, *35*
Jonasson, J., Rosengren, E., Waldeck, B. 147, *158;* 189, 192, *197*
— Rutledge, C. O. 225, *233*
— see Carlsson, A. 227, 229, 230, *232*
— see Dahlström, A. 22, *34*
— see Häggendal, J. 100
Jonsson, G., see Corrodi, H. 20, *34*
Jost, A., see Margolies, F. 217, *218*
Jouvet 143
Junggren, U., see Carlsson, A. 224, *232*

Kalsner, Nickerson 166
Kanno, T., Douglas, W. W. 114, *125*
— see Douglas, W. W. 114, *125*
Kapeller, K., Mayor, D. 4, 14, *16;* 20, 21, *35;* 42, 59, *71*
Karlsson, J.-O. 23, *35*
— Sjöstrand, J. 31, *35*
— see Sjöstrand, J. 23
Katz, B. 91, 95, *97;* 112, *126;* 152, *158*
— Miledi, R. 121, 124, *126*
— see Del Castillo, J. [46] 85, *88*
Kaufmann, S., Friedmann, S. 22, *35*
— see Levin, E. Y. 50, *56;* [28] 80, *87*
Kaye, M. P., see Jellinek, M. M. P. 137, *141*
Kety, S. S. 221
— see Javoy, F. 243, *247*
— see Thierry, A. M. 245, *248*
Khairallah, see Peach Bumpus 209
Khan, M. 178
— Mantegazza, P., Piccinini, F. *184*
Kidaka, H., see Nagatsu, T. [32] 80, *87*
Kidwai, A. M., Ochs, S. 23, *35*
Kielly, R. K., see Kielly, W. W. 189, *197*
Kielly, W. W., Kielly, R. K. 189, *197*
Kirpekar, S. M., Cervoni, P. 115, *126*
— Misu, Y. 114, *126*
Kirshner, A. G., Kirshner, N. 61, *71;* [17] 79, *87*
— see Kirshner, N. 45, 46, *55;* [7, 11] 79, *86*
Kirshner, N. 42, 43, 56, 57, 75, 78, [25, 27] 80, *87;* 89, 90, 97, 144, 148, *158;* 159, 188, *197;* 200, 220, 273, 274, 281
— Goodall, McC. 216, *218*
— Holloway, C., Smith, W. J., Kirshner, A. G. 45, 46, *55;* [11] 79, *86*
— Sage, H. J., Smith, W. G., Kirshner, A. G. [7] 79, *86*
— — — [8] 79, *86;* 121, *126*
— Viveros, O. H. 115, 122, *126*
— see Duch, D. S. [29] 80, *87*
— see Kirshner, A. G. 61, *71;* [17] 79, *87*
— see Sage, H. J. 47, 49, 53, *56;* [6] 79, *86*
— see Smith, W. J. *56;* 61, *72;* [15] 79, *86*
— see Viveros, O. H. 42, 46, 51, 54, *56;* 58, *72;* [34, 35, 36] 80, 81, *87;* [41, 44, 45] 82, 83, *88;* 115, 121, 122, *127*
Koch-Weser, J. 202, *208*
Koe, B. K., see Weissman, A. 245, *248*
Koenig, H. L., Droz, B. 15, *16*
Kopin, I. J., Breeze, G. R., Krauss, K. R., Weise, V. K. 124, *126;* 241, *247*
— see Glowinski, J. 245, *247*
— see Hertting, G. 223, *233*
— see Rosell, S. 100, 102, *109;* 121, *126;* 206, *209*
— see Sedvall, G. C. 133, *141;* 212, *219;* 245, *248*
— see Thoa 212, 220
Kordon, C., Glowinski, J. 245, *248*
Korn, E. D. 15, *16*
Korr, I. M., Wilkinson, P. N., Chornock, F. W. 12, *16*
Kottegoda, S. R. 169, *185*
Krakoff, L., see De Champlain, J. 138, *141*
Krasney, J. A., Paudler, F. T., Hogan, P. M., Lowe, R. F., Youmans, W. B. 202, *208*
— — Smith, D. C., Davis, L. D., Youmans, W. B. 202, *208*
Krauss, K. R., see Kopin, I. J. 124, *126;* 241, *247*
Kreutzberg, G. 31, *35*
Krnjevic, K., Miledi, R. 121, *126*
Kroneberg, G. 1, 2, 3
Krop, S., see McNamara, B. 169, *185*

Kuff, E. L., see Schneider, W. C. 12, 15, *17*
Kuriyama, H., see Burnstock, G. 120, *125*
Kuzuya, H., see Nagatsu, T. [32] 80, *87*

Lacasse, G., see Belleau, B. 148, *156*
Laduron, P., Belpaire, F. 20, 22, *35;* 36, 43, 46, 50, 51, *56;* [57] 85, *88;* 221
Lagercrantz, H. see Stjärne, L. 122, *127;* 193, *197*
Lagerstedt, S., see Hillarp, N.-A. [9] 79, *86*
Landsberg, L., Axelrod, J. 215, *218*
Lang, K., Siebert, G., Estelmann, T. 32, *35*
Langer, S. Z. 265, *267;* 268, 269
— Trendelenburg, U. 165, *165*
Larrabee, M. G., see Bronk, D. W. 94, *96*
Leach, C. S., Lipscomb, H. H. 217, *218*
Leaders, F. E. 178, *185*
— Long, J. P. 178, *185*
— see Chiang, T. S. 178, *184*
Lee, K. S., see Schwartz, A. 195, *197*
Lee, W. C., Shidemann, F. E. 169, *185*
Lehmann, J. 231, *233*
Levenberg, B., see Levin, E. Y. 50, *56;* [28] 80, *87*
Levi-Montalcini, R., Angeletti, P. U. 130, 137, *141*
— Booker, B. 130, *141*
Levin, E. Y., Levenberg, B., Kaufman, S. 50, *56;* [28] 80, *87*
Levitt, M., Spector, S., Sjoerdsma, A., Udenfriend, S. 140, *141;* 212, *218*
Levy, J. V. 193, *197;* 199
Lightman, Iversen 166
Lindmar, R. 180, 181, 182, *185*
— Löffelholz, K., Muscholl, E. 114, *126;* 152, *158;* 169, 175, 177, 178, 182, *185;* 262, *267*
— Muscholl, E. 148, *158;* 161, 162, *165;* 168, 169, 170, 182, *185*
— see Löffelholz, K. 175, *185*
Lindner, E., see Schöne, H. 189, *197*
Lindqvist, M., see Carlsson, A. 224, 225, 226, 227, 229, 230, *232;* 252, 254, *255*
— see Häggendal, J. 245, *247*
Ling, G. M., see Malik, K. U. 152, *158;* 183, *185*
Lipscomb, H. H., see Leach, C. S. 217, *218*
Lisander, B., see Folkow, B. [48] 85, *88;* 91, 92, 95, *97;* 100, 101, 104, 107, *108;* 113, 115, 117, 118, *125;* 155, *157*
Lishajko, F. 144, *158*
— see Euler, U. S. v. 24, *34;* 103, *108;* 123, *125;* 144, 145, 146, 147, 148, 150, 152, *157;* 188, 189, 192, 193, *196*
— see Stjärne, L. 121, *127*
Livett, B. G. 58
— Geffen, L. B., Austin, L. 40, *41;* 58, *71*
— — Rush, R. A. 24, *35;* 52, *56;* [58] 85, *88*
— see Austin, L. [33] 80, *87*
— see Geffen, L. B. 54, *55;* 58, 68, *71*
Ljung, B. 93, *97*
— see Häggendal, J. 100
— see Johansson, B. 94, *97;* 99
Löffelholz, K. 169, 171, 182, 183, *185*
— Lindmar, R., Muscholl, E. 175, *185*
— Muscholl, E. 114, *126;* 168, 176, 177, 181, *185*
— see Lindmar, P. 114, *126;* 152, *158;* 169, 175, 177, 178, 182, *185;* 262, *267*
Long, J. P., see Leaders, F. E. 178, *185*
Lowe, R. F., see Krasney, J. A. 202, *208*
Lubinska, L. 4, 14, *16*
Ludden, C. T., see Stone, C. A. 140, *141*

MacDonald, G. J., see Wilcken, D. E. L. 150, *158*
— see Wilcken, P. E. L. 195, *197*
MacIntosh, Emmelin 280
Macon, Sokoloff, Glowinski, J. 245
Maître, see Muscholl 76
Makinose, H., see Balzer, H. 193, *196*
Malamed, S., Poisner, A. M., Trifaro, J. M., Douglas, W. W. [24] 80, *87;* 122, *126*
Malawista, S. E., Bensch, K. G., Sato, H. 32, *35*
Malek, G., see Belleau, B.
Malik, K. U. 186
— Ling, G. M. 152, *158;* 183, *185*
Malmfors, T. 101, 110, *108*
— Sachs, C. 140, *141*
— see Carlsson, A. 225, *232*
— see Corrodi, H. 252, *255*
— see Häggendal, J. 20, *35;* 117, *125*
Mangnall, D., see Banks, P. 22, *33*
Mantegazza, P., Naimzada, M. K. 154, *158*
— see Khan, M. *184*
Margolies, F., Roffi, J., Jost, A. 217, *218*
Marley, E. 75
— Paton, W. D. M. [51] *88;* 115, 116, *126*
— Prout, G. I. [50] *88;* 115, *126*
Maronde, R., see De Quattro, V. [37] 81, *87;* 214, *218;* 220
Marsha l, E., Stirling, G. S., Tait, A. C., Todrick, A. 223, *233*
Martel, R., see Belleau, B. 148, *156*
Martin, A. R. [47] 85, *88*

Massani, Z. M., Finkielmann, S., Worcel, M., Agrest, A., Paladini, A. C. 207, *208*
— see Finkielmann, S. 207, *208*
Masuoka, D., see Hamberger, B. 224, *233*
Mayor, D., see Banks, P. 22, *33;* 47, 51, 52, *55;* 67, *71;* 121, *124*
— see Kapeller, K. 4, 14, *16;* 20, 21, *35;* 42, 59, *71*
McGiff, J. C., Burns, R. B. P., Blumenthal, M. R. 179, *185*
McKay, E. A., see McNamara, B. 169, *185*
McKenzie, G. M., Szerb, J. C. 242, *248*
McNamara, B., Krop, S., McKay, E. A. 169, *185*
Melamed, J., Trujillo-Cenóz, O. 12, *16*
Menard, M., see Belleau, B. 148, *156*
Merlis, J. K. *267*
Merrillees, N. C. R. 93, *97*
Metcalf, J. C., Seeman, P., Burgen, A. S. V. 192, *197*
Michaelson 129
Middleton, H. H., see Middleton, S. 169, *185*
Middleton, S., Oberti, C., Prager, R., Middleton, H. H. 169, *185*
— see Hoffmann, F. 169, *184*
Miledi, R., see Katz, B. 121, 124, *126*
— see Krnjevic, K. 121, *126*
Misu, Y., see Kirpekar, S. M. 114, *126*
Mitchell, J. F. 238, *248*
Moerman, E. J., see De Potter, W. P. 54, *55;* 67, 68, *71;* 73, 75, 121, 122, *125*
Molinoff 37
Moore, K. E., see Carr, L. A. 242, *247*
Morré 17, 18
Mori, S. 14, *16*
Mueller, R. A. 212
— Thoenen, H. 138
— — Axelrod, J. [38, 39] 82, *87;* 133, 140, *141;* 212, 213, 214, 215, 216, *218*
— Willard, Axelrod, J. 214
— see Axelrod, J. 215
— see Thoenen, H. 28, *35;* [40] 82, *88;* 140, *141;* 212, 213, 214, 215, 216, *219*
Munoz, E., Nachbar, M. S., Schor, M. T., Salton, M. R J. 49, *56*
Musacchio 221
Muscholl, E. 110, 168, *185;* 186, 187, 200, 209, 210, 268, 281
— Maître 76
— see Hukovic, S. 114, *125;* 168, 177, *184;* 202, *208*
— see Lindmar, P. 114, *126;* 148, 152, *158;* 161, 162, *165;* 168, 169, 170, 175, 177, 178, 182, *185;* 262, *267*
— see Löffelholz, K. 114, *126;* 168, 175, 176, 177, 181, *185*
Myers, R. D., see Feldberg, W. 231, *233;* 258, *266*

Nachbar, M. S., see Munoz, E. 49, *56*
Nachmansohn, D. 152, *158*
Nagatsu, T. [31] 80, *87*
— Kuzuya, H., Kidaka, H. [32] 80, *87*
— see De Quattro, V. [37] 81, *87;* 214, *218;* 220
Nahmod, V. E., see Finkielmann, S. 207, *208*
Naimzada, M. K., see Mantegazza P. 15,4, *158*
Nairn, R. C. 62, *71*
Nasmyth, P. A. 148, *158*
Nedergaard, Beran 186
Neff, N. H., see Costa, E. 244, *247*
Newton 77
Nickerson, M., see Dontas, A. S. 212, *218*
— see Kalsner 166
Niedorf, H. R., see Blümcke, S. 14, *16*
Nigh, C. A., see Jellinek, M. M. P. 137, *141*
Nilson, B., see Hillarp, N.-A. [9] 79, *86;* 144, *157*
Norberg, K.-A. 25, 26, *35*
Normann, T. C. [42] 82, *88*
Novikoff, A. B., Goldfisher, S. 11, *16*
Nybäck, H., Sedvall, G. 244, *248*

Oberti, C., see Middleton, S. 169, *185*
Ochs, S. 23
— Sabri, M. I. 23, *35*
— — Johnson, J. 23, *35*
— see Kidwai, A. M. 23, *35*
Ohad, I., see Amsterdam, A. [43] 82, *88*
Oka, M., see Izumi, F. 195, *197*
Oliver, A. P., see Bloom, M. F. E. *266*
Ostberg, A., see Geffen, L. B. 40, *41;* 59, *71*
Owen, D. A. A., see Day, M. D. 150, *157*
Owman, C., see Falck, B. 65, *71*
Owman, Ch., Sjöstrand, N. O. 153, *158*

Pacha, W., see Salzmann, R. 100, 105, *109*
Paladini, A. C., see Finkielmann, S. 207, *208*
— see Massani, Z. M. 207, *208*
Palay, S. L. 14, *16;* 19
Palm, D. 73, 188, 198, 199, 200, 234
— see Grobecker, H. 189, 190, 191, 192, 193, 194, *197*
— see Holtz, P. 188, *197*
— see Philippu, A. 189, *197*
Paton, D. M. 75
— Gillis, C. N. 206, *209*

Paton, W. D. M., see Marley, E. [51] *88;* 115, 116, *126*
Paudler, F. T., see Krasney, J. A. 202, *208*
Peach Bumpus, Khairallah 209
Pellegrino de Iraldi, A. 2, 4, 42
— De Robertis, E. 4, 5, 7, 11, 12, *16;* 23, 24, *35; 41*
— Gueudet, R. 12, *16*
— Jaim Etcheverry, G. 14, *16*
— Rodriguez de Lores Arnaiz 5, 11, 13
— see De Robertis, E. 12, *16*
Perron, Y. G., see Belleau, B. 148, *156*
Persson, H., Waldeck 250
— see Ehinger, B. 114, *125*
Persson, N.-Å., see Euler, U. S. v. 150, *157*
Petersen, O. H., see Poulsen, J. H. 114, *126*
Pfeiffer, A., see Philippu, A. 24, *35*
Pfenninger, K., Sandri, C., Akert, K., Eugster, C. H. 155, *158*
Philippu, A. 76, 77, 160, 187, 198, 201, 234, 235, 236, 250, 258, 267, 268, 273
— Heyd, G., Burger, A. 258, 259, 260, 261, 262, 264, 265
— Palm, D., Schümann, H. J. 189, *197*
— Pfeiffer, A., Schümann, H. J. 24, *35*
— Przuntek, H. 258, 262, *267*
— see Burger, A. 146, *156;* 188, *196*
— see Przuntek, H. 258, *267*
— see Schümann, H. J. 53, *56;* 147, *158*
Piccinini, F., see Khan, M. *184*
Piezzi, see Rodriguez 18
Pillai, P. A., see Weiss, P. 14, *17*
Pitts, R. F., see Bronk, D. W. 94, *96*
Pletscher, A., see Berneis, K. H. 61, *71*
— see Da Prada, M. 254, *255*
Poisner, A. M., Trifaro, J. M., Douglas, W. W. [23] 80, *87;* 122, *126*
— see Douglas, W. W. [1, 2, 3] 78, *86;* 121, *125*
— see Malamed, S. [24] 80, *87;* 122, *126*
— see Trifaro, J. M. [22] 79, *87*
Porter, C. C., see Stone, C. A. 140, *141*
Porter, K. R., see Tilney, L. G. 15, *17*
Posner, H., see Weil-Malherbe, H. 189, *197*
Potter, L. T. 121, *126*
— Axelrod, J. 52, *56;* [56] 85, *88;* 258, *267*
— see Axelrod, J. 148, *156*
Poulsen, J. H., Petersen, O. H. 114, *126*
Prager, R., see Middleton, S. 169, *185*
Prout, G. I., see Marley, E. [50] *88;* 115, *126*
Przuntek, H., see Philippu, A. 258, 262, *267*
— Philippu, A. 258, *267*

Rand, M. J., see Burn, J. H. 114, *124;* 147, 152, 155, *156;* 169, 179, *184*
Reger, J. F., see Breemen, V. L. 14, *16*
Richards 143
Richards, J. G., see Tranzer, J. P. 133, 136, *142*
Richardson, H. L., Graupner, K. I., Richardson, M. E. 195, *197*
Richardson, J. A., Woods, E. F. 169, *185*
Richardson, K. G. 23, *35*
Richardson, M. E., see Richardson, H. L. 195, *197*
Rinaldi, F. V., see Everett, G. M. 227, *233*
Ritchie, J. M., see Armett, C. J. 153, *156*
Robinson, G. A., see Sutherland, E. W. 144, 145, *158*
Robinson, I. M., see Bhagat, B. 182, *184*
Robinson, P. M., see Burnstock, G. 114, *125;* 152, *156*
Rodriguez, Piezzi 18
Rodriguez de Lores Arnaiz, see Pellegrino de Iraldi, A. 5, 11, 13
Roffi, J., see Margolies, F. 217, *218*
Roos, B. E. 252, 254, *255*
— Sjöström, R. 231, *233*
— see Andén, N. E. 252, 254, *255*
Rosell, S., Kopin, I. J., Axelrod, J. [63] 100, 102, *109;* 121, *126;* 206, *209*
Rosenblueth, A., see Cannon, W. B. 130, *141*
Rosengren, E., see Ehinger, B. 114, *125*
— see Jonasson, J. 147, *158;* 189, 192, *197*
Ross, G., see Wurtmann, R. J. 217, *219*
Rossum, J. M. van 173, *185*
Rubin, R. P., see Douglas, W. W. [1, 2, 3] 78, *86;* 114, 121, *125*
Rush, R. A. 58
— see Geffen, L. B. 20, *34;* 39, 40, *41;* 54, 58, *71*
— see Livett, B. G. 24, *35;* 52, *56;* [58] 85, *88*
Rutledge, C. O., see Jonason, J. 225, *233*

Sabri, M. I., see Ochs, S. 23, *35*
Sachs, C., see Malmfors, T. 140, *141*
Sage, H. J., Smith, W. J., Kirshner, N. 47, 49, 53, *56;* [6] 79, *86*
— see Kirshner, N. [7, 8] 79, *86;* 121, *126*
Salmoiraghi, G. C., see Bloom, M. F. E. *266*

Salton, M. R. J., see Munoz, E, 49, *56*
Salzmann, R., Pacha, W., Taeschler, M., Weidmann, H. 100, 105, *109*
Sampson, S. R., see Douglas, W. W. 114, *125*
Sanchez-Garcia, P., see Furchgott, R. F. 164, *165*
Sandler, M., see Youdim, M. B. H. 195, *198*
Sandri, C., see Pfenninger, K. 155, *158*
Sato, H., see Malawista, S. E. 32, *35*
Schiefthaler, T., see Hertting, G. 136, 137, *141;* 206, *208*
Schild, H. O. 114, *126*
Schildkraut 234
Schlote, W., Hager, H. 12, *17*
Schmitt, F. O. 24, 29, *35*
Schneider, F. H., Smith, A. D., Winkler, H. 49, 53, 54, *56;* 61, *71;* 75 [19] 79, *87*
— see Blaschko, H. 45, 54, *55;* [18] 79, *87;* 121, *124*
Schneider, W. C., Kuff, E. L. 12, 15, *17*
Schöne, H. 188, *197*
— Lindner, E. 189, *197*
Schor, M. T., see Munoz, E. 49, *56*
Schramm, M., see Amsterdam, A. [43] 82, *88*
Schümann, H. J. 1, 2, 121, *126;* 144, *158;* 159, 187, 199, 202, 210, 211, 270, 273, 275, 276, 277, 281
— Philippu, A. 53, *56;* 147, *158*
— see Burger, A. 146, *156;* 188, *196*
— see Grobecker, H. 193, *197*
— see Holland 274
— see Philippu, A. 24, *35;* 189, *197*
— see Starke, K. 202, 203, 204, 207, *209*
Schwartz, A., Lee, K. S. 195, *197*
Sedvall, G. C., Kopin, I. J. 133, *141;* 212, *219*
— Weise, U. K., Kopin, I. J. 245, *248*
— see Nybäck, H. 244, *248*
Seeman, P. M. 188, 190, 192, *197;* 198, 200
— Bialy, H. S. 190, 192, *197*
— see Metcalf, J. C. 192, *197*
Serck-Hanssen, G. 49, 55, *56;* 122, *126*
— see Helle, K. B. 49, 52, *55*
Shelanski, M. L., see Wisniewski, H. 32, *36*
Shidemann, F. E., see Lee, W. C. 169, *185*
Shore, P. A., see Giachetti, A. 161, *165*
Shorey, C. D., see Wilcken, D. E. L. 150, *158;* 195, *197*
Siebert, G., see Lang, K. 32, *35*
Sigg, E. B. 223, *233*
— Soffer, L., Gyermek, L. 223, *233*
Silver, M., see Blaschko, H. 45, 54, *55;* [18] 79, *87;* 121, *124*
Sjoerdsma, A., see Gordon, R. 212, *218*
— see Levitt, M. 140, *141;* 212, *218*
— see Spector, S. 212, *219*
Sjöstrand, J., Karlsson, J.-O. 23
— see Karlsson, J.-O. 31, *35*
Sjöstrand, N. O. 88, 152, *158;* 159, 209
— see Owman, Ch. 153, *158*
Sjöström, R., see Roos, B.-E. 231, *233*
Smith, see Järlfors 24
Smith, A. D. 49, *56;* 58, 61, *71;* [55] 85, *88;* 152, *158*
— Winkler, H. 45, 46, *56;* 60, *71;* [13, 16] 79, *86*
— see Blaschko, H. 45, 54, *55;* [12, 18] 79, *86; 87;* 121, *124*
— see De Potter, W. P. 54, *55;* 67, 68, *71;* 73, 75, 121, 122, *125*
— see Schneider, F. H. 49, 53, 54, *56;* 61, *71;* 75, [19] 79, *87*
Smith, D. C., see Krasney, J. A. 202, *208*
Smith, P. 75
Smith, W. G., see Kirshner, N. [7, 8] 79, *86*
Smith, W. J., Kirshner, N. *56;* 61, *72;* [15] 79, *86*
— see Sage, H. J. 47, 49, 53, *56;* [6] 79, *86*
— see Kirshner, N. 45, 46, *55;* [11] 79, *86;* 121, *126*
Snipes, R. L., see Tranzer, J. P. 133, 136, *142*
Snyder, see Axelrod, J. 37, 249
Snyder, S. H., Green, A. I., Hendley, E. D. 242, *248*
Soffer, L., see Sigg, E. B. 223, *233*
Sohal, R. S., see Sun, S.-C. 195, *197*
Sokoloff, see Macon 245
Sotelo, J. R., see De Robertis, E. 4, *16*
— see Estable, C. 4, 12, *16*
— see Wettstein, R. 12, *17*
Spector, S., Gordon, R., Sjoerdsma, A., Udenfriend, S. 212, *219*
— see Gordon, R. 212, *218*
— see Levitt, M. 140, *141;* 212, *218*
Spiegel, H. E., see Dengler, H. J. 223, *233*
Sporrong, B., see Ehinger, B. 114, *125*
Starke, K., Werner, U., Schümann, H. J. 202, 203, 204, 207, *209*
Stavorski, J. M., see Stone, C. A. 140, *141*
Stein, L., see Wise, C. D. 245, *248*
Stirling, G. S., see Marshall, E. 223, *233*

Stjärne, L. 112, 113, 115, 121, *126;* 127, 128, 129, 144, *158;* 192, *197;* 198, 200, 201, 220, 256, 268, 275, 276, 278, 281
— Euler, U. S. v. 161, *165*
— Hedqvist, P. 116, 117, 118, 119, 120, *126*
— — Bygdeman, S. [49] 85, *88;* 92, *97; 109; 126*
— — Lagercrantz, H. 122, *127*
— Lagercrantz, H., Hedqvist, P. 193, *197*
— Lishajko, F. 121, *127*
— see Euler, U. S. v. 24, *34;* 103, *108;* 189, 192, *196*
Stone, C. A., Porter, C. C., Stavorski, J. M., Ludden, C. T., Totaro, J. A. 140, *141*
Strieder, N., see Winkler, H. 53, *56*
Sulser, F., Dingell, J. V. 223, *233*
Sun, S.-C., Sohal, R. S., Colcolough, H. L., Burch, G. E. 195, *197*
Sutherland, E. W., Robinson, G. A., Butcher, R. W. 144, 145, *158*
Szerb, J. C., see McKenzie, G. M. 242, *248*

Taeschler, M., see Salzmann, R. 100, 105, *109*
Tait, A. C., see Marshall, E. 223, *233*
Tait, A., see Yates, C. M. 226, *233*
Talesnik, J., see Hoffmann, F. 169, *184*
Taugner, G., Hasselbach, W. 146, *158;* 188, *197;* 270
Taylor, A. C., see Weiss, P. 14, *17*
Tenen, S. S., see Weissman, A. 245, *248*
Terry, R. D., see Wisniewski, H. 32, *36*
Thierry, A. M., Blanc, G., Glowinski, J. 242, 245, 246, *248*
— Fekete, M., Glowinski, J. 245, *248*
— Javoy, F., Glowinski, J., Kety, S. S. 245, *248*
— see Javoy, F. 243, *247*
Thoa, Kopin 212, 220
Thoenen, H. 37, 41, 76, 77, 109, 142, 143, 209, 212, 269, 279
— Häfely, W., Gey, K. F., Hürlimann, A. 130, *141*
— Hürlimann, A., Haefely, W. 100, *109*
— Mueller, R. A., Axelrod, J. 28, *35;* 36, [40] 82, *88;* 140, *141;* 212, 213, 214, 215, 216, *219*
— Tranzer, J. P. 133, 134, 135, 140, *141;* 212, *219*
— see Haefely, W. 117, *125;* 182, *184*
— see Häusler, G. 138, *141;* 175, 178, *184*
— see Mueller, R. A. [38, 39] 82, *87;* 133, 138, 140, *141;* 212, 213, 214, 215, 216, *218*
— see Tranzer, J. P. 21, 24, *35;* 130, 133, *141;* 136, *142*
Thomas 89
Thornburg, W., De Robertis, E. 4, *17*
Tilney, L. G., Porter, K. R. 15, 17, *17*
Titus, E. O., see Dengler, H. J. 223, *233*
Todrick, A., see Marshall, E. 223, *233*
— see Yates, C. M. 226, *233*
Totaro, J. A., see Stone, C. A. 140, *141*
Tramezzani, J., see Wurtmann, R. J. 217, *219*
Tranzer, P. 37, 130
— Thoenen, H. 21, 24, *35;* 130, 133, *141*
— — Snipes, R. L., Richards, J. G. 133, 136, *142*
— see Thoenen, H. 133, 134, 135, 140, *141;* 212, *219*
Trendelenburg, U. 130, 138, *142;* 147, 150, *158;* 161, 164, 166, 168, 173, 180, *185;* 186, 210, 221, 268, 269
— see Drasköczy, P. R. 161, *165*
— see Langer, S. Z. 165, *165*
Trifaro, J. M., Poisner, A. M., Douglas, W. W. [22] 79, *87*
— see Malamed, S. [24] 80, *87;* 122, *126*
— see Poisner, A. M. [23] 80, *87;* 122, *126*
Trujillo-Cenóz, O., see Melamed, J. 12, *16*

Udenfriend, S., see Gordon, R. 212, *218*
— see Levitt, M. 140, *141;* 212, *218*
— see Spector, S. 212, *219*
Ungerstedt, U., see Andén, N. E. 255, *255*
— see Carlsson, A. 242, *247*
— see Fuxe, K. 224, *233;* 242, *247*
Uretsky, Iversen 143

Van Anderson, E., see Breemen, V. L. 14, *16*
Vaz Ferreira, A., see De Robertis, E. D. P. 44, 53, *55;* [20] 79, 80, *87*
Vesell, E., see Wurtmann, R. J. 217, *219*
Viveros, O. H. 78, 220
— Arqueros, L., Connett, R. J., Kirshner, N. 46, 51, 54, *56;* [36] 81, *87;* [45] 82, *88*
— — Kirshner, N. 58, *72;* [34, 35] 80, 81, *87;* [41, 44] 82, 83, *88;* 115, 121, 122, *127*
— see Duch, D. S. [29] 80, *87*
— see Kirshner, N. 115, 122, *126*
Vogt, M. 237, *248;* 265, *267*

Vogt, M., see Iggo, A. 28, *35;* 36, 212, *218*

Waldeck, B., see Carlsson, A. 24, *34;* [26] 80, *87;* 103, *108;* 144, 145, 148, *156;* 188, 189, *196;* 252, *255*
— see Jonasson, J. 147, *158;* 189, 192, *197*
— see Persson 250
Wechsler, W., Hager, H. 12, *17*
Weidmann, H., see Salzmann, R. 100, 105, *109*
Weil-Malherbe, H., Posner, H. 189, *197*
Weinberg, N. L., see Belleau, B. 148, *156*
Weiner 220
Weiner, N., see Alousi, A. 212, *218*
Weise, U. K., see Kopin, I. J. 241, *247;* 245, *248*
Weise, V. K., see Kopin, I. J. 124, *126*
Weiss, P. 23, 24, 29, *35*
— Hiscoe, H. 20, *36*
— Taylor, A. C., Pillai, P. A. 14, *17*
Weissman, A., Koe, B. K., Tenen, S. S. 245, *248*
Werdinius, B., see Andén, N. E. 252, 254, *255*
Werner, U., see Starke, K. 202, 203, 204, 207, *209*
West, W. L., see Bhagat, B. 182, *184*
Wetzstein, R. [21] 79, 80, *87*
Wettstein, R., Sotelo, J. R. 12, *17*
Whitby, L. G., see Axelrod, J. 223, *232*
— see Hertting, G. 223, *233*
Wiegand, R. G., see Everett, G. M. 227, *233*
Wilcken, P. E. L., Brender, D., MacDonald, G. J., Shorey, C. D., Hinterberger, A. 195, *197*
— — Shorey, C. D., Mac Donald, G. J. 150, *158*
Willard, see Mueller, R. A. 214
Wilkinson, P. N., see Korr, I. M. 12, *16*
Wilson, A. B. 182, 183, *185*
— see Birmingham, A. T. 149, *156;* 182, 183, *184*
Winkler 270
Winkler, H., Ziegler, E., Strieder, N. 53, *56*
— see Blaschko, H. [12] 79, *86*
— see Schneider, F. H. 49, 53, 54, *56;* 61, *71;* 75, [19] 79, *87*
— see Smith, A. D. 45, *56;* [13, 16] 79, *86*
— see Smith, A. G. 60, *71*
Wise, C. D., Stein, L. 245, *248*
Wisniewski, H., Shelanski, M. L., Terry, R. D. 32, *36*
Withrington, P. G., see Davies, B. N. 117, *125;* 153, *157*
Wolinsky, H., Glagov, S. 137, *142*
Wolstencroft, J. H., see Bradley, P. B. 262, *266*
Wood 43
Woods, E. F., see Richardson, J. A. 169, *185*
Worcel, M., see Finkielmann, S. 207, *208*
— see Massani, Z. M. 207, *208*
Wurtmann, R. J. 215, 217, *219*
— Axelrod, J. 215, 217, 218, *219*
— — Tramezzani, J. 217, *219*
— — Vesell, E., Ross, G. 217, *219*
Wysokowski, J., see Carlsson, A. 224, *232*

Yates, C. M., Todrick, A., Tait, A. 226, *233*
Yoshida, H., see Izumi, F. 195, *197*
Youdim, M. B. H., Sandler, M. 195, *198*
Youmans, W. B., see Krasney, J. A. 202, *208*

Zaimis, E. 195, *198*
Ziegler, E., see Winkler, H. 53, *56*
Zieher, L. M., see Jaim Etcheverry, G. 12, *16*

Subject Index

accumulation
— drugs 193
— enzymes 22
— storage granules 20—21
acetylcholin 100, 112, 122, 152—154, 170, 176, 186, 219, 238, 260—262
ACTH 215—217
S-adenosylmethionine 216
adrenal medulla (see also medulla)
— antibodies for chromogranin prepared from 59—60
— content of chromogranin 61
— differences to sympathetic nerves in exocytosis 121—123
— model system for neurotransmitter secretion 113—115
— quantal or graded secretion 115
— release of chromogranin 53, 58
— reserve capacity 122
— secretion of catecholamines and dopamin-β-hydroxylase 78—88, 113, 273
— synthesis of NA 212
adrenalin resynthesis 89
adrenergic blocking agents 148, 192
adrenergic nerves
— degeneration 133
— function 20, 29
— identification 133
adrenergic transmitter (see also transmitter)
— effects of drugs 147—151, 168—186
— release 100—109, 168—186
— scheme for transmission 154—155
aggregated particles 60
alkaloids 29, 32
all-or-none phenomenon 189
amine storage granules 20—36
— feed back mechanism 28
— synthesis 24
— transport 29
amine uptake mechanisms 223—233
amino acids
— accumulation 23
— radioactive 23
— transport 23
amphetamine 147, 239f.
analysis of protein molecular weight 23
angiotensin 202—211
antigen
— chromogranin 46, 59, 60
— dopamin-β-hydroxylase 59—60
ATP 32, 144—145, 151—152, 188, 198, 200
— ratio to catecholamines 51, 61, 78, 273
ATP-ase 188—189
atropin 169, 178, 187, 221, 262f.
autoinhibition 173
axon
— adrenergic 24, 29
— axonal transport 20—36, 39
— compression 11, 15 (see also ligature)
— intraganglionic 29
— ligation 14, 70
— membrane 124
— regeneration 12—14
— submicroscopic changes 4
— terminals 39
— unligated 23, 40
axoplasm 8, 12
— axoplasmic transport 29
— biochemical studies 11—12
— fractionation of axoplasmic particles 11, 18
— morphological changes 9, 14

biochemical studies
— catecholamines 89, 212—219, 254
— fractions of axoplasmic particles 12
— Golgi membranes 11
blockade of the membrane pump 223 to 233
blood
— brain barrier 142—143, 223f.
— flow 101—102, 104
— pressure 28, 142, 212
brain 223—233, 252—255
bretylium-like action of DMPP 182

calcium 61, 68, 106—107, 113—114, 128, 152, 260—262, 281
carbachol 54
cardiostimulation 178
catecholamines (see also noradrenalin and transmitter)
— amount per splanchnic impuls 115
— cellular mechanisms of secretion 121
— distribution by histochemical techniques 65—66
— effect of drugs on uptake and release 144—158

catecholamines
— effect of metabolic factors on uptake and release 144—158
— fluorescence 62—64
— hypothalamus 258—267
— postsynaptic receptors 255
— quantal release 113
— ratio to ATP 51, 61, 78, 273
— ratio to dopamin-β-hydroxylase 273
— relationship to chromogranins 53, 58, 70
— release by drugs 187—198, 260—262
— secretion 78—88, 273
— synthesis 89, 212—219, 254
— turnover 122
catechol O-methyl transferase 164
cellular mechanisms of CA secretion 121
central nervous system 237—248, 266
chemical sympathectomy 130—142
chlorpromazine 188f.
cholinergic-adrenergic interactions 178
"cholinergic link" hypothesis 178—179
cholinergic transmission 95, 100, 131
cholinomimetic drugs 168—186
chromaffin cells 142
chromaffin granules 51
chromatography
— chromogranin 47
— paper 265—266
— proteins from vesicles 61
chromogranin
— chromogranin-like protein 53
— distribution 51, 61—62
— electrophoretic mobility 53, 61
— granule membranes 49
— immunohistochemical localisation 58—72
— immunological assay 46—49
— immunological properties 45—56
— isolation methods 45
— nervous tissue 51
— part of the vesicle membrane 56—57
— preparation 59—61
— preparation of antigen 47, 59—60
— protein binding of NA 61
— relationship to catecholamines 53, 58, 70
— released by nerve impulses 58—72
— released from adrenal medulla 53
— released from sympathetic nerve endings 54
— reserpine 221
— secretion 53, 58, 79, 127
— structural homology in various animals 53
— subcellular localisation 66, 67
— turn over rate 73
circulation 107
cocaine 109, 166—167
colchicine
— binding capacity of neurotubules 18
— effect on the transport of granules 29
compartmentalized transmitter granules 95, 123, 272
compression (see also ligature)
— axons 11, 15
— various nerve types 42
coronary flow 177, 205
cross reaction 52
cytochrome oxidase 198
cytoplasmic concentration of NA 161 to 163

degeneration of adrenergic nerve endings 133
denervation 214
— preganglionic 29
dense core vesicles 18, 21, 23, 37 (see also vesicles)
— depletion by reserpine 41
— electron microscopic observations 60
— non adrenergic nerves 42
desimipramin 101, 242
dexamethasone 217
distribution
— catecholamines by histochemical methods 65—66
— chromogranins 51, 61—62
DMPP (1,1-dimethyl-4-phenyl-piperazine) 169—170, 175, 177, 260—262
— "bretylium-like" actions 182—183
— non nicotinic actions 181—183
— "tyramin-like" actions 181—182
DNA 32
dopa 227f.
dopa-decarboxylase 22, 68, 231
dopamin 240f.
dopamin-β-hydroxylase 22, 43, 47, 50, 54, 80—81
— chromogranin 54, 68
— secretion from adrenal medulla 78—88, 273
dopamin-β-oxidase
— as a marker for NA storage vesicles 37
— soluble and insoluble 115
double ligature 22
double membrane 15, 17, 18, 40
drugs
— accumulation 193
— α-adrenergic blocking 100, 103, 192, 198
— β-adrenergic blocking 192, 198
— γ-adrenergic blocking 198
— analgesics 225
— antidepressants 225, 242f.

— antihistamines 225
— cholinomimetic 168—186
— effect on adrenergic nerve stimulation 147—151
— effect on amine uptake 223—233
— effect on axonal transport 20—36
— effect on hypothalamus 258—267
— effect on synthesis and release 240f.
— ganglionic blocking 268
— membrane effects 188—198
— neuroleptics 244
— nicotinic 175
— psychotropic 148, 192
— treatment of storage vesicles 84

effector cell 103
electrical stimulation 67—68
electronmicroscope-autoradiography 43
electronmicroscopic observations 5—11, 21, 40, 60, 131—132, 190—191, 194 to 195
electrophoresis of proteins from vesicles 61
electroshoc 221
ergotamin 104—105
exocytosis 44, 53—54, 70—77, 121—123, 270ff.
extragranular stores 150, 160

false transmitter 76, 130, 134
feed back mechanism
— influence on the transmitter release 103—106, 119
— intraneural 243, 254
— production of amine storage granules 28
— resynthesis of transmitter 155
— transsynaptic 254
— tyrosine hydroxylase 212
fixation techniques 24, 42—43
fluorescence microscopy 25
fractionation of axoplasmic particles 12

ganglion
— properties of granules 37
gap
— junctional 117
— synaptic 69, 98—99, 103, 275
glucose 200, 230
glucose-6-phosphatase 18
glutaraldehyde-dichromate-reaction 11—12, 43
glutaraldehyde-osmiumtetroxide-reaction 21, 23, 43
Golgi complex 9, 271—272
— thiamine-pyrophosphatase content 11
— relationship with neurotubules 14
granules (see also vesicles)
— binding of transmitter 151—152
— compartmentalization 95
— differences between medullary and nervous 145—146
— extragranular stores 150
— feed back mechanism 28
— ganglionic 37
— life span 127—128
— NA content 92, 128
— origin 22—26
— quantal packet 98
— solubisation of granule constituents 50
— storage 20—36, 128, 143
— synthesis 24—27, 274
— transmitter granules 112
— transport 22, 27, 29
— turnover 26
— uptake of NA 145

heart
— effect of angiotensin on NA release 202—209
hexamethonium 159, 171, 177, 182
histochemical changes
— fluorescence methods 20, 24, 224, 228—229
histochemical technique 42
hormonal control 212—219
5-hydroxydopamine 37, 130
6-hydroxydopamine 82, 130—142, 212 to 213
— blood brain barrier 142—143
— storage in granules 143
5 hydroxytryptamine 186, 227, 254
hypertension 138, 142, 214
hypothalamus
— catecholamine release 258—267

imipramine 242
immunofluorescence 24, 61—64
immunohistochemical methods 51, 61—64
immunological properties of chromogranin 45, 61
impulse frequency 111
inactivation of intraneural mechanisms 165
induction
— enzyme 221
— transsynaptic 212f., 219f.
insulin 81—82, 89, 274
interstitial space 93
intraneuronal
— feedback mechanism 243, 254
— forms of monoamine storage 244f.
— mechanisms of inactivation 165
inulin 240
iproniazid 40, 228—229

key cells 98—99

ligature (see also compression)
— accumulation of storage granules 20—21
— axon 11—12, 18, 70, 221
— delayed double 27
— double 22
lipoprotein 15
lysolecithin 270

macromolecular mechanism 15
maximum capacity of NA compartment 127
medulla (see also adrenal medulla)
— perfusion 53
— release of chromogranin 53
— similiarity with CA storing particles 54
membrane
— accumulation of drugs 193
— chromogranin 56—57
— double (unit) membrane 15, 17, 18, 40
— effect of drugs 188—198
— exocytosis 123, 124, 271
— granules 49, 84, 190
— membranal uptake 165
— neuronal cell 161—163
— neurotubules 17
— postjunctional potential 112
— pump 100—101, 109—111, 223f.
— secretion 79—80
— smooth endoplasmic reticulum 17
— stimulation 82
— transport 188
metabolic factors
— on transmitter uptake and release 144—147
microsomes 37, 66—67, 195, 258f.
— varicosity 106
microtubule (see also tubule and neurotubule)
— number in adrenergic axons 19
— transport of storage granules 31
migration, proximo-distal 26
mitochondria 145, 194, 200, 258f.
mitosis 29, 32
model of transmitter release 106
monoamine
— intraneuronal storage and release 244f.
— oxidase 20, 161—162, 164, 195
— oxidase inhibitor 40, 151, 228—229, 240, 267
— turnover 237
monomolecular layer 256
morphology of axoplasm 9, 14
multimolecular packets 112
muscarinic inhibition of NA release 175 to 181

nerve
— amount of CA per splanchnic impulse 115
— chromogranin release 54
— differences between adrenal medulla and sympathetic nerves 114—115
— impulses 39, 115
— particulate fractionation of splenic nerves 52
— postganglionic sympathetic 39
— supramaximal stimulus 117—119
— sympathetic terminals 40
neuroleptic drugs 252—255
neurons
— adrenergic 142
— content of noradrenergic 69—70
— dopaminergic 142
— localisation of chromogranins 61, 66 to 67
— membrane 163
— monoamine 252—255
— neurogenic stimulation 83
— neuronal control 212—219
— tonic motoneurons 95, 272
neurotubule (see also tubule and microtubule)
— binding capacity of colchicine 18
— mother tubule of dense core vesicle 23
— number after ligation 19
— relationship with Golgi complex 14
— secretory activity 14
nicotinic
— drugs 175
— non nicotinic actions of DMPP 181 to 183
— receptors 173
— release mechanisms 169—174
noradrenaline (see also catecholamines and transmitter)
— accumulation after ligation 20—21, 40
— axonal transport 39
— binding on chromogranin 61, 123
— content after double ligature 22
— content per varicosity 97
— content of one granule 92
— cytoplasmic concentration 161—163
— depletion after 6-hydroxydopamine 133—134, 137
— effect of angiotensin 202—211
— inactivation 100—101
— injection into coeliac ganglion 40
— metabolites 265—267
— muscarinic inhibition of release 175 to 181
— nicotinic release mechanisms 169—174
— quantal secretion 121
— radioactive labelled NA 40, 258f.

— release after phenoxybenzamine 92, 101
— release after stimulation 67—68, 102
— reserpine resistant pool 42
— resynthesis 89
— saturation of uptake 161
— secretion per nerve stimulus 117
— spontaneous release from hypothalamus 258—262
— storage 26
— subcellular distribution 259
— supranormal levels 27, 221
— synthesis 22, 26, 110, 212
— turnover 26, 73, 127
— uptake and output 93, 103, 145—147, 166

operon 220
osmium tetroxid 5
overflow 227—231

pargyline 162—163
perfusion
— adrenal gland 275
— heart 202—208
— hypothalamus 258—262
perikaryon 9, 23, 272
perinuclear zone 25
peripheral resistance 103—104
periphery 29
phenoxybenzamine 92, 100—101, 109, 212—213
phenylethanolamine N-methyl-transferase 79, 212—219
l-phenylephrine 164
phospholipids 54, 129
pool, "storage and function" 245, 249, 279
potassium permanganate fixation 21
precursor of monoamines 237—248
prenylamine 188f.
preparation of chromogranin 59—61
properties of chromogranin 45
prostaglandin 107, 119, 152—154, 256
protein
— chromatography 61
— chromogranin like 51, 53
— electrophoresis 61
— release from nerves 67—68
— synthesis
— — in nerve endings 15, 36
— — in neurons 28
— — from labelled aminoacids 23
— transport 40
protriptyline 224f.
proximo-distal migration 26
psychotropic drugs 148, 192, 242f.

quantal aspects of secretion 78—88, 91 to 97, 112—113, 116, 122, 155, 272
quantal packets 91, 94, 98

radioactive
— amino acids 23
— precursors of monoamines 237—248
α-receptor blockers 102—105
refilling of vesicles 123
release
— atropine 187
— catecholamines from hypothalamus 258—267
— model 106
— monoamines 237—248
— muscarinic inhibition 175—181
— nicotinic mechanisms 169—174
— NA after electrical stimulation 67—68
— phenoxybenzamine 92, 101
— "points" 106
— proteins from nerves 67—68
— stimulation 67—68, 102
reserpine 20, 24, 27, 41—42, 81, 135, 149—150, 159, 161—162, 181, 188f., 198f., 212—213, 277
reserve capacity from adrenal medulla 122
respiratory chain 145, 159

S-adenosylmethionine 216
sandwich technique 62
saturation of uptake 161
scheme for adrenergic transmission 154 to 155
sciatic nerve 5, 25, 221
second messenger 144—145
secretion
— adrenal medulla 115, 122
— catecholamines 78—88, 122
— compartments 123
— graded 112—127
— neurotransmitter 116—121
— vesicles after stimulation 83
serotonin 143
sieve chromatography of chromogranins 47, 61
soma 58, 70, 272
— formation of synaptic vesicles 39—40
— somatopetal transport 219
splanchnic nerve impulse 115
stabilising effect 193
stereospecificity 161—162
stimulation
— hypothalamus 258—267
— nerve 149, 175—178
— neurogenic 81—82
— perfused adrenal glands 80
— post- and preganglionic 159

stimulation
— prolonged 119
supraphysiological 117
— sympathetic 102, 131—132, 138, 203f.
storage
— amines in granules 20—36
— effect of reserpine 37
— granules transport 22—23, 28
— intraneuronal 244f.
— various forms 250
striatum 238f.
subcellular
— fractions 69
— localisation of chromogranins 66—67
subthreshold
— potentials 94
— release 99
supramaximal nerve stimulus 117—119
surface tension 190—192, 198—199
sympathectomy
— chemical 130—142
— immuno 130, 136—137
sympathetic ganglia 24, 61
sympathetic nerve endings
— difference to adrenal medulla in exocytosis 121
— release of chromogranin 54
synapses
— cerebral cortex 14
— motor end-plate 14
— olfactory bulb 14
synaptic
— gap 69, 98—99, 103, 275
— vesicles 112
synaptosomes 266
synthesis
— amine storage granules 24
— ^{3}H-amines 240f.

tetrabenazine 25
thiamine-pyrophosphatase 11, 15
— in Golgi membranes 11
— in smooth endoplasmic reticulum 18
thymoleptic drugs 225—227
transmitter (see also noradrenalin and catecholamines)
— action of acetylcholin and prostaglandins 152—154
— binding to storage granules 151
— discrete packages 58
— effect of tyramine 147
— false 76, 130, 134
— feed back mechanism 103—106
— granules 112
— inhibitors on uptake and release 144
— metabolic inhibitors 145
— multimolecular packets 112
— quantal release 91—97
— recapture 117
— release 68, 72—73, 100—109, 144 to 147, 168—186
— secretion compartments 123, 272
— storage in vesicles 12
transport
— active 193
— axoplasmic 29
— catecholamines 188
— effect of drugs 20—36
— enzymes 22
— proximo-distal 20—22
— rate 28
— somatofugal 71
— storage granules 22, 28—29
transsynaptic induction 212f., 219f., 254
trihydroxyphenol 140
Triton X-100 49, 52
tubule (see also neurotubule and microtubule)
— membrane 17
— number in adrenergic axons 19
— structure 9
turnover
— catecholamines 122
— chromogranin 73
— 5-hydroxytryptamine 254
— monoamines 237
— noradrenalin 26, 73, 127
— vesicles 95
tyramine 147, 181, 187
— "tyramine-like" action of DMPP 181—182
^{3}H-tyrosine 241f.
tyrosine-hydroxylase 22, 28, 36, 133, 140, 160, 212—219, 248

ultramorphological changes
— by hydroxydopamin 133, 143
— by psychotropic drugs 194—195
uncouplers 145—146
unit
— bundle 93, 118
— membrane 15, 17—18, 40
uptake kinetics 146, 223—233

varicosity 92—93, 97, 106
vasopressin 205—206
vesicles 4—17 (see also granules)
— axonal transport 4
— connection of typical neurotubules with vesicles 9, 14
— damage by reserpine 37
— dense core vesicles 18
— electron microscopic observations 5, 60

— elementary neurosecretary 42
— glutaraldehyde dichromate reaction 12
— granulated 12
— life span 70
— membranes after stimulation 82
— origin 4, 42, 69—70
— quantal unit 83, 91
— recharging 70
— refilling 123
— soluble content after stimulation 82
— storage 12, 37
— synaptic 39, 112
— — half life 42
— turnover 95
— ungranulated 12
vinblastine 29, 32—33

www.ingramcontent.com/pod-product-compliance
Ingram Content Group UK Ltd.
Pitfield, Milton Keynes, MK11 3LW, UK
UKHW051003220926
475831UK00003B/101
* 9 7 8 3 6 4 2 4 6 2 4 3 6 *